Climate Change and Current Challenges for Landscapes and Cultural Heritage

Climate Change and Current Challenges for Landscapes and Cultural Heritage

Editors

Jan K. Kazak
Katarzyna Hodor
Magdalena Wilkosz-Mamcarczyk

MDPI

Basel • Beijing • Wuhan • Barcelona • Belgrade • Novi Sad • Cluj • Manchester

Editors

Jan K. Kazak
Wrocław University of Environmental and Life Sciences
Wrocław, Poland

Katarzyna Hodor
University of Technology
Kraków, Poland

Magdalena Wilkosz-Mamcarczyk
University of Agriculture in Krakow
Kraków, Poland

Editorial Office
MDPI
St. Alban-Anlage 66
4052 Basel, Switzerland

This is a reprint of articles from the Special Issue published online in the open access journal *Land* (ISSN 2073-445X) (available at: https://www.mdpi.com/journal/land/special_issues/landscape_cultural_heritage).

For citation purposes, cite each article independently as indicated on the article page online and as indicated below:

Lastname, A.A.; Lastname, B.B. Article Title. *Journal Name* **Year**, *Volume Number*, Page Range.

ISBN 978-3-0365-8504-8 (Hbk)
ISBN 978-3-0365-8505-5 (PDF)
doi.org/10.3390/books978-3-0365-8505-5

Cover image courtesy of Jan K. Kazak

Contents

Jan K. Kazak, Katarzyna Hodor and Magdalena Wilkosz-Mamcarczyk
Climate Change and Current Challenges for Landscapes and Cultural Heritage
Reprinted from: *Land* **2022**, *11*, 2323, doi:10.3390/land11122323 1

Gabriel Silva Dantas, Ildikó Réka Nagy and Pedro Brizack Nogueira
Implementation of Green Infrastructure in Existing Urban Structures: Tracking Changes in Ferencváros, Budapest
Reprinted from: *Land* **2022**, *11*, 644, doi:10.3390/land11050644 5

Gül Aktürk and Hannah Fluck
Vernacular Heritage as a Response to Climate: Lessons for Future Climate Resilience from Rize, Turkey
Reprinted from: *Land* **2022**, *11*, 276, doi:10.3390/land11020276 17

Bo Li, Yue Wang, Tong Wang, Xiaoman He and Jan K. Kazak
Scenario Analysis for Resilient Urban Green Infrastructure
Reprinted from: *Land* **2022**, *11*, 1481, doi:10.3390/land11091481 37

Izabela Krzeptowska-Moszkowicz, Łukasz Moszkowicz and Karolina Porada
Urban Sensory Gardens with Aromatic Herbs in the Light of Climate Change: Therapeutic Potential and Memory-Dependent Smell Impact on Human Wellbeing
Reprinted from: *Land* **2022**, *11*, 760, doi:10.3390/land11050760 57

Anna Staniewska
Gardens of Historic Mental Health Hospitals and Their Potential Use for Green Therapy Purposes
Reprinted from: *Land* **2022**, *11*, 1618, doi:10.3390/land11101618 79

Ning DongGe, Jinbiao Yan, Peilin Liu, Martin van den Toorn and Albert Fekete
Historical Water Management Strategies—Case Study of Traditional Villages in Southern China, Hunan Province
Reprinted from: *Land* **2022**, *11*, 2107, doi:10.3390/land11122107 103

Monika Trojanowska
Climate Change Mitigation and Preservation of the Cultural Heritage—A Story of the Municipal Park in Rumia, Poland
Reprinted from: *Land* **2022**, *11*, 65, doi:10.3390/land11010065 117

Yingzhuo Zhang, Haoran Yin, Lianqi Zhu and Changhong Miao
Landscape Fragmentation in Qinling–Daba Mountains Nature Reserves and Its Influencing Factors
Reprinted from: *Land* **2021**, *10*, 1124, doi:10.3390/land10111124 135

Andrzej Długoński, Diana Dushkova and Dagmar Haase
Urban Cemeteries—Places of Multiple Diversity and Challenges. A Case Study from Łódź (Poland) and Leipzig (Germany)
Reprinted from: *Land* **2022**, *11*, 677, doi:10.3390/land11050677 155

Albert Fekete and Máté Sárospataki
Baroque Gardens in Transylvania: A Historic Overview
Reprinted from: *Land* **2022**, *11*, 949, doi:10.3390/land11060949 177

Wojciech Korbel, Filip Suchoń and Marta Łapuszek
Water Dams of the Krakow Fortress: Potential of a Vanishing Heritage
Reprinted from: *Land* **2021**, *10*, 1273, doi:10.3390/land10111273 . . . . **195**

Eliza Szczerek
The Problem of Densification of Large-Panel Housing Estates upon the Example of Cracow
Reprinted from: *Land* **2021**, *10*, 1359, doi:10.3390/land10121359 . . . . **215**

land

MDPI

Editorial

Climate Change and Current Challenges for Landscapes and Cultural Heritage

Jan K. Kazak [1,*], Katarzyna Hodor [2] and Magdalena Wilkosz-Mamcarczyk [3]

1 Institute of Spatial Management, Wrocław University of Environmental and Life Sciences, Grunwaldzka 55, 50-375 Wrocław, Poland
2 Chair of Landscape Architecture, Faculty of Architecture, Cracow University of Technology, Warszawska 24, 31-155 Kraków, Poland
3 Department of Spatial Management and Landscape Architecture, Faculty of Environmental Engineering and Land Surveying, University of Agriculture in Krakow, Balicka 253 c, 30-198 Kraków, Poland
* Correspondence: jan.kazak@upwr.edu.pl

Citation: Kazak, J.K.; Hodor, K.; Wilkosz-Mamcarczyk, M. Climate Change and Current Challenges for Landscapes and Cultural Heritage. *Land* **2022**, *11*, 2323. https://doi.org/10.3390/land11122323

Received: 26 November 2022
Accepted: 30 November 2022
Published: 18 December 2022

Publisher's Note: MDPI stays neutral with regard to jurisdictional claims in published maps and institutional affiliations.

Landscapes and cultural heritage have faced many challenges over time, including modifications and degradation that appear with time, overlap with other challenges not previously observed, and the influence of cities and management [1–3]. One of the most visible and globally discussed challenges is climate change [4–6]. Therefore, there is an urgent need to launch initiatives to tackle climate change and other current challenges from the perspective of landscape and cultural heritage protection. This Special Issue presents selected papers of studies conducted in the relation to the 28th Conference on the Series of Garden Art and Historical Dendrology titled "Climate Change and Current Challenges for Landscapes and Cultural Heritage" that was held on 28 and 29 October 2021. This collection serves as a platform for the exchange of experiences among researchers from different scientific domains. It focuses on an important discussion related to the changing climate around the world, and identifies the current problems and challenges in maintaining and preserving the cultural heritage of cities, villages, and open spaces. Raising this issue may significantly increase awareness of the benefits associated with the protection of historical heritage sites, as well as the vulnerability and hazards of these monuments [7–9]. This is especially important considering that, in regions dominated by traditional and historical methods of managing natural resources, local communities' acceptance of improvements in the field of modern renewable energy sources and climate protection is relatively low [10]. This links to tangible cultural heritage, which includes cultural landscapes, historical buildings and gardens, archaeological sites, and historical sites. It is worth noting that such heritage plays an important role in economic, touristic, and recreational development, bringing significant benefits at social, environmental, and economic levels. Urban greenery also plays a significant role in the context of urban areas. Considering that social activity and the cocreation of common spaces in large cities are among the most important needs of residents, not discounting projects related to mobility infrastructure, the issues of preserving and using green areas for recreation and relaxation are particularly relevant [11].

Climate change affects regions around the world and is associated with changes in average temperature, climate patterns, and extreme weather events (e.g., storms, floods, and heat waves). Moreover, an additional factor is the development of urban areas, which contributes to lowering the quality of water, increasing the number of impermeable surfaces, and creating urban heat islands. As the reports of IPCC [12] and ICOMOS [13] show, unfavorable phenomena may also threaten the survival of cultural heritage in the future. ICOMOS has been working diligently to improve the integration of culture and heritage into climate research in general, and the work of the IPCC in particular, and the International Meeting on Culture, Heritage, and Climate Change represents a significant step in this direction. The current state of discussion in this domain can be ascertained from the scientific outcomes of this meeting [14]; however, in this discussion, there were no representatives from Central Europe, which highlights the importance of our conference and this Special Issue. Moreover, there are shortcomings in the policy of adaptation to climate change for heritage, and shortages of technical guidelines for individual countries, which may contribute to the loss of valuable landscapes in coming years. As stated in the March 2021 European Cultural Heritage Green Paper, *"we firmly believe therefore that cultural heritage is a vector for achieving the long-term vision and policy goals of the European Union, including the European Green Deal. Cultural heritage is not just about preserving our past—it is about shaping our future"*. Taking this into account, this Special Issue focuses on three main pillars:

- Green infrastructures and gardens responding to climate change and improving human health [15–19];
- The protection of natural heritage in the context of climate change [20,21];
- Problems and challenges of natural and cultural heritage in terms of landscape [22–25].

Our expectation is that the collection of interdisciplinary research will add real value to this scientific domain. This approach should generate the attention of experts from various professions and highlight the need to employ complex strategies when developing sustainable cities and regions.

Author Contributions: J.K.K., K.H. and M.W.-M. contributed equally to this work. All authors have read and agreed to the published version of the manuscript.

Acknowledgments: The cooperation under XXVIII Conference in the Series of Garden Art and Historical Dendrology, X International Edition entitled "Climate Change and Current Challenges for Landscapes and Cultural Heritage" and the Special Issue on "Climate Change and Current Challenges for Landscapes and Cultural Heritage" in Land was supported by scientific activity conducted within the Leading Research Group: Sustainable Cities and Regions at the Wrocław University of Environmental and Life Science, the Chair of Landscape Architecture at the Cracow University of Technology and the Department of Spatial Management and Landscape Architecture at the University of Agriculture in Krakow.

Conflicts of Interest: The authors declare no conflict of interest.

References

1. Wieteska-Rosiak, B. Real estate sector in the face of climate change adaptation in major Polish cities. *Real Estate Manag. Valuat.* **2020**, *28*, 51–63. [CrossRef]
2. Rząsa, K.; Ciski, M.; Ogryzek, M. Application of GIS tools in spatial distribution modeling of historical monuments. *Geomat. Environ. Eng.* **2019**, *13*, 61–71. [CrossRef]
3. Bazan-Krzywoszańska, A.; Mrówczyńska, M.; Skiba, M.; Sztubecka, M. Sustainable urban development on the example of the housing deveopment of Zielona Góra (Poland), as a response to the climate policy of the European Union. In Proceedings of the 10th International Conference on Environmental Engineering (ICEE 2017), Vilnius, Litwa, 27–28 April 2017.
4. Zięba, Z.; Dąbrowska, J.; Marschalko, M.; Pinto, J.; Mrówczyńska, M.; Leśniak, A.; Petrovski, A.; Kazak, J.K. Built Environment Challenges Due to Climate Change. *IOP Conf. Ser. Earth Environ. Sci.* **2020**, *609*, 012061. [CrossRef]
5. Belčáková, I.; Świąder, M.; Bartyna-Zielińska, M. The Green Infrastructure in Cities as A Tool for Climate Change Adaptation and Mitigation: Slovakian and Polish Experiences. *Atmosphere* **2019**, *10*, 552. [CrossRef]
6. Tokarczyk-Dorociak, K.; Walter, E.; Kobierska, K.; Kołodyński, R. Rainwater Management in the Urban Landscape of Wroclaw in Terms of Adaptation to Climate Changes. *J. Ecol. Eng.* **2017**, *18*, 171–184. [CrossRef]
7. Ćmielewski, B.; Sieczkowska, S.; Kościuk, J.; Bastante, J.M.; Wilczyńska, I. UAV LiDAR Mapping in the Historic Sanctuary of Machupicchu: Challenges and Preliminary Results: Part 1. *Wiadomości Konserw.—J. Herit. Conserv.* **2021**, *67*, 159–170.
8. Rącka, I.; Palicki, S. Influence of Urban Renewal on the Assessment of Housing Market in the Context of Sustainable Socioeconomic City Development. In *EAI Endorsed Transactions on Smart Cities*; Springer: Cham, Switzerland, 2016; pp. 1–9.
9. Bieda, A. Urban renewal and the value of real properties. *Stud. Reg. Lokal.* **2017**, *69*, 5–28. [CrossRef]
10. Furmankiewicz, M.; Hewitt, R.J.; Kapusta, A.; Solecka, I. Climate Change Challenges and Community-Led Development Strategies: Do They Fit Together in Fisheries Regions? *Energies* **2021**, *14*, 6614. [CrossRef]
11. Kołat, K.; Furmankiewicz, M.; Kalisiak-Mędelska, M. What Are the Needs of City Dwellers in Terms of the Development of Public Spaces? A Case Study of Participatory Budgeting in Częstochowa, Poland. *Int. J. Environ. Res. Public Health* **2022**, *19*, 5171. [CrossRef] [PubMed]

12. IPCC. *Climate Change 2022: Impacts, Adaptation and Vulnerability. Working Group II Contribution to the Sixth Assessment Report of the Intergovernmental Panel on Climate Change*; IPCC: Geneva, Switzerland, 2022.
13. ICOMOS. *The Future of Our Pasts: Engaging Cultural Heritage in Climate Action*; ICOMOS: Paris, France, 2019.
14. Morel, H.; Megarry, W.; Potts, A.; Arikan, Y.; Brondizio, E.; Cassar, M.; Flato, G.; Hosagrahar, J.; R.Jigyasu, R.; Mason-Delmotte, V.; et al. *Global Research and Action Agenda on Culture, Heritage, and Climate Change*; ICOMOS: Paris, France, 2022.
15. Dantas, G.S.; Nagy, I.R.B.; Nogueira, P.B. Implementation of Green Infrastructure in Existing Urban Structures: Tracking Changes in Ferencváros, Budapest. *Land* **2022**, *11*, 644. [CrossRef]
16. Aktürk, G.; Fluck, H. Vernacular Heritage as a Response to Climate: Lessons for Future Climate Resilience from Rize, Turkey. *Land* **2022**, *11*, 276. [CrossRef]
17. Li, B.; Wang, Y.; Wang, T.; He, X.; Kazak, J.K. Scenario Analysis for Resilient Urban Green Infrastructure. *Land* **2022**, *11*, 1481. [CrossRef]
18. Krzeptowska-Moszkowicz, I.; Moszkowicz, Ł.; Porada, K. Urban Sensory Gardens with Aromatic Herbs in the Light of Climate Change: Therapeutic Potential and Memory-Dependent Smell Impact on Human Wellbeing. *Land* **2022**, *11*, 760. [CrossRef]
19. Staniewska, A. Gardens of Historic Mental Health Hospitals and Their Potential Use for Green Therapy Purposes. *Land* **2022**, *11*, 1618. [CrossRef]
20. Trojanowska, M. Climate Change Mitigation and Preservation of the Cultural Heritage—A Story of the Municipal Park in Rumia, Poland. *Land* **2022**, *11*, 65. [CrossRef]
21. Zhang, Y.; Yin, H.; Zhu, L.; Miao, C. Landscape Fragmentation in Qinling–Daba Mountains Nature Reserves and Its Influencing Factors. *Land* **2021**, *10*, 1124. [CrossRef]
22. Długoński, A.; Dushkova, D.; Haase, D. Urban Cemeteries—Places of Multiple Diversity and Challenges. A Case Study from Łódź (Poland) and Leipzig (Germany). *Land* **2022**, *11*, 677. [CrossRef]
23. Fekete, A.; Sárospataki, M. Baroque Gardens in Transylvania: A Historic Overview. *Land* **2022**, *11*, 949. [CrossRef]
24. Korbel, W.; Suchoń, F.; Łapuszek, M. Water Dams of the Krakow Fortress: Potential of a Vanishing Heritage. *Land* **2021**, *10*, 1273. [CrossRef]
25. Szczerek, E. The Problem of Densification of Large-Panel Housing Estates upon the Example of Cracow. *Land* **2021**, *10*, 1359. [CrossRef]

land

MDPI

Article

Implementation of Green Infrastructure in Existing Urban Structures: Tracking Changes in Ferencváros, Budapest

Gabriel Silva Dantas [1,*], Ildikó Réka Báthoryné Nagy [1] and Pedro Brizack Nogueira [2]

[1] Department of Urban Planning and Urban Green Infrastructure, Institute of Landscape Architecture Urban Planning and Garden Art, Hungarian University of Agriculture and Life Sciences, 1118 Budapest, Hungary; bathoryne.nagy.ildiko.reka@uni-mate.hu

[2] Department of Exact Sciences, State University of Feira de Santana, Feira de Santana 44036-900, Brazil; peunogueira@gmail.com

* Correspondence: dantassgabriel@gmail.com; Tel.: +36-30-113-3678

Abstract: Understanding the resilience of urban forms as a latent force that drives a place's physical characterization and social cohesion is essential for defining successful adaptive processes of pre-existing urban fabrics. Budapest's ninth district (Ferencváros) is an outstanding example of transforming a complex historical urban context, which underwent renovation strategies guided by maintaining and enhancing essential morphological elements. Courtyards have great relevance in conditioning the well-being in areas of high occupational density, especially in terms of accessibility to urban green infrastructure. In the case of Ferencváros, they were reframed to add new layers of use and to improve territorial integration by unifying smaller private courtyard unities into more extensive communal areas, creating a comprehensive urban green network, preserving urban heritage, and increasing green coverage. This study assesses how this recent re-urbanization phenomenon is related to political changes in a post-socialist city. The conjuncture found in Ferencváros is unique, yet it can be applied in other similar contexts. The methodology applied to this study is supervised classification for the quantitative analysis of remote-sensing image data with GIS software assistance—a procedure rarely applied in medium-scale urban analysis. However, it was verified to be precise and effective in tracking morphological changes. The preliminary results indicate a significant intensification in greenery in the urban pattern, especially in the core areas of the blocks: the courtyards. After the intervention, green areas became more predominant, cohesive, and articulated.

Keywords: courtyard; urban green infrastructure; urban pattern; urban renewal; supervised classification

Citation: Dantas, G.S.; Nagy, I.R.B.; Nogueira, P.B. Implementation of Green Infrastructure in Existing Urban Structures: Tracking Changes in Ferencváros, Budapest. *Land* **2022**, *11*, 644. https://doi.org/10.3390/land11050644

Academic Editors: Jan K. Kazak, Katarzyna Hodor and Magdalena Wilkosz-Mamcarczyk

Received: 25 March 2022
Accepted: 26 April 2022
Published: 27 April 2022

1. Introduction

More than two-thirds of the population of the countries of the European Union currently live in urban areas. Despite the growing effort to include the urban development agenda as a mainstream element of the bloc's Cohesion Policy, it was only in the period between 2007 and 2013 that the EU established it as one of its main development guidelines. At first, the European Commission began to investigate the problems to a conceptual degree and act on an experimental basis, which later also resulted in intergovernmental cooperation agreements in the field of urban development in favor of cohesion [1].

On a global level, this was a response to a major paradigm shift: the advent of the intensification of the globalization process, which boosted the strengthening of international ties, aiming to achieve economic and social prosperity. However, the trend towards globalization occurs unequally in different contexts. The developing world countries would have more opportunities to access exponential growth, while facing difficulties in income distribution and expanding the city's infrastructure network. Around the world, in Asian and South American cities, or even in post-Soviet centers, two eminent development patterns can be identified: peripheral growth (both office spaces and houses) and the participation of the private sector in infrastructure provision [2]. Despite the similarities, the socioeconomic particularities of each region outlined the conditions for the occurrence of this process. Little has been studied regarding the reverberation of this phenomenon in post-socialist urban centers, especially concerning adaptive measures for their inclusion in a global urban network [3].

This scenario was also asymmetric among the countries of the European Union, which motivated the intensification of isolation and urban segmentation [1]. Post-Soviet countries, such as Hungary, only joined the political–economic bloc later, and faced challenges left by the communist legacy, such as deficiencies in urban infrastructure, management of mass-housing estates, and environmental problems [4].

With the end of the communist regime, the urban centers previously inserted in this socio–spatial conjuncture went through an initial population shrinkage, marginalization, and isolation [5]. Public policies played a fundamental role in structuring the privatization operation and introducing the market-based economy. Cities underwent, therefore, an acute adaptive procedure through a situation of disturbance of the political and social order [6]. On the one hand, it was essential to create mechanisms to ensure urban resilience, both in the socioeconomic scope and morphologically, by preserving urban patterns and social aspects, making cities more attractive and competitive [7]. On the other hand, investments in green infrastructure to promote urban resilience did not follow the same tendency in most eastern European cities. The case of Ferencváros stands out in this context, especially for the centralized coordination between public and private interests to rebuild a sizeable urban stretch.

Environmental issues are one of the most significant drawbacks to urban development, considering the interdependence between social, economic, and environmental dimensions for achieving sustainable urban development [8]. In cities with infrastructural limitations, such as Budapest, the obstacles in promoting social cohesion are even more significant. For this reason, mitigating accessibility to green infrastructure can be a tool for territorial integration and a consequent increase in urban resilience [9].

Vienna is often an analogous example at the urban-structure level when establishing a comparative framework with Budapest. These cities share numerous similarities due to their intertwined historical processes. Despite the gradual disconnection they have gone through since World War II, recently, both have faced a common issue: urban decay and increasing urban voids [10]. Notwithstanding their divergent political conditions, between 1970 and 1990, several projects targeting the renovation of central areas emerged in those cities.

One of the biggest obstacles in observing changes in cities is the lack of detailed and model-ready morphological data at the urban scale. In Vienna, GIS data analysis proved to be efficient in obtaining the morphological heterogeneity across the urban landscape, which implies the possibility of using this method to track changes in the infrastructure of Budapest [11].

The IX District of Budapest is an example of the requalification of an emptied historic area with heterogeneous territorial occupation, marked by the existence of urban voids (initially occupied by small- and medium-sized industries) [12]. The voids generated by restructuring the regional industrial production system contributed to the acceleration of the urban decay process and territorial fragmentation, making it difficult for the population to remain in the area [13]. Historic urban morphology, traditionally found in eastern central European cities, is also found in Ferencváros. Before the intervention, most buildings had typological characteristics such as continuous and aligned facades and individual courtyards [14].

In this intervention, the restructuring of the courtyard system was essential for implementing an extensive system of green infrastructure in the region, shaping the intensification of green areas in the densely occupied urban fabric. This action took into account morphological elements characteristic of the site and had a relevant impact on increasing urban cohesion.

2. Materials and Methods

The investigation follows a case-study approach, relying on research on evidence of satellite image processing and map analysis to identify variations in the morphological structure and the green infrastructure over time. The methodology was drawn on qualitative and quantitative analysis of land-use and land-cover (LULC) mapping, highlighting the transformations that occurred in the polygonal of the study in historical periods remarkable in terms of variations in the pattern of urban development [15]. The analysis was carried out in three different periods, in 2000, 2011, and 2021, in order to obtain parameters to establish a comparative framework. The input dates were defined based on milestones in the change in conduct in the management of the urban domain and migration movements—especially regarding Ferencváros—and the availability of material for investigation.

The study area comprises a specific region within the entirety of the IX District of Budapest and is located between the geographic coordinates 47°28′48.82″ north latitude and 19°04′38.91″ east longitude and has an extension of 71.5403 hectares. The database used is composed of digital images from orbital sensors made available by Google Earth Pro on its image catalog by NASA, and were processed in an IACS-compatible environment, on geographical information system (GIS). The software used in this analysis was ArcGis version 10.1, using the ArcToolBox tools extension, with components designed for supervised classification such as Create Sign-nature, Filter statistics, and Maximum Likelihood Classification [16].

The classes selected to carry out the supervised classification consider artificial and natural elements that constitute the urban landscape. Land-cover classes are identified as natural earth resources (e.g., forests, water, bogs, marshlands), while land-use classes are considered artificial areas (e.g., agriculture, roads, cities) [15]. The classes used for the supervised classification are vegetation, exposed soil, street, and building. These classes are essential to identify and follow the pattern of urban development in the region, highlighting demolitions, new buildings, streets and pedestrian paths, urban voids, and green areas over time. Twenty sample pixels for each of the classes mentioned before were obtained, thus enabling measurement on different dates.

Due to the relatively low resolution of the images accessed to perform the procedure, the results found present a degree of inaccuracy. In addition to that, clouds and shadows are expected in optical remote-sensor images, decreasing the precision of the analysis. The occlusion of features is another limiting factor, which reduces the available useful area of the image, compromising the quantitative analysis [17]. Those factors also made it unmanageable to obtain images for classification before the 2000s, which were freely available, and with a satisfactory resolution.

Although some urban areas, mainly in Asia and North America, have been the object of scientific study with supervised classification, little material is produced on this subject in Eastern European cities. Furthermore, this methodology is commonly found in studies covering large territorial portions, often aiming to measure the growth of urban areas towards rural or natural environments, for example [18]. In the case of this present case study, the supervised classification is applied to identify changes in the urban morphology in a segment of the ninth district of Budapest, generating evidence of the renovation operation performed in this region.

3. Results

The Ferencváros requalification project differs from other initiatives in Budapest in this field, as it was set up to enhance urban resilience through the maintenance and improvement of morphological elements of structural importance to the urban fabric, such as the original layout of blocks, continuous facades, overall height, and plot size (or use of architectural elements that symbolize the original individuality of each of them, even when multiple sites were joined for the development of new development) [19]. However, the courtyards are the most striking feature in the definition and spatial articulation in the region. These elements were restructured, enabling new layers of use.

The courtyards, initially independent, were primarily devoid of greenery or had poor and fragmented green areas with restricted access to the building's residents [20]. The restructuring proposed the unification of these elements, also encompassing the urban voids and creating a mesh of public or semi-public pedestrian crossings. The LULC maps produced from satellite images were used in this study as a tool to define the impacts of this intervention on spatial conformation.

3.1. Social–Political Conjuncture and the Designation of Input Satellite Images

The first restructuring projects of Frencváros date from the late 1970s and were conceived under modern guidelines of spatial elaboration and articulation, foreseeing the demolition of most pre-existing historic buildings and emphasizing the creation of new roads, green areas, and buildings for residential and community use [16]. The project underwent revisions in the 1980s, making it more adaptable to local circumstances, as shown in Figure 1. The decision to reconfigure the city's outskirts emerged from the increasing densification of its agglomeration belt, with the migratory movement from the countryside to the city, accelerating the demand for rapid housing construction [21]. Despite the extensive planning, few changes were implemented in the area during this period.

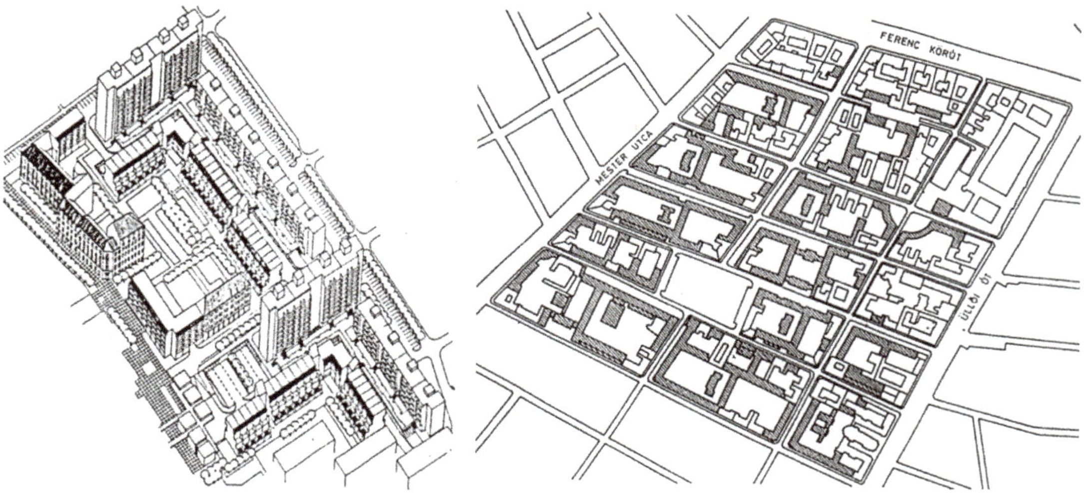

(a) (b)

Figure 1. (**a**) Scheme of a requalification plan from the late 1970s for the Middle Ferencváros. Most of the historical buildings were set to be demolished and replaced by large prefab residential blocks [10]; (**b**) Revised master plan from 1982 to 1983. The original street network and the layout of the urban blocks are redesigned to bear wider streets and more extensive communal courtyards. For the first time, the intention to implement a public square was indicated in the south–central region of Middle Ferencváros. New buildings are hatched [14].

The political transition process in Hungary began in the early 1990s, but it was only around the 2000s that national and international investments in real estate and public infrastructure expanded [22]. The process of urban sprawl and suburbanization started in the late 19th century. This escalated in the mid-20th century in Budapest and impacted the densification of the Ferencváros region, a trend that continues to rise. Table 1 [16] indicates that the urban sprawl movement grew significantly between 2001 and 2011.

Table 1. Continuous urban sprawl process in Budapest even after the communist era.

Year/Region	1990	2001	2011
Proper Budapest	2,016,000	1,775,000	1,729,040
Agglomeration Belt	567,000	672,000	805,848
Total Agglomeration	2,583,000	2,447,000	2,534,888

Furthermore, as mentioned earlier, between 2007 and 2013, the EU established urban development as one of the most relevant factors for its cohesion policy. This decision boosted more evident transformations in the city's urban structure, especially in areas with significant transformative potential and subject to real estate investments, as occurred in the analyzed district [23]. For the reasons mentioned, the year 2011 was selected as one of the landmarks for the elaboration of this research. Finally, the third period selected for sampling was the year 2021, seeking to obtain more recent data that portray the current situation found in the place.

3.2. Urban Design and the Implementation of Green Infrastructure

At the local level in Budapest, the end of the communist regime implied the return to self-governance in Budapest, accordingly inferring the two-tier administrative system and the subsequent shift in decision-making from the city to the district level. This scenario of increasing the individuality and competencies of the districts has resulted in their ability to reformulate their social and housing policies, making them capable of launching urban requalification projects [24].

In the global policies scope, the set of new Sustainable Development Goals (SDG), established in the post-2015 Development Agenda in September 2015, is defined in the General Assembly of the United Nations [25]. Among the guidelines mentioned in the document, SDG 11.7 specifies the following for public spaces:

> *"By 2030, provide universal access to safe, inclusive and accessible, green and public spaces, particularly for women and children, older persons and persons with disabilities".*

Regarding accessibility to green and leisure areas, Ferencváros presented deficiencies. The coefficients of accessibility to those areas in the district until the 2000s (when urban intervention was not yet consolidated) were below 9 m^2 per capita within 15 min of walking distance stated by the World Health Organization [26]. To achieve these parameters, the system of integrated green courtyards was designed, in addition to the implementation of green elements at specific points on the streets—a measure that made the core and the edges of the blocks greener, also improving the landscape conditions of the place.

To meet the indicators of the World Health Organization and to achieve a better ratio between population density and accessibility to green open spaces, a new compact urban park [27] was created from the demolition of some poorly conserved buildings and empty plots in the south–central region of the polygonal, as shown in Figure 2. The Kerekerdő park is one of the most significant public elements for the configuration of the green grid in the area, being a (public) confluence point for the green paths [28]. The same is true of Ferenc tér, a pre-existing square in the north–central region that also plays the role of a significant urban green infrastructure element.

Among the benefits achieved by implementing a comprehensive green system are strengthening social relations, increments in connectivity, improvements in urban cohesion, and an increase in local economic activities, leading to the resilience of this urban territory [9]. Ensuring accessibility to green open spaces is a possible response to a healthier urban environment [29]. In that perspective, the presence of green spaces, pedestrian paths, and leisure equipment was accomplished by redesigning former brownfields and residential plots.

(a) (b)

Figure 2. (**a**) Process of establishing the Kerekerdő park from the agglutination of empty plots and the demolition of residential buildings [14]; (**b**) Kerekerdő park already in use in the early 2000s. The public space was designed to prioritize the integration of the territory and promote accessibility to green areas and playgrounds [14].

3.3. Supervised Image Classification for Tracking Urban Transformations

As shown in Figure 3, the supervised classification performed in the satellite image of the 2000s reveals a territory still lacking a comprehensive urban green infrastructure. The green elements are presented in a fragmented and diluted way, evidencing the typological characteristics of a historical urban fabric ascended from closed blocks and buildings developed around relatively narrow and poorly lit courtyards, devoid of green components capable of significantly impacting the landscape composition and the quality of life for local inhabitants [30].

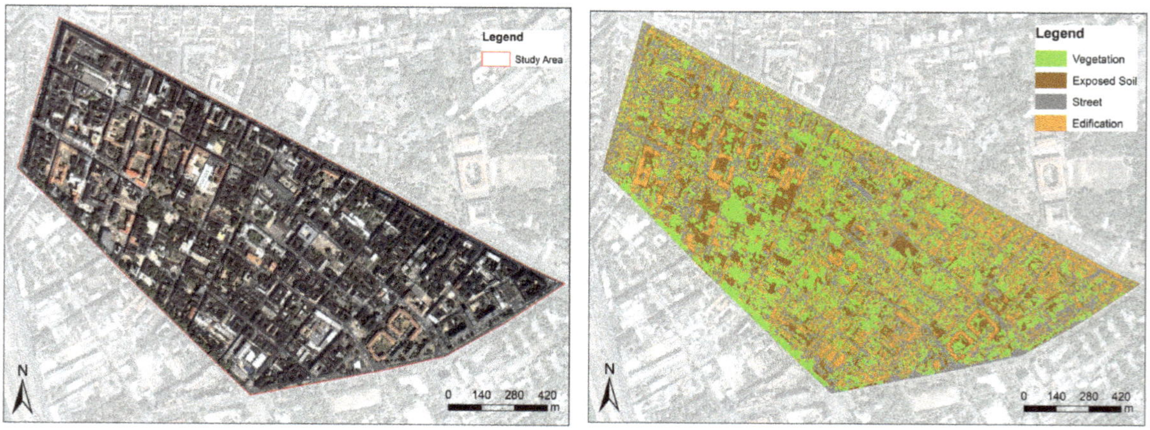

(a) (b)

Figure 3. (**a**) Satellite image from the 2000s. It is possible to notice that most buildings have typological characteristics common to historic structures, which reverberate in the dense urban condition, despite the existent industrial voids.; (**b**) The supervised classification reveals urban fragmentation and the deficiency of green infrastructure in the area.

There are also high rates of exposed soil, indicating, at this time, the prevalence of urban voids characteristic of deactivated industrial facilities [12]. These idle areas were mainly concentrated in the central portion of the study polygonal since the edges (better served by the public transport network and important mobility hubs), already in this period, were mostly occupied by residential and commercial buildings.

In 2011, it was already possible to visualize the grid of the green infrastructure defined in the urban requalification project, as indicated in Figure 4. The reduction in exposed soil areas is notorious, as are the growth of areas occupied by buildings. This period is marked by the coexistence of several new residential developments with old historic buildings in a poor state of conservation, designated for demolition, but most of which had not yet been demolished or rebuilt—except for a set of buildings located in the vicinity of Kerekerdő park [31], in the south–central region of the polygonal. In this area, there is a marked predominance of exposed soil.

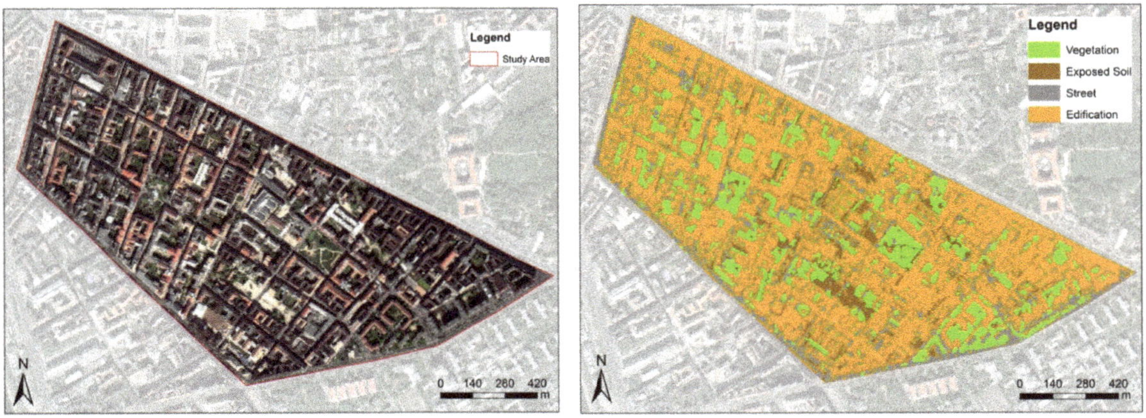

(a) (b)

Figure 4. (**a**) Satellite image from 2011. It is clear that the intervention has already reached a certain degree of consolidation, and the redesign of some urban blocks has already been completed, especially in the northern region; (**b**) The supervised classification indicates the first significant advances toward urban cohesion and green infrastructure begins to emerge.

Despite the prominent involvement of the private sector in the reconstruction process, this was a project guided by the demands arising from the local public administration. Contrary to most districts of Budapest after the political transition, the local government was able to continue the Middle Ferencváros renovation project, as this area was officially designated as an "urban rehabilitation site" [16]. For this reason, the stages of development of the intervention were defined primarily from the perspective of social conflicts, and aimed to address solutions to the most urgent infrastructural deficiencies. The city designated the priority areas as "centers of gravity", and the green infrastructure would establish the articulation between them. The implementation of Kerekerdő park, for example, was one of the measures adopted to stimulate the development of one of the regions of the district with the most significant deficit in infrastructure, landscape conditions, and socio–spatial cohesion [31].

Based on the 2021 satellite image analysis, visible in Figure 5, it is possible to observe the consolidation of the greenspace system. The intensification of linear and compact green elements, implemented in previous stages of the intervention in the planned urban fabric, is noted. At the same time, simultaneously, new shared courtyards were also established. At this stage, the proportion between green components and built area is more balanced than in previous stages of the requalification process, with approximately twice as many areas occupied by buildings as green areas, as indicated in Table 2.

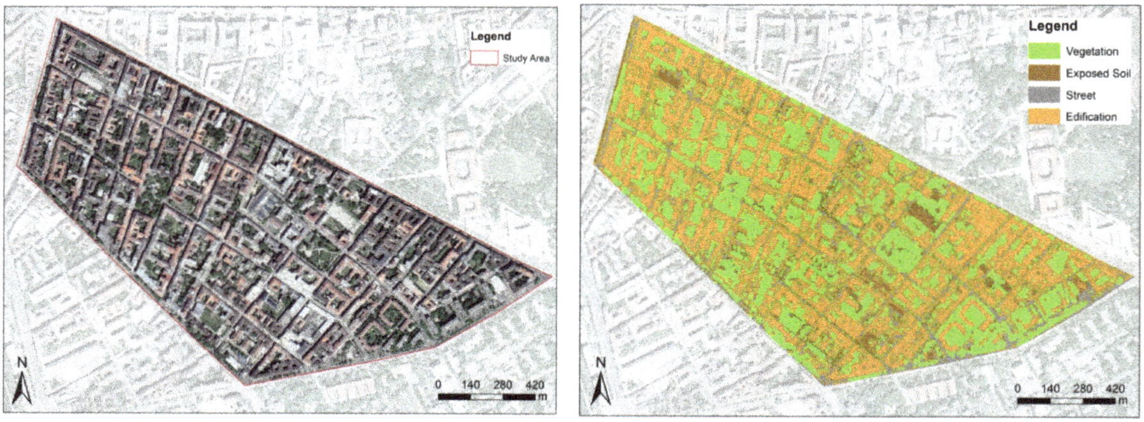

(a) (b)

Figure 5. (**a**) Satellite image of 2021. Based on the morphological analysis, it is clear that the intervention has reached its stage of maturity; (**b**) implementation of green infrastructure is almost complete, and there are few areas marked as exposed soil.

Table 2. The total area occupied by each of the respective classes over time. Results were obtained with supervised classification.

Class Name	2000	2011	2021
Exposed soil	9,193,263	3,964,979	5,005,173
Street	26,787,694	13,280,933	11,864,784
Building	25,055,177	51,467,673	41,295,866
Vegetation	17,627,973	10,004,571	20,546,853

The green areas are also more equidistantly distributed throughout the territory, bringing the conditions currently existing in the district closer to the parameters established by the World Health Organization concerning accessibility to urban public green areas [26]. However, there is still a denser strip of occupation where the urban green infrastructure system still did not reach its full potential, east of Kerekerdő park. This is one of the last "centers of gravity" among the priority areas for the requalification implementation. There is still a relatively high percentage of historic buildings that have not undergone renovation or demolition interventions in this area.

4. Discussion

Changes in the social–political context of post-socialist cities caused challenges in terms of their ability to persevere as resilient urban forms, leading them to the need to adapt to the new model of social and spatial structuring. In this context, international public policies, mediated by the European Union, were designed to assist in the development of these cities, aiming at cohesion and reducing the gap—especially in terms of infrastructure and environmental management—that exists between eastern and western European cities [6].

Parallel to these events, the dynamics of suburbanization and urban sprawl also guided cities such as Budapest to a scenario of social and physical fragmentation (also in terms of landscape conditions), particularly on the fringes of large metropolitan areas such as the situation found in Ferencváros [32]. In this sense, green infrastructure is an essential component in the viability of sustainable urban planning, facilitating social cohesion and supportive social networks, enhancing equity, and the development of social capital and promoting a healthier environment [33].

The partial or total demolition of some historic buildings during the renovation process raises questions about urban heritage preservation and urban morphology. In that regard, Kropf (2018) [34] states the following:

> *"Configuration is an arrangement of parts, and a type is a configuration with a degree of modularity and integration as a cultural habit. The type is a configuration that is or has been actively reproduced. While each example of a type might be slightly different, the configuration remains the same."*

Following that perception, demolition performed in some specific points of the district and the reconfiguration of the historic courtyards, despite being antagonistic to the preservation of urban heritage and its morphological characteristics, allowed for the revival of the area's configuration. This process happened through the typological re-constitution, providing cohesion to the fragmented urban territory and boosting its resilience [35].

The results reported in this paper show that the urban requalification of Ferencváros was efficient in improving territorial cohesion through a gradual reduction in urban voids (represented mainly by the exposed soil class) and the implementation of an extensive system of green elements. It is relevant to underline that the year 2011, as an intermediate period of evolution of the urban condition, presents classes with percentages of land use and occupation different from the general trend seen between 2000 and 2021 (Table 2). This movement is justified by the fact that this was a transition period, pointing to the coexistence of sections already restructured with others where the action was still in progress. The co-consolidation of the transformations was only confirmed in the outcome of the investigation carried out in the most recent image.

Furthermore, the supervised classification of satellite images has proved to be a robust method for analyzing the gradual reconstruction activity. The LULC maps produced with the help of GIS tools provide a detailed report on the development of the different layers that combined composed the urban fabric over time in the narrowly selected periods [36]. Nevertheless, the relatively low resolution of the images used for the study resulted in higher misclassification rates because of the spectral similarity, mainly between the exposed soil and the building materials. The use of higher resolution images (produced by Sentinel-2, for example) and the employment of Synthetic Aperture Radar (SAR) to extract textural features could be possible solutions to this issue [37].

As Hammerberg et al. (2018) demonstrated in Vienna, the use of remote-sensing imagery to determine morphological features on an urban scale can be beneficial for tracking urban structure and precisely evaluating the performance of cities at the environmental level [11]. The classes analyzed in the study proved to be substantially sufficient in the stratification of land-use information to capture the heterogeneity of the urban landscape morphology. Similar results were reported by Forget et al. (2018) through the supervised classification based on manually digitized training samples to frame the heterogeneity of sub-Saharan African urban areas. The automated approach could reach classification effectiveness similar to that of manual sampling strategies [37]. In this case study, the authors identified some inaccuracies related to the attributes of the openly available aerial images (difficulties also found in the study developed in Ferencváros).

The supervised classification as a method assisted in the correlation of the development strategies adopted by the public–private initiative with the remarkable historical events that defined the project's guidelines. Nevertheless, the impossibility of obtaining older images of free access conditioned the analysis to periods in which the intervention had already started, limiting the comparative framework concerning previous periods.

Author Contributions: Conceptualization, G.S.D.; methodology, G.S.D.; software, G.S.D. and P.B.N.; validation, G.S.D. and I.R.B.N.; formal analysis, G.S.D.; investigation, G.S.D.; resources G.S.D.; data curation, G.S.D.; writing—original draft preparation, G.S.D.; writing—review and editing, G.S.D.; visualization, G.S.D. and P.B.N.; supervision G.S.D. and I.R.B.N.; project administration, G.S.D.; funding acquisition, I.R.B.N. All authors have read and agreed to the published version of the manuscript.

Funding: This research received no external funding and the APC was funded by the Hungarian University of Agriculture and Life Science.

Informed Consent Statement: Not applicable.

Data Availability Statement: Publicly available datasets were analyzed in this study. This data can be found here: [https://earth.google.com/] (accessed on 10 March 2022).

Conflicts of Interest: The authors declare no conflict of interest.

References

1. Cotella, G. Chapter 7—The Urban Dimension of EU Cohesion Policy. In *Territorial Cohesion. The Urban Dimension*; Springer: Berlin/Heidelberg, Germany, 2019; pp. 133–151.
2. Di Palma, F.; Amato, F.; Nolè, G.; Martellozzo, F.; Murgante, B. A SMAP supervised classification of landsat images for urban sprawl evaluation. *ISPRS Int. J. Geo-Inf.* **2016**, *5*, 109. [CrossRef]
3. Richardson, H.W.; Bae, C.H.C. (Eds.) *Globalization and Urban Development*; Springer Science & Business Media: Berlin/Heidelberg, Germany, 2005.
4. Van Kempen, R.; Vermeulen, M.; Baan, A. (Eds.) *Urban Issues and Urban Policies in the New EU Countries*; Ashgate: Aldershot, UK, 2005.
5. Hirt, S.; Stanilov, K. *Twenty Years of Transition: The Evolution of Urban Planning in Eastern Europe and the Former Soviet Union, 1989–2009*; United Nations Human Settlements Programme: Nairobi, Kenya, 2009.
6. Hirt, S. The Post-Public City: Experiences from Post-socialist Europe. In *102nd Annual Conference of the Association of Collegiate Schools of Architecture*; Association of Collegiate Schools of Architecture: Washington, DC, USA, 2014.
7. Murray, M. Spatial Restructuring on a Global Scale: Enclave Urbanism and the Fragmentation of Urban Space. In *The Urbanism of Exception: The Dynamics of Global City Building in the Twenty-First Century*; Cambridge University Press: Cambridge, UK, 2017; pp. 93–146. [CrossRef]
8. Elkington, J. *Cannibals with Forks: The Triple Bottom Line of 21st Century*; Capstone: Oxford, UK, 1997; ISBN 1-900961-27-X.
9. Pauleit, S.; Ambrose-Oji, B.; Andersson, E.; Anton, B.; Buijs, A.; Haase, D.; Elands, B.; Hansen, R.; Kowarik, I.; Kronenberg, J.; et al. Advancing Urban Green Infrastructure in Europe: Outcomes and reflections from the GREEN SURGE project. *Urban For. Urban Green.* **2019**, *40*, 4–16. [CrossRef]
10. Lichtenberger, E. Urban decay and urban renewal in Budapest and Vienna, a comparison. *Acta Geograph. Lovan.* **1994**, *34*, 469–477.
11. Hammerberg, K.; Brousse, O.; Martilli, A.; Mahdavi, A. Implications of employing detailed urban canopy parameters for mesoscale climate modelling: A comparison between WUDAPT and GIS databases over Vienna, Austria. *Int. J. Climatol.* **2018**, *38*, e1241–e1257. [CrossRef]
12. Locsmándy, G. Városmegújítás a Ferencvárosban. In *Városrehabilitáció és Társadalom*; Tamás, E., Ed.; Hungarian Academy of Sciences: Budapest, Hungary, 2005.
13. Reed, B. Shifting from 'sustainability' to regeneration. *Build. Res. Inf.* **2007**, *35*, 674–680. [CrossRef]
14. Locsmándi, G. Large-Scale Restructuring Processes in the Urban Space of Budapest. In *Urban Models and Public-Private Partnership*; Springer: Berlin/Heidelberg, Germany, 2011; pp. 131–212.
15. Alhassan, V.; Henry, C.J.; Ramanna, S.; Storie, C. A deep learning framework for land-use/land-cover mapping and analysis using multispectral satellite imagery. *Neural Comput. Appl.* **2020**, *32*, 8529–8544. [CrossRef]
16. Oliveira, G.; Martins, V.; Fernandes Filho, E.I.; Xavier, F. Classificação supervisionada como ferramenta para avaliação da dinâmica do uso e cobertura do solo. In *Enciclopédia Biosfera*; Centro Cientifico Conhecer: Jadaia, Brazil, 2014; p. 10.
17. Chen, D.; Stow, D. The effect of training strategies on supervised classification at different spatial resolutions. *Photogramm. Eng. Remote Sens.* **2002**, *68*, 1155–1162.
18. Keuchel, J.; Naumann, S.; Heiler, M.; Siegmund, A. Automatic land cover analysis for Tenerife by supervised classification using remotely sensed data. *Remote Sens. Environ.* **2003**, *86*, 530–541. [CrossRef]
19. Levy, A. Urban morphology and the problem of the modern urban fabric: Some questions for research. *Urban Morphol.* **1999**, *3*, 79–85.
20. Sugár, V.; Talamon, A.; Horkai, A.; Kita, M. Assessment of rehabilitation possibilities in case of budapest jewish quarter building stock. *Int. J. Archit. Environ. Eng.* **2017**, *11*, 386–392.
21. Kocsis, J. Patterns of urban development in Budapest after 1989. *Hung. Stud.* **2015**, *29*, 3–20. [CrossRef]
22. Drahokoupil, J. *Globalization and the State in Central and Eastern Europe: The Politics of Foreign Direct Investment*; Routledge: Abingdon-on-Thames, UK, 2008.
23. Meerow, S.; Newell, J.P.; Stults, M. Defining urban resilience: A review. *Landsc. Urban Plan.* **2016**, *147*, 38–49. [CrossRef]
24. Kovacs, Z.; Wiessner, R.; Zischner, R. Urban Renewal in the Inner City of Budapest: Gentrification from a Post-socialist Perspective. *Urban Stud.* **2013**, *50*, 22–38. [CrossRef]
25. Pafi, M.; Siragusa, A.; Ferri, S.; Halkia, M. *Measuring the Accessibility of Urban Green Areas: A Comparison of the Green ESM with Other Datasets in Four European Cities*; JRC Technical Reports, European Commission; Publications Office of the European Union: Luxembourg, 2016.

26. WHO Regional Office for Europe. Second Meeting of the WHO City Action Group on Healthy Urban Planning. Healthy Urban Planning and Transport. In Proceedings of the WHO Meeting, Milan, Italy, 23–24 May 2002; WHO Regional Office for Europe: Copenhagen, Denmark, 2002.
27. Nordh, H.; Hartig, T.; Hagerhall, C.M.; Fry, G. Components of small urban parks that predict the possibility for restoration. *Urban For. Urban Green.* **2009**, *8*, 225–235. [CrossRef]
28. Tzoulas, K.; Korpela, K.; Venn, S.; Yli-Pelkonen, V.; Kaźmierczak, A.; Niemela, J.; James, P. Promoting ecosystem and human health in urban areas using Green Infrastructure: A literature review. *Landsc. Urban Plan.* **2007**, *81*, 167–178. [CrossRef]
29. Kaplan, S. The restorative benefits of nature: Toward an integrative framework. *J. Environ. Psychol.* **1995**, *15*, 169–182. [CrossRef]
30. Benkő, M.; Germán, T. Crime prevention aspects of public space renewal in Budapest. *J. Place Manag. Dev.* **2016**, *9*, 191–209. [CrossRef]
31. Balogh, P.I.; Takács, D. The significance of urban open spaces and green areas in urban property developments. In *First International Conference "Horticulture and Landscape Architecture in Transylvania" Agriculture and Environment Supplement*; Sapientia Hungarian University of Transylvania: Cluj-Napoca, Romania, 2011; Volume 110, p. 121.
32. McMahon, E.T.; Benedict, M.A. Green infrastructure. *Plan. Comm. J.* **2000**, *37*, 4–7.
33. Barton, H.; Grant, M.; Mitcham, C.; Tsourou, C. Healthy urban planning in European cities. *Health Promot. Int.* **2009**, *24* (Suppl. S1), i91–i99. [CrossRef]
34. Kropf, K. *The Handbook of Urban Morphology*; Wiley: Chichester, UK, 2017.
35. Aksözen, M.; Hassler, U.; Kohler, N. Reconstitution of the dynamics of an urban building stock. *Build. Res. Inf.* **2017**, *45*, 239–258. [CrossRef]
36. Rwanga, S.S.; Ndambuki, J.M. Accuracy assessment of land use/land cover classification using remote sensing and GIS. *Int. J. Geosci.* **2017**, *8*, 611–622. [CrossRef]
37. Forget, Y.; Linard, C.; Gilbert, M. Supervised classification of built-up areas in Sub-Saharan African cities using Landsat imagery and openstreetmap. *Remote Sens.* **2018**, *10*, 1145. [CrossRef]

land

MDPI

Article

Vernacular Heritage as a Response to Climate: Lessons for Future Climate Resilience from Rize, Turkey

Gül Aktürk [1,*] and Hannah Fluck [2]

1 Department of Architecture, Delft University of Technology (TU Delft), 2628 BL Delft, The Netherlands
2 Historic England, The Engine House, Swindon SN2 2EH, UK; hannah.fluck@historicengland.org.uk
* Correspondence: g.akturk@tudelft.nl; Tel.: +31-(0)617285407

Abstract: Vernacular heritage is undergoing rapid changes caused by the effects of the changing climate, such as loss of lands, biodiversity, building materials, integrity, traditional knowledge, and maladaptation. However, little is known about the causes of deterioration in vernacular heritage sites under changing climate and landscape conditions from a user perspective. This paper provides insights into the perceptions of local people on climate change and how it has changed the landscape in the Fındıklı district of Rize in the Eastern Black Sea area of Turkey. The study proposed analyzing vernacular architecture as a heritage category for localizing the management of climate change impacts using field survey, on-site observations, and unstructured interviews with local people. The results of the shared concerns regarding the changing climate and landscapes from a local perspective evoke the use of narratives as a tool for local authorities to include local communities in building resilience of cultural heritage to climate change.

Keywords: climate resilience; vernacular heritage; climate narratives; climate adaptation; climate stories

Citation: Aktürk, G.; Fluck, H. Vernacular Heritage as a Response to Climate: Lessons for Future Climate Resilience from Rize, Turkey. *Land* **2022**, *11*, 276. https://doi.org/10.3390/land11020276

Academic Editors: Jan K. Kazak, Katarzyna Hodor and Magdalena Wilkosz-Mamcarczyk

Received: 17 January 2022
Accepted: 9 February 2022
Published: 11 February 2022

1. Introduction

Though vernacular is a linguistic term, it represents the "architectural language of the people with its ethnic, regional, and local dialects" [1]. The simplicity of this folk, peasant, humble architecture is a product of local builders and artisans that extends beyond native construction materials, techniques, and details. Vernacular architecture responds to local conditions, particularly the climate, everyday living, crafts, and culture. Its legacy is represented by traditional houses from the pre-industrial period that were built by laypeople [2]. Although vernacular architecture is majorly prominent within its traditional building systems, the term "vernacular heritage" is used in this paper to describe its value as not only architectural heritage but also the value of its surrounding landscape and the intangible assets attached to it.

As a fundamental part of cultural landscapes, the vernacular efforts of buildings, bridges, countryside dwellings, farmhouses, cottages, mills, and kilns are recognized as vernacularly built heritage according to the ICOMOS Charter on Built Vernacular Heritage in 1999 [3]. Through this holistic approach, vernacular heritage helps us to understand traditional landscapes as a whole and encompasses buildings, land use and cover, vegetation, biodiversity, water sources, and artisanship.

Local climate has an inextricable role in forming vernacular heritage, as do other physical, social, economic, and cultural determinants. Rebuilt, restored, and adapted, vernacular settlements have evolved with the changing climate, cultural practices, community aspirations, and a gradual influx in modernization and urbanization. Reflecting our contemporary lifestyle in the construction technique hinders the preservation of local architectural identity and its valuable heritage [4]. As a result, successive generations have forgotten the symbiotic relationship between "people", "land", and "nature".

Vernacular heritage fosters a creative way for people to connect themselves with their culture, customs, and surroundings. The transmission of local knowledge, both formal and informal, helps generations to safeguard their vernacular heritage. Since the rise of industrialisation, the use of "modern" materials and techniques has become widespread in the maintenance of vernacular heritage [5]. Traditional construction systems were replaced by contemporary construction elements that are easier to access, quicker, and cheaper to build than vernacular buildings. Constructing vernacular buildings requires traditional knowledge, skills, and money to obtain once-surrounding materials such as timber and stone. Due to these constraints, local communities prefer the use of modern materials and techniques in the reconstruction of their vernacular buildings. However, these interventions are often poorly integrated into historic buildings [6] and cause rapid deterioration, including thermal discomfort [7]. In many cases, vernacular heritage is less energy-demanding and more environmental friendly than modern construction systems [8]. With today's environmental concerns, low-cost techniques and environmentally friendly and energy efficient solutions such as those used in vernacular buildings are in demand once again [9]. Although it is an underappreciated form of cultural heritage, it has been highly valued among local communities and local tourists. Recently, the return of people to their hometowns and the increasing interest in village and organic lifestyles has helped to boost local economies. Thus, the preservation of these buildings is significant for the flourishing of the local, city, and regional economies.

Vernacular heritage may have a role in responding to climate, but it is also vulnerable to its effects in a wider social, economic, and cultural context. Climate change multiplies the risks of disasters [10] on vernacular heritage [11] and its interdependencies on agriculture [12], infrastructure, livelihoods, forestry, and traditional knowledge with several other pressures from natural disasters [13] and anthropic interventions [2]. Abandonment [14], crop failures, desertification, maladaptation, and loss of knowledge and intangible values [15] are among the risks observed in vernacular landscapes [2]. Although local vernacular heritage may not be able to respond to present climatic conditions anymore, climate stories of various sites can disclose resilient places and communities for the future [16].

Scholars have covered vernacular architecture as a type of heritage under the concepts of sustainability and disaster-resiliency inspired by its construction in favor of the local context, e.g., [17]. However, the acceleration of climate change has increased the existing vulnerabilities of vernacular heritage. It has brought "climate resilience" of the existing settlements to the attention of practitioners and scholars. In this study, the term climate resilience is regarded as "the ability of a social or ecological system to absorb disturbances while retaining the same basic structure and ways of functioning, the capacity for self-organization and the capacity to adapt to stress and change" [18]. Vernacular heritage conveys traditional knowledge and practices to improve the resilience of local communities in the face of climate change-related disasters [19,20].

Local knowledge and skills are not only important sources for the proper maintenance of historic buildings [21], it is also important for building climate resiliency in both heritage sites and new settlements. It is argued that the climate resilience of heritage assets can foster both traditional preservation and transformation by allowing flexibility for the continuity of the place meaning and heritage values [22,23]. Furthermore, the integration of people's attitudes, behaviours, and values has been suggested as a more just and inclusive way to consider disaster risk management and climate resilience in local communities [24]. Systematic literature reviews focusing on climate change and cultural heritage reveal major gaps in facilitating "traditional knowledge for preservation and adaptation" and potentially for building climate resiliency [25,26]. Local knowledge and insights, which hold invaluable information regarding these low-cost adaptation strategies [27], are often underestimated in policy and practice [28]. In this regard, local environmental knowledge [29] together with scientific contributions such as climate modelling and predictions, may complement climate action and adaptation policies at the regional and national level.

The perceptions, insights, and statements of the locals living in these communities may highlight the know-how knowledge, failures, and success stories of vernacular heritage from past to present. Marcy Rockman stated that "Every place has a climate story" [30], emphasizing the importance of analyzing "narratives" and "stories" of local people. The place-based insights of local people can highlight how past societies interacted with their environment [30,31] to plan vernacular settlements and the present experiences of local communities due to changing climate and landscapes [32]. Communicating the local climate knowledge of the recent past and present impact of climate change on vernacular heritage can improve the response of local communities to future threats. However, there has been little research regarding the impacts of the changing climate and vernacular landscapes from a user perspective. Thus, there is a need to understand the local user perspective of how they emerged in relation to local climate, how they are changing due to changing climate, and the climate-resilient features of these settlements.

The case of Fındıklı in Rize, Turkey, illustrates vernacular heritage as the production of past climate knowledge as well as a legacy that is at risk due to climate change. The villages from major riverside areas are majorly prone to heavy rainfalls, floods, and landslides, among other risks. This study, using the field survey data from the selected site, aims to reveal the insights of local people to the localized impacts of climate change and their response to these changes in the vernacular heritage context. The results of the analysis of the narratives from local residents on climate change can be embedded into collaborative works by local stakeholder aiming to build climate resilience.

Literature: Local People's Knowledge and Perceptions on Changing Climate and Landscape

An overwhelming amount of literature focuses on the local perceptions of climate change and its associated risks as evidence for climate change vulnerability and impact assessments. The integration of this "insider knowledge" with climate modelling can further validate the results better during impact and adaptation studies [33]. The qualitative analysis of the perceptions that local people have regarding the changing climate in the case of the Peruvian Andes reveals that there is a conflict between local knowledge and scientific knowledge as well as the knowledge that is shared among the stakeholders involved [34]. The translation of local and indigenous knowledge is recognized as the "downscaling of climate adaptation" [34]. By uncovering their perceptions, insights, and experiences, indigenous and local communities reclaim their voice as it pertains to their heritage, landscape, and the global concern for the changing climate [35]. Uncovering local awareness of the effects of climate change and the analysis of these perceptions aims to bridge the gap between the stringent climate targets set by national policies and community-led local actions.

The analysis of local narratives [36] often brings up a complex relationship between climate change, landscape changes, and livelihoods [37], though it is not specifically mentioned. For instance, in the context of rural communities, the respondents of a survey in southeastern Zimbabwe reveal landscape changes occurring at the local scale due to forest and vegetation losses [38]. While heritage loss is inevitable and recognized by local communities [39], a great deal of local knowledge regarding the specific context of these heritage sites and their growing transformation may serve as an invaluable source for building climate resiliency [40]. Documentation covering changes in the clime in landscape are often published from the lens of local stakeholders [41] and should not be overlooked when generating knowledge on how to build climate resiliency [42]. Thus, it is particularly important to understand how heritage assets are managed, monitored, and interpreted along with their surroundings [43]. Embracing diverse voices, stories, memories, and narratives on the challenges that heritage assets face can activate local participation in terms of strengthening adaptation efforts [44].

2. Methodology

2.1. Case Study Area

The focus of the present study is Fındıklı, a small coastal town and district in the province of Rize in the Eastern Black Sea region of Turkey (Figure 1). This area is particularly relevant when considering the relationship between climate resilience and vernacular heritage for two reasons: (1) the province of Rize experiences frequent rainfall, severe floods, and landslides; (2) Fındıklı is rich in vernacular heritage [2].

Figure 1. The map produced using ArcGIS software shows the location of the Fındıklı district and boundaries of the city of Rize. Source: Aktürk, G., and Hauser, S. J. (2021).

The province of Rize has the highest annual total rainfall in Turkey, receiving over 2300 mm each year [45]. The annual relative humidity level in the province was higher than 78 percent in 2018, and therefore, there have no drought months reported (Figure 2). In the future, it is expected that the frequency of floods and landslides will increase following excessive rainfall events, some of which may exceed 250 mm, such as those that took place between 2016–2018 (Figure 3), due to the acceleration of the impacts of climate change.

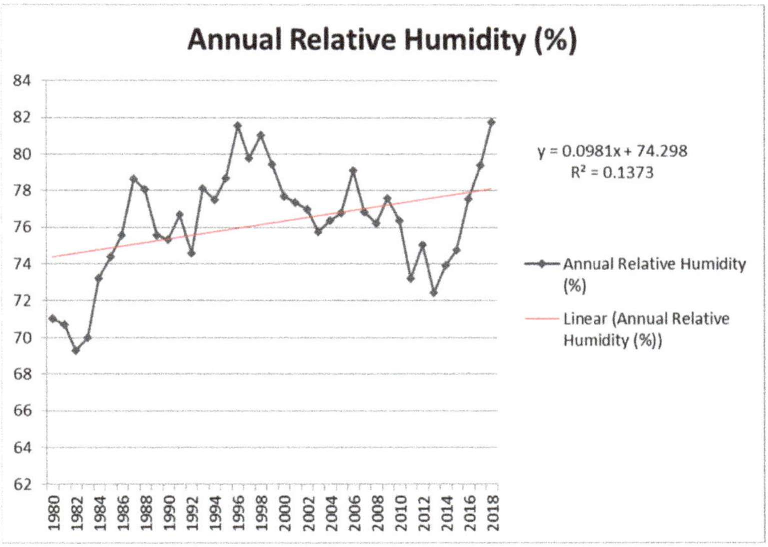

Figure 2. Annual relative humidity (%) in the city of Rize between 1980 and 2018 from 17,040 stations in the city centre, which is at latitude 41.0400, longitude 40.5013, and altitude 3 m. The monthly data were obtained from the 11th Regional Directorate of Meteorology in the province of Trabzon and were analyzed annually by the first author.

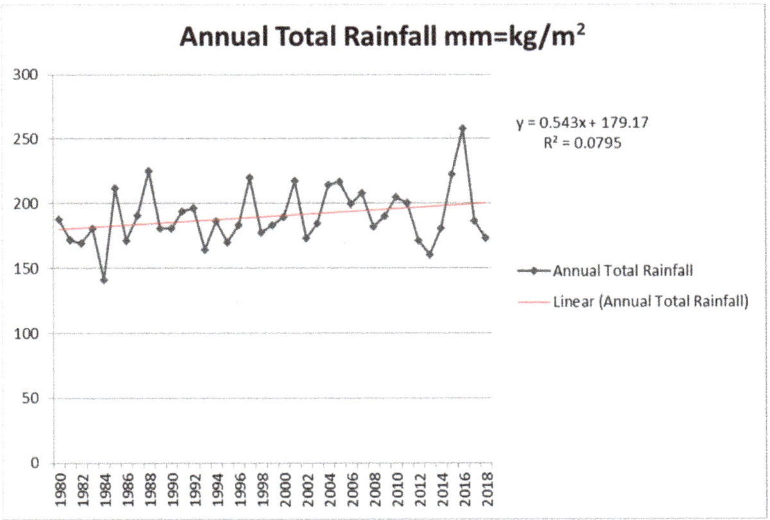

Figure 3. Annual total rainfall (mm = kg/m^2) in the city of Rize between 1980 and 2018 17,040 stations in the city centre, which is latitude 41.0400, longitude 40.5013, and altitude 3 m. The monthly data were obtained from the 11th Regional Directorate of Meteorology in the province of Trabzon and were analyzed annually by the first author.

The regional climate varies greatly from the coast to the inland areas of the region. The coast has a typical oceanic climate, with high humidity and rainfall. There area has a mild climate, with warm, humid summers and cool and damp winters. In contrast, there is a transition from an oceanic to continental climate in the hinterlands. Summers are warm and dry, whereas the winters are cold and humid.

The Fındıklı district in Rize has a population of 16,678 [46] and covers an area of 409 km^2 [47]. Due to its narrow coastal strip, settlements are widespread in the hinterlands as well as in the valleys and expand towards the Kaçkar mountains. Fındıklı has 23 villages and 8 neighbourhoods. The Sümer, Arılı, and Çağlayan rivers have shaped the plains and valleys. Most of the vernacular heritage is concentrated along the Çağlayan and Arılı valleys. The topography of the selected villages of Çağlayan, Beydere, Hara, and Gürsu varies (Figure 4). For instance, the altitude of Çağlayan village is 252 m, whereas the village of Beydere has the highest elevation, with an altitude of 508 m [48].

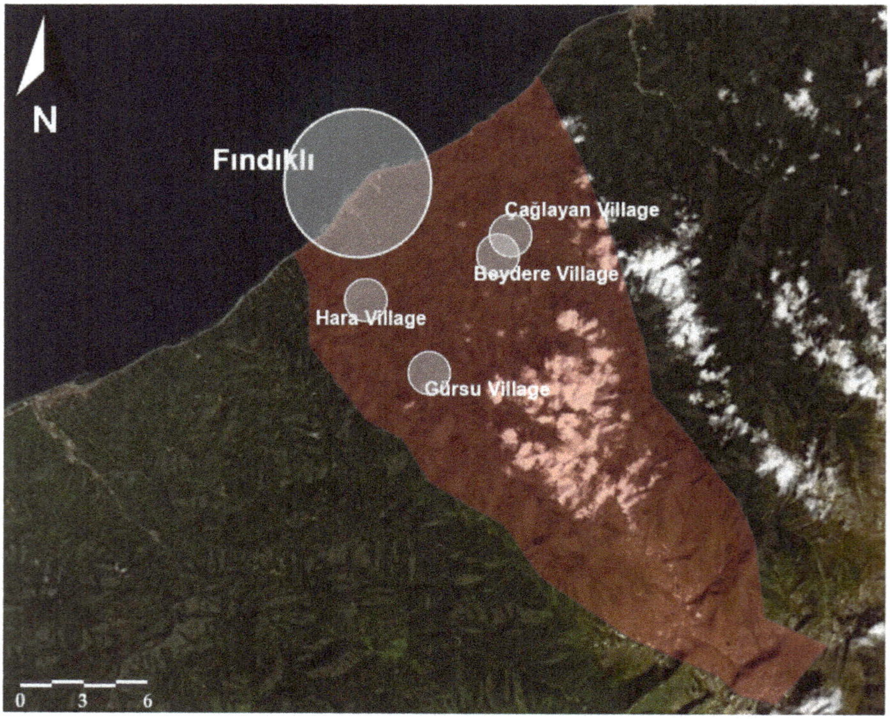

Figure 4. The boundaries of the Fındıklı district and the locations of the interviews conducted in Fındıklı created in ArcGIS software Source: Created by the first author.

Villagers have access to farmlands, with the local economy being dependent agriculture and animal husbandry. Recently, due to financial help from European Union projects, the locals in Fındıklı have switched to organic farming, which is also encouraged by local institutions for farmers and female workers for the development of sustainable farming practices. The main cultivation activities include tea, hazelnut, corn crops as well as fruit growing, fisheries, and beekeeping [49]. Villagers store nuts, corn, and grains in timber cupboards in their storage houses, which stand on four to six timber pillars. Locals often search for additional sources of income in addition to harvesting tea, such as the production of kiwi, citrus fruits, and Concord grapes [50], though the changing climate has forced them to find other sources of agricultural products.

Aside from the topographic setting, the rivers in the Fındıklı district have also played a significant role in the formation of forests that are rich in spruce and chestnut trees, which were used by locals for the construction of houses in the hinterlands. Wood carving and timber artisanship have become traditional in this region due to lumbering, timber trading, and the availability of rich forests in the area. Wood carving is used for making boats and for building storage houses, decorations, and furniture [51]. Along with agricultural activities, locals in this area have acquired ironworking skills for making reap hooks, slings, and knives in order to collect the crops. Coppersmithing, weaving, stonemasonry, wattle-making, and quilting are among other types of traditional knowledge that are at risk of being lost [2].

The vernacular landscapes in the area are shaped by the natural heritage of the region, e.g., Sarıçam [52] and Ihlamurlu [53] forestry; the tangible heritage of stone bridges, mills, fountains, mansions, and timber mosques [49]; and the intangible heritage of wooden carvings [51]. Site selection, the layout of pedestrian access, the location and orientation of houses, the design of building elements, and building typologies are closely connected to the local environmental context. In relation to the natural surroundings, the buildings were structured at a variety of scales, including grand mansions, village houses, and storage houses with fruit gardens and land holdings as a part of the cultural landscape. Privately owned small land holdings include building plots, arable land, and livestock as part of the rural landscapes.

Two main types of construction: stacking and framing, have emerged from the local needs applied to different contexts [54,55]. While timber and stone masonry systems are under the first category, the latter is the main focus of this article, which comprises four different types of construction systems, including (1) timber (Figure 5a), (2) stone-infilled timber-framed (Figure 5b), (3) amulet filling (Figure 5d), and (4) mixed construction systems (Figure 5c) [56], though the stone-infilled timber-framed is the most common type of construction not only in this region, but also in Fındıklı. The main differences between these types are the filling system and frame warping [57].

Figure 5. (**a**) Timber buildings elevated on stone in Hara village; (**b**) stone-infilled building in Çağlayan village; and (**c**) mixed construction on a building in Hara village. The stone and stone-infilled technique on the ground floor and the plastered façade of the first floor is an example of the çakatura technique. Source: Photograph by the first author taken in July 2019. (**d**) Amulet-filled building in the district of Pazar in Rize. Source: https://karadeniz.gov.tr//konut-ahmet-sakir-efendi-konagi/#prettyPhoto (accessed on 8 February 2022).

Timber and masonry construction systems that do not use studs are common in the region in forested areas at high altitudes, providing flexibility for building disassembly and relocating to another location [58]. However, the overexploitation of timber in nearby forests has led locals to prefer stone-infilled timber-framed construction systems, also known as eye-filling [59], over timber buildings. A system that emerged after cell-filling houses, "muskali dolma", also known as amulet filling, is a timber-framed construction technique where houses are built with triangular-shaped timber frames filled with small pieces of stone and mortar. It is more durable to side forces compared to cell-filled buildings [60].

A more recent construction technique known as "çakatura" differs from "muskali dolma", as the façade is plastered when this technique is used [58]. This type of construction system begins to decay faster than other systems. The facade is plastered with lime to prevent the timber from being exposed to the air [61], but the plaster wears away due to rainfall, meaning that it can no longer protect the timber. When at least two of four systems are used in one building, it is known as mixed construction [59]. Although there are rare examples of buildings combing all of these construction methods, stone-infilled timber buildings are the most preferable.

Stone-infilled timber buildings are built with stones that are 5–6 cm in width and 23–25 cm in length that are then placed in timber frames. Mortar covers the spaces in between the frames and stones to weatherproof them and to prevent rain and wind from penetrating the building [62]. They are generally built as two floors, in which the ground is used as barn and the first floor is dedicated to the living space. The ground floor is located partially under the slope that covers half of the surface of the upper floor. However, Çaglayan village is an exception, as the topography is rather flat, meaning that the ground floor usually covers the entire area of the first floor. In some cases, there is a walking path between the slope and the rear façade of the building.

Due to thick snow cover in the past, the roofs of these buildings are supported with timber thrusts and beams. Although the roofs are not high-pitched, the roofs were built in a way to bear the weight of snow. "Hartama"—a type of traditional roof cladding—is often used, and this material is made of 1 cm thick timber from fir, spruce (Picea orientalis in Latin), or oak or trees 2 cm thick timber from chestnut trees [63]. The board-on-board roof is installed vertically. During roof construction, the four-ridge structure is preferable to three-ridged roofs due to their resistance to wind and snow. The roof eaves extend to 150 cm, particularly in Çağlayan village, to prevent the decay of the façade because of the rainfall. Because of the substantial space between the roof and the windows, the risk of receiving rainwater in the interior is relatively low.

These traditional houses have two entrances: one main entrance that is elevated with quarter-circle stone steps and another located on the opposite side next to the toilet. The main entrance usually faces the north or north-east direction to avoid strong winds from northwest and west, which usually bring precipitation. Therefore, the façade facing the prevalent wind direction is usually built with stone [64]. The other entrace was traditionally used by the landlord to clean himself up after working on the land before presenting himself to the household or the guests.

There are several implications of climate-induced disasters on vernacular heritage. To provide an example, Turkey is facing climate migration, particularly among seasonal farm workers, due to the slow-onset effects of climate change [65]. Climate change puts pressure on the depopulation of rural areas; as a result, rural populations are forced to leave their lands, food sources, and houses. Scarcity and loss of lands, water resources, and crops may lead to abandonment and, following that, to the deterioration of vernacular landscapes. Anthropic influences such as changes in land-use and -cover in the hinterlands in Rize aggravate climate-induced risks on vernacular landscapes along with the rapid urbanization and planning of hydro-electrical power plants.

2.1.1. Data Collection

This paper draws on a series of unstructured interviews with local people and on-site observation notes undertaken by the first author during January and July 2019 [51]. The first fieldwork in the selected case area was conducted for a week in January 2019, while a second site visit was carried out for two weeks in July 2019 [51]. In total, 16 unstructured interviews were conducted face to face with 14 individuals from four different villages and with one person from the administrative centre of the district, Ref. [51] as shown in Figure 4. As emphasized earlier, the reason behind the selection of these villages is that the majority of the vernacular sites are located in these areas.

The interviewees were selected using the snowball sampling technique [66]. The sample size was defined by the saturation point [67], which is described as "the point in coding when you find that no new codes occur in the data" [68]. In-depth interviews were carried out on site, mostly at the historic houses, and by walking through the lands. During the fieldwork, observational notes were taken based on guided tours around the sites.

Table 1 demonstrates that the interviewees were mainly senior citizens and included five females and nine males who aged from 40 to 84 years old (Table 1) [51]. The majority of the sample consisted of historic homeowners, as the aim of the present paper is to analyze the resilience of these buildings from the perception of its end users. Due to the low number of skilled artisans, only two artisans were selected for the interview process. Eight were retirees, and four used to be teachers. Interviews were held conversationally (rather than semi-structured) and were based on the narratives of local people, including their perceptions, stories, and concerns [69]. This provided flexibility and spontaneity to the interviewees, as it allowed them to choose the order and nature of the narrative. It also allows a deeper understanding of the reality and the situation. This method enabled locals from different educational, cultural, and socio-cultural backgrounds to participate in the study. The aim of this method is to understand the effects of climate change on vernacular heritage from the statements of local users, mainly those who own vernacular buildings. As occupants of these buildings, they experience the challenges of the managing of their lands, buildings, and crops firsthand.

Table 1. Characteristics of interviewees. Source: adapted from "Remembering traditional craftsmanship: conserving a heritage of woodworking" by Gül Aktürk.

Interviewee	Gender	Relevancy	Profession	Villages	No. of Interviews	Date
C.K.	Male	Homeowner	Retired/teacher	Hara	3	3 July 2019
F.H.	Female	Homeowner	Retired	Çağlayan	1	12 January 2019
G.A.	Female	Homeowner	Retired	Hara	1	3 July 2019
H.Ş.	Male	Homeowner	Retired/teacher	Çağlayan	1	30 June 2019
M.A.	Male	Farmer	Retired	Çağlayan	1	12 January 2019
Ş.S.	Male	Stone mason	Constructor	Fındıklı centre	1	14 January 2019
S.T.	Female	Homeowner	Housewife	Gürsu	1	2 July 2019
Ş.Ö.	Male	Homeowner	Land registry	Çağlayan	1	6 July 2019
S.Ş.	Female	Homeowner	Retired	Çağlayan	1	11 January 2019
Y.G.	Male	Homeowner	Retired	Çağlayan	1	5 July 2019
Y.Y.	Male	Project Manager	Teacher	Fındıklı centre	1	11 January 2019
A.S.	Male	Stone mason	Constructor	Beydere	1	11 January 2019
B.U.	Female	Homeowner	Retired/teacher	Çağlayan	1	10 January 2019
T.H.	Male	Homeowner	Retired	Gürsu	1	2 July 2019

The interviews ranged from 45 min to two hours [51]. All interviews were audio and video recorded after receiving permission from the interviewees, and they were then transcribed [51]. The transcriptions were translated by the first author from Turkish to English while maintaining the meaning of local terms [51].

2.1.2. Data Analysis

Qualitative data analysis software (Atlas.ti version 8) was used to interpret the data after the coding of each interview. The verbatim interview transcripts and field notes were analyzed through inductive or "data-driven" thematic coding as the list of codes derived from the explanations provided by the locals. The six steps of this method include (a) familiarizing oneself with the data, (b) generating the initial codes, (c) searching for themes, (d) reviewing themes, (e) defining themes, and (f) writing-up the findings [70]. In the choice of coding, attention is given to the themes or concepts, practices, and context. Key themes and codes were generated to track some of the repeated notions, experiences, and patterns in understanding the effects of climate change and its associated risks to vernacular heritage sites. The initial coding revealed that the statements follow a chronological order of the events. This resulted in the emergence of three specific themes, including (1) past stories, (2) current problems, and (3) future threats.

Following that, six codes were created based on the inductive analysis of the recurring issues and patterns that could be observed in the content. The data from the interviews were qualitatively and quantitatively analyzed. The number of mentions were quantified and provided in numbers (n) and percentages (%). Drawing on field surveys undertaken by the first author, we explored the resilience of vernacularly built heritage to the changing climate and how communities are responding to the changes affecting vernacular heritage. While the local challenges regarding vernacular heritage preservation are not only limited to climate change effects, the interviews were analyzed through the themes of the impacts that climate change has on vernacular heritage and its resilience to the changing climate for the purposes of this paper.

3. Results

This section analyzes three major themes and six codes. The findings highlight that the current problems (n = 33, 49%) are the most frequently mentioned themes (Table 2), whereas past knowledge (n = 17, 25%) and future threats (n = 17, 25%) were equally quoted. This is due to the loss of past climate stories by the successive generation.

Table 2. The table shows the themes, codes, example quotations, and number of quotations. Note that the number of quotations shows the number of times the quotes were mentioned.

Themes	Codes	Example Quotations	No. of Quotes
Past Stories	Past climate	"It is snowing less now compared to the past. In the past, the rain was more excessive, but there is still rain otherwise the tea crops would not grow."	9
	Past settlements	"In the past, they used to build a house above the cornfield . . . "	8
Current Problems	Flooding	"When the rivers flooded, no one is around."	14
	Landslide	"Here is a landslide zone."	16
	Rainfall	"But what rain, I have not seen anything like. It again rained like this last year."	3
Future threats	Climate resilience	"The native tiles are water-resistant and durable."	17

The knowledge of the past was equally phrased with climate-resilience (n = 17, 25%). Within the coding for the current problems, the least mentioned issue was rainfall (n = 3, 4%), whereas the code for climate resilience within future threats was mentioned the most (n = 17, 25%).

3.1. Past Knowledge

Past knowledge shapes the life cycle of a building which starts by choosing a site, followed by construction, and finally, the reuse of the building. In the past, Fındıklı was not preferred for settlement, particularly the shoreline of the district. Despite of the marine culture at the coastline, residing at the seaside was not common. Regarding the coastal settlements, Y.Y.—the project manager for the EU-funded project "Training Masters for Rural Built Heritage in the Eastern Black Sea Region"— emphasized that "there was a risk of malaria due to a bite of a swamp mosquito as it was a wetland in the past." Referring to the 1600s, Y.G. reflected on that:

> *"Back then, Fındıklı was a swamp. In other words, no one would have settled in the city centre due to the mosquitoes and swamps ... Our grandfather ... came here ... built his home there ... on the hill across here ... "*

The rural settlements and village houses along the river valleys are scattered and are mostly located on top of the slopes with long and narrow strips of land comprising tea plants. The new settlements, which have been built below the cornfields, are more prone to floods due to their proximity to the rivers. On the contrary, historic buildings, which are located along the river valleys, were built on one and a half metres of foundation walls to elevate the house. C.K. noticed the role of animal husbandry and farmland in the site selection, which was repeated by Ş.S.:

> *"Here the houses are distant ... In the past, the mansions were built on top of the cornfield so that the rainwater carried the scat of animals down to the slope and fertilized the land."*

The reasons behind this settlement pattern are in close proximity to water sources, provide exposure to solar radiation in the morning, allow farmland to be manged easily, face the landscape. When dealing with the micro-climate, local communities were self-reliant when selecting building sites. Unlike today, strong attachment to the land indicated the management of private-owned lands at the time.

The past climatic conditions of the district according to the locals reveal that the snowfall and cover was greater than it is today. It appeared that the local climate also determined the land use, land tenure, and cropping pattern as well as the vernacular settlements, as T.H. noted:

> *"When I saw the first great snowfall in our village in 1948, there was three metres high snow cover. Two metres and a half. It did not snow in the last two years, including this year. When there is no snowfall, it does nothing good to agriculture."*

In terms of collective neighborhood relations, H.S. pointed out "1.5–2 m high snow would cover the village, and villagers would help each other out to shovel the snow on the roofs with the help of stairs." Because the roofs in this area are not high-pitched, it is questionable how the current three to four ridged low-pitched roof structures could carry the snow load. The timber beams are wide enough to carry the snow weight. Having experienced this firsthand, T.H. answered:

> *"Timber trusses hold the roof. In addition, there are timber beams that stand next to each other to hold the roof tiles. Poles under the trusses carry the full weight. I know with the help of the neighbor that the snow used to be cleared from the roof when the snow is very heavy. The neighbors would step in the clearance of snow from the rooftop."*

Past knowledge of the climate and settlements reveals a lot about the climate resilience of vernacular heritage to a changing landscape. However, the number of the statements on this theme indicates that the design of these buildings was not merely arbitrary. In response to the harsh climate, the locals managed their lands and settlements and worked as a community to deal with the challenges of it. Learning-by-doing and shared knowledge advanced the coping strategies of locals in the past.

3.2. Current Problems

More than half of the interviewees mentioned flooding, landslides, and extreme rainfall as major challenges induced by climate change, whereas the remaining interviewees linked these events to human intervention and natural disasters. Among the current problems (n = 33, 49%), landslides (n = 16, 24%) accounted for the most. Following that, flooding (n = 14, 21%) was mentioned. Issues related to rainfall were reported the least (n = 3, 4%).

3.2.1. Flooding

Flooding mostly happens in the months of March–April and May due to the melting of the snow on top of the hills and mountains. According to the statement of F.H., historic building owners do not stay in their houses when floods occur.

Ş.S. explained that deadliest flood occurred 20 years ago and destroyed a house in the village above, where a mother with two kids lived, was swept away. An 11-year-old boy who was dragged away in a flood in 2018 has still not been found. Ş.Ö. explained the life-threatening event:

> *"There was a flood 50 years or 45 years ago but the most serious one was in 2016 ... The damage in the Beydere village was more severe. A child was drowned."*

According to B.U., the Çağlayan river was similarly flooded, and the road to the village was closed. In the first visit to the field, the locals repeated that Beydere village was a landslide zone, but the observations during the second site trip revealed that even though the Çağlayan village was still at a relatively low risk, there is now a higher risk of the river flooding. Today, there is a flood defense constructed by the Hydraulic Works team along the Çağlayan river, although the area is relatively safer compared to the other villages. On the positioning of Caglayan village as a disaster area, H.Ş. described:

> *"The land facing the rear façade eroded twice. We were affected by it. It eroded two years ago, and the rear façade degraded twice. North facade ... There are a lot of landslides happening here and this why this area of Caglayan village is announced as a landslide zone ... An eanormous storage house has gone down below. There are a few storage houses that went as such, some houses in Aslandere and Beydere villages were damaged by floods ... The upper villages were affected more severely."*

Aside from the mansions, the storage houses, which are made of a lighter material, drifted away in the floods and were destroyed by landslides. According to the statements of Y.G. and Ş.S., the floods caused the collapse of the biggest storage house in the village with a double door in the region. M.A. detected the loss of vernacular heritage:

> *"There was a beautiful old storage house on the side. It [referring to the flooding] damaged the storage house too ... They came to us while they [referring to the owner of the storage house] escaped from the disaster, they sheltered here ... "*

River flooding occurs frequently in these villages, thus putting rural vernacular buildings at high risk. It is also noted that the risk classification of the villages may change over time along with the severity of the events. The number of the areas that are at risk of flooding continues to increase and the homeowners are not prepared for the anticipated risks of flooding.

3.2.2. Landslides

Beydere village, which has a more elevated topography, concurrently appeared in the mentions of locals as the area that has been affected by floods and landslides the most, resulting it being named a disaster region due to the frequency of these events. There are very few original remnants of these historic buildings in this village, and the remaining ones have been extensively modified (Figure 6). Following statements on Beydere village, C.K. mentioned Karaali village as a landslide zone where historic buildings were damaged.

Figure 6. The hazard area of Beydere village with mostly new settlements. Source: Photograph by the first author.

Public places, such as the school in the village of Çağlayan, was eroded by the landslide in 2018 according to B.U., and the land was detached. Ş.Ö. recognized the damage of it on his site:

"The land there was eroded. It shook the storage house but didn't demolish it. However, this time, it destroyed the trees ... There was a landslide at the back of my house, even though it is not a tea field ... "

However, even though most of the houses survived these disasters, the rear facades of some of the houses have become degraded due to landslides (Figure 7). Particularly, in the case of the house of Ş.Ö., the main door faces the rear façade, the location of the house that experiences the most damage during landslides. He reacted surprisingly:

"Couldn't these be thought of when these structures were built? The storage house is 250–200 years old why did they build it here? If I knew, I would dismantle it and rebuild it in another place ... I sometimes think whether I should change the position of the main door this way?"

Most of these buildings are not orientated to face the slopes where landslides pose a threat; however, the house of Ş.Ö. is an exception. This building was rebuilt with the remaining materials of a historic building that had been damaged by bombs during the Russian invasion of the region from 1916 to 1918. Afterwards, successive generations lost the original building knowledge and experiences of this house built 250 years before. The differences in generational construction practices present the future threats to local construction knowledge and experience with the local environment.

Among other changes, the locals observed that there are anthropic interventions to the landscape, which may have contributed to the frequency and occurrence and climate-risk disasters such as landslides. One major impact is deforestation for tea plantation, one of the favoured crops in the region. Forested areas with abundant trees prevent the risk of landslides and rockfalls. Rockfall is only referred to once by Y.H. Facing a slope of 70 degrees perpendicular to the rear façade of his house, this façade was built with stone to prevent the risk of rockfall rather than a landslide.

Figure 7. Site visit of a house in Çağlayan village in Fındıklı on 12th January in 2019. Maladaptation example of rebuilding the eradicated rear façade with briquette wall after the landslide in 2018. Source: Photograph by the first author.

Today, the rear façades of the village houses are surrounded by retaining walls. In the past, there was no need to construct a retaining wall behind these houses, as the walking pathway behind the buildings did not exceed 1 and a half metres. Together with the opening of the roads to vehicles between the houses and the slopes, villagers also cleared the slopes either for tea plantation or for the parking of personal vehicles through deforestation. This led to a gradual increase in the risk of landslides. In contrast, some locals have experienced landslides in the areas where there was no deforestation or tea plantations. M.A. recalled the event:

> *"Landslide happened this way. It is not because it rained a lot here. Here, a natural water discharge came and drained the water there. The excessive water maybe came from the sea as a hose . . . , so the rainwater damages something along the way. But in this case, there was no place to accumulate water."*

H.Ş. claimed that in the case of these events, the Disaster and Emergency Management Presidency (AFAD) helps communities to recover. They document and report the damage to the buildings. If the retaining wall is already built by the house owner, they do not receive reimbursement from them for it, meaning that they are not responsible for the construction of a retaining wall. This incentive is only for the preservation of historic buildings if they are damaged by landslides.

3.2.3. Rainfall

B.U. mentioned the local term of "rotten month" for the month of July, when the area receives excessive rainwater. This means that the area is rather damp; therefore, the month of July is known to be the rotten month. The dampness and changes in temperature also causes crop failures. As a part of this integrated historic environment, Y.G. noted that pears do not grow; similarly, H.Ş. mentioned that cherries used to grow in the past but that now only a few grow. Although some of the locals emphasized the effects of chemical fertilizers and the construction of dams as external factors, they could not deny the effects of the changing climate. In relation to the croplands, S.Ş. exemplified:

> *"Now even the crops are not growing. The environment is decaying. We do organic farming with pigeon manure, but the weather pattern has changed. For example, orange drops timelessly, but it should not fall from its tree so early. It has just matured. We produced the orange in 2017 but it is not ripening this year...But now the fruits do not mature enough and fall from the trees earlier."*

The locals who were born and raised in the area are well-aware of their local climate. However, incidents in the area have recently proven that climate change is showing its impacts more seriously. Ş.Ö. echoed the event:

> *"We took the car and waited inside the car in the school garden. Rain falls, then the rain level goes up to 25 cm as if a movie is directed ... It was raining in the past too, but I have not seen anything like that. Now that we experienced it, we are afraid."*

C.K.'s house experienced degradation due to rainfall, and the eaves of the roof could not protect the façade from the rainwater. The roof eaves extend to 150 cm in the Çağlayan village, but his is house located in Hara village and has 100 cm eaves, which were hit by the rainwater, leading to the decay of the facades of the upper floor.

As it is closely related to the local climate conditions, the severity and frequency of rainfall are not observed closely by the locals. The locals emphasized the damage of extreme rainfall on the historic houses that had been abandoned. If historic buildings are neglected by the mansion owners, then a drop of water can destroy it according to C.K. Even though some of these historic buildings have been renovated, they are beginning to decay once again. He further suggested:

> *"Both this part of the façade and wooden windows are decayed ... Çakatura type of construction technique could not survive."*

3.3. Future Threats

The stories from the individuals in these areas highlight the severity of the issue in cultural heritage context that are underscored by climate policies. For instance, the Minister of Environment, Urban Planning, and Climate Change in Turkey, Murat Kurum, announced the Regional Climate Action Plan for the Black Sea region on 12 July 2019 [71]. He explained the active role of NGO's and universities in mitigating the effects climate change. The 15 actions to be taken concerning the city of Rize inter alia include several practices in the building sector. The most relevant to cultural heritage is the 13th article on encouraging the use of local materials in construction for climate resiliency [71]. One important implication of the decision to use local materials and techniques is a legal exemption from any type of fees or taxes during house construction [71].

The eaves of the historic buildings extend between 80 cm to 180 cm to keep the facade secure from precipitation and rainfall. Local roofing tiles, known as Ottoman tiles, make a resilient roof covering. Y.Y. indicated that the native tiles are more water-resistant and durable than modern replacements. Sheet, ondulin, and European tiles are not accepted for the restoration of the listed historic buildings. The native tiles were made slightly wider so that they can cover the roof more efficiently than any other types of tiles. One major difficulty is renovating the roofs of the village houses, as it costs more money.

The main stakeholders of disaster management only focus on post-recovery, whereas pre-disaster management falls on the shoulders of the locals. Governmental institutions such as the Hydraulic Works team reveal issues related to anthropic activities, whereas the locals provide more insight into climate change. Engaging with local issues such as the mismanagement and maladaptation of vernacular heritage under climate change can teach us a lot about tackling climate change in a broader sense.

4. Discussion and Conclusions

This paper examines climate as a factor that shapes climate change as a challenge to vernacular heritage sites by assessing the perceptions of local people in Fındıklı, Rize. The results reveal that landslides pose the greatest risk to vernacular heritage as well as flooding. Locals are not well-prepared for the increased damage caused by flooding, landslides, and rainfall. The findings from this study overlap with the results from those reporting damages on the landscape scale but reveal more lost and damaged sites [2].

Climate narratives and stories of indigenous practices enable access to local knowledge regarding climate resilience and can be taken as a methodological approach to identify the risks and damages on vernacular heritage globally. Narratives of climate change and heritage through the landscape can link different temporalities, people, and places to encourage more ground-up approaches when managing these sites [44]. Therefore, there is a need for more ethnographic and storytelling methods to identify local narratives on the present challenges related vernacular heritage and to incorporate them in climate change policies [72]. Much of the data mentioned by the members of the local communities in heritage sites could be gathered, sorted, translated, and analyzed to support the disaster risk management database. While the method used in this paper is a traditional ethnographic method used to collect the field survey data, digital and visual ethnography methods in the form of video, film, and photography also contribute to the field significantly.

The preservation of cultural heritage sites prioritizes the documentation of single buildings, and inventories lack oral stories, small-scale structures, and landscapes. Although building risk assessment provides significant information on possible decay, the vulnerability of the cultural heritage sites should be perceived in a larger management context. Consequently, cultural heritage sites cannot be disentangled from their setting. Oral stories of climate perceptions from local communities were found to be complementary to the spatial analysis of vulnerable heritage sites [2]. According to the result of a ArcGIS analysis in a similar study which discusses disaster prone-sites in the context of the case area, only 3 out of 58 vernacular heritage sites were located in disaster-prone areas [2]. These interviews highlighted more sites that have been damaged and lost during the floods and landslides. Therefore, the use of oral stories can assist climate risk assessments.

While climate-induced disasters can extend to regional, national, and global borders, disasters become much more localized relatively sooner than the risks imposed by climate change [73]. The Localized impacts of climate change may differ in various other geographical locations; however, there are similar patterns, challenges, and concerns over the local implications of the impacts of climate change. Thus, it is important to understand the impacts on heritage assets and knowledge at the local scale and how people are responding to these disasters. The use of the narrative approach when considering heritage sites in changing climates is applicable and transferable to other geographical locations.

Through apprenticeship, vernacular buildings and the building knowledge that has been transformed through observation, information by word, and replication can be handed down. The construction knowledge from the past shows generational differences in vernacular practices. Every generation interprets its own vernacular heritage in its own time in response to technological advancements, financial, social, cultural, and individual needs. In today's circumstances, changes in the local climate and environmental conditions should be systematically addressed for maintaining, preserving, and adapting the vernacular heritage.

Manufacturing and industrialisation made the traditional knowledge of vernacular landscapes less important even though it creates the strongest connection between nature and culture. Traditional craftsmanship and practical skills go hand in hand. When manufacturing imposes a threat on the availability of timber masters, the timber making skill will be lost. of the preferences that local people have for modern global practices over low-tech solutions in vernacular settlements degrade the authenticity of these buildings [74].

The knowledge of vernacular heritage goes beyond understanding the material, construction technique and extends to the culture and lifestyles of the users. The characteristics of these houses are their modularity, flexibility, adaptability, transformability, and reusability in an unfixed nature. The local knowledge of building-resilient constructions can reduce the effects of new exposures and disasters.

The documentation of vernacular heritage should not only focus on capturing the authentic state of the cultural asset but also the later interventions, which are essential in understanding the maladaptation or best practices of climate adaptation. Cultural heritage is not fixed; thus, heritage professionals should embrace constant change. The proactive (preventative) or anticipatory approaches in climate adaptation allows for more time to prepare for the effects of climate change. Reactive adaptation planning responses take place after disasters hit. Thus, it is often suggested that heritage professionals should move away from reactive to proactive planning during adaptation, increasing the climate-resiliency of historic buildings in the future [75].

Decision-making processes need a combination of top-down and bottom-up approaches. The cultural heritage knowledge network and its interdependencies relies on the engagement and inclusion of users, artisans, local administrators, and other beneficiaries. The investigation of indigenous practices from underrepresented communities will enable access to specific climate knowledge and adaptation practices. The value of this knowledge should not be undermined during the creation of climate policies, as they mention incentives for the use of local construction materials and techniques.

Author Contributions: Conceptualization, G.A. and H.F.; methodology, G.A.; formal analysis, G.A. and H.F.; investigation, G.A.; resources, G.A.; writing—original draft preparation, G.A.; writing—review and editing, H.F.; visualization, G.A. All authors have read and agreed to the published version of the manuscript.

Funding: This research received no external funding.

Institutional Review Board Statement: Not applicable.

Informed Consent Statement: Not applicable.

Conflicts of Interest: The authors declare no conflict of interest.

References

1. Oliver, P. *Built to Meet Needs: Cultural Issues in Vernacular Architecture*; Routledge: London, UK, 2006.
2. Aktürk, G.; Hauser, S.J. Detection of Disaster-Prone Vernacular Heritage Sites at District Scale: The Case of Fındıklı in Rize, Turkey. *Int. J. Disaster Risk Reduct.* **2021**, *58*, 102238. [CrossRef]
3. ICOMOS. *Charter On the Built Vernacular Heritage*; ICOMOS: Charenton-le-Pont, France, 1999.
4. Al-Mohannadi, A.; Furlan, R.; Major, M. A Cultural Heritage Framework for Preserving Qatari Vernacular Domestic Architecture. *Sustainability* **2020**, *12*, 7295. [CrossRef]
5. Tawayha, F.A.; Braganca, L.; Mateus, R. Contribution of the Vernacular Architecture to the Sustainability: A Comparative Study between the Contemporary Areas and the Old Quarter of a Mediterranean City. *Sustainability* **2019**, *11*, 896. [CrossRef]
6. Gil-Piqueras, T.; Rodríguez-Navarro, P. Tradition and Sustainability in Vernacular Architecture of Southeast Morocco. *Sustainability* **2021**, *13*, 684. [CrossRef]
7. Rashid, M.; Ara, D.R. Modernity in tradition: Reflections on building design and technology in the Asian vernacular. *Front. Archit. Res.* **2015**, *4*, 46–55. [CrossRef]
8. Dabaieh, M.; Maguid, D.; El-Mahdy, D. Circularity in the New Gravity; Re-Thinking Vernacular Architecture and Circularity. *Sustainability* **2022**, *14*, 328. [CrossRef]
9. Patidar, S.; Raghuwanshi, B. Vernacular to modern in the search of sustainable development. *ITU J. Fac. Archit.* **2016**, *13*, 115–126. [CrossRef]
10. ICOMOS Climate Change and Cultural Heritage Working Group. *The Future of Our Pasts: Engaging Cultural Heritage in Climate Action*; ICOMOS: Paris, France, 2019.
11. Aktürk, G. The Rural Landscape as Heritage in Turkey Under Changing Climate//Le Paysage Rural Turque, un Patrimoine Soumis au Changement Climatique. In Proceedings of the Rural Heritage—Landscapes and Beyond/Patrimoine Rural: Paysages et au-delà, Marrakesh, Morocco, 17 October 2019; Volume 4. [CrossRef]
12. Cillis, G.; Statuto, D.; Picuno, P. Vernacular Farm Buildings and Rural Landscape: A Geospatial Approach for Their Integrated Management. *Sustainability* **2020**, *12*, 4. [CrossRef]
13. Shinde, K.A. Disruption, resilience, and vernacular heritage in an Indian city: Pune after the 1961 floods. *Urban Studies* **2016**, *54*, 382–398. [CrossRef]

14. Güler, K.; Kâhya, Y. Developing an approach for conservation of abandoned rural settlements in Turkey. *ITU J. Fac. Archit.* **2019**, *16*, 97–115. [CrossRef]
15. Aktürk, G.; Lerski, M. Intangible cultural heritage: A benefit to climate-displaced and host communities. *J. Environ. Stud. Sci.* **2021**, *11*, 305–315. [CrossRef]
16. Fluck, H.; Dawson, M. Climate Change and the Historic Environment. *Hist. Environ. Policy Pract.* **2021**, *12*, 263–270. [CrossRef]
17. Dipasquale, L.; Mecca, S.; Özel, B.; Ovalı, P.K. *Resilience of Vernacular Architecture. VERSUS: Heritage for Tomorrow, Vernacular Knowledge for Sustainable Architecture*; Firenze University Press: Firenze, Italy, 2014; pp. 65–73.
18. IPCC. Climate Change 2007: Impacts, Adaptation, and Vulnerability. *Contribution of Working Group II to the Fourth Assessment Report of the Intergovernmental Panel on Climate Change*. 2007. Available online: https://www.ipcc.ch/report/ar4/wg2/ (accessed on 6 June 2021).
19. UNESCO. Climate Change and World Heritage. Available online: https://whc.unesco.org/en/climatechange (accessed on 6 June 2021).
20. UNESCO. Vanuatu's Traditional Architecture Makes a Community More Resilient in the Face of Climate Change-Related Disasters. Available online: https://en.unesco.org/news/vanuatus-traditional-architecture-makes-community-more-resilient-face-climate-change-related (accessed on 24 June 2021).
21. Ahmadreza Shirvani, D.; Massimo, S.; Ilenia, P. Climate Change Challenges to Existing Cultural Heritage Policy. *Sustainability* **2019**, *11*, 5227. [CrossRef]
22. Seekamp, E.; Jo, E. Resilience and transformation of heritage sites to accommodate for loss and learning in a changing climate. *Clim. Chang.* **2020**, *162*, 41–55. [CrossRef]
23. Aktürk, G.; Dastgerdi, A.S. Cultural Landscapes under the Threat of Climate Change: A Systematic Study of Barriers to Resilience. *Sustainability* **2021**, *13*, 9974. [CrossRef]
24. Ghahramani, L.; McArdle, K.; Fatoric, S. Minority community resilience and cultural heritage preservation: A case study of the gullah geechee community. *Sustainability* **2020**, *12*, 2266. [CrossRef]
25. Fatoric, S.; Seekamp, E. Are Cultural Heritage and Resources Threatened by Climate Change? *A Systematic Literature Review. Clim. Change* **2017**, *142*, 227–254. [CrossRef]
26. Orr, S.A.; Richards, J.; Fatorić, S. Climate Change and Cultural Heritage: A Systematic Literature Review (2016–2020). *Hist. Environ. Policy Pract.* **2021**, *12*, 434–477. [CrossRef]
27. Jigyasu, R. Does Cultural Heritage Make More Resilient Cities? Available online: https://www.urbanet.info/does-cultural-heritage-make-more-resilient-cities (accessed on 6 June 2021).
28. Nakhaei, M. Facing Climate Change: The Importance of Protecting Earthen Heritage Traditional Knowledge. In Proceedings of the ICOMOS GA2020—6ISCs Joint Meeting, Sydney, Australia, 1–10 October 2020.
29. García, B.M. Resilient Cultural Heritage for a Future of Climate Change. *J. Int. Aff.* **2019**, *73*, 101–120.
30. U.S National Park Service. Cultural Resources Climate Change Strategy. 2016. Available online: https://www.nps.gov/subjects/climatechange/culturalresourcesstrategy.htm (accessed on 6 June 2021).
31. Graves, D.; Villano, E.; Cooper, C. Expanding the Narratives: How Stories of Our Past Can Help Inspire Our Response to the Climate Crisis. *Hist. Environ. Policy Pract.* **2021**, *12*, 292–312. [CrossRef]
32. Harvey, D.C.; Perry, J. *The Future of Heritage as Climates Change: Loss, Adaptation and Creativity*; Routledge: London, UK, 2015; p. 286.
33. Becken, S.; Lama, A.K.; Espiner, S. The cultural context of climate change impacts: Perceptions among community members in the Annapurna Conservation Area, Nepal. *Environ. Dev.* **2013**, *8*, 22–37. [CrossRef]
34. Weber, A.; Schmidt, M. Local perceptions, knowledge systems and communication problems around the climate change discourse—Examples from the Peruvian Andes. *Erdkunde* **2016**, *70*, 355–366. [CrossRef]
35. Lin, Y.-R.; Tomi, P.; Huang, H.; Lin, C.-H.; Chen, Y. Situating Indigenous Resilience: Climate Change and Tayal's "Millet Ark" Action in Taiwan. *Sustainability* **2020**, *12*, 10676. [CrossRef]
36. Aktürk, G. Learning from the Climate Narratives of Cultural Heritage. Available online: https://www.globalheritage.nl/news/learning-from-the-climate-narratives-of-cultural-heritage (accessed on 19 June 2021).
37. Lyons, I.; Hill, R.; Deshong, S.; Mooney, G.; Turpin, G. Protecting what is left after colonisation: Embedding climate adaptation planning in traditional owner narratives. *Geogr. Res.* **2020**, *58*, 34–48. [CrossRef]
38. Musakwa, W.; Mpofu, E.; Nyathi, N.A. Local Community Perceptions on Landscape Change, Ecosystem Services, Climate Change, and Livelihoods in Gonarezhou National Park, Zimbabwe. *Sustainability* **2020**, *12*, 4610. [CrossRef]
39. DeSilvey, C.; Harrison, R. Anticipating Loss: Rethinking Endangerment in Heritage Futures. *Int. J. Herit. Stud.* **2020**, *26*, 1–7. [CrossRef]
40. Holtorf, C. Averting loss aversion in cultural heritage. *Int. J. Herit. Stud.* **2015**, *21*, 405–421. [CrossRef]
41. Megarry, W.; Hadick, K. Lessons from the Edge: Assessing the impact and efficacy of digital technologies to stress urgency about climate change and cultural heritage globally. *Hist. Environ. Policy Pract.* **2021**, *12*, 336–355. [CrossRef]
42. Altschuler, B.; Brownlee, M. Perceptions of climate change on the Island of Providencia. *Local Environ.* **2016**, *21*, 615–635. [CrossRef]
43. Berenfeld, M.L. Planning for Permanent Emergency: "Triage" as a Strategy for Managing Cultural Resources threatened by Climate Change. 2015. Available online: http://www.georgewright.org/321berenfeld.pdf (accessed on 6 June 2021).

44. Ferraby, R.; Powlesland, D. Heritage and landscape change: Recording, archiving and engaging with photogrammetry on the Jurassic Coast World Heritage Site. *Proc. Geol. Assoc.* **2019**, *130*, 483–492. [CrossRef]
45. Rize İl Tarım ve Orman Müdürlüğü. Coğrafi Yapı. Available online: https://rize.tarimorman.gov.tr/Menu/12/Cografi-Yapi (accessed on 8 April 2020).
46. TUİK. ADKS Sonuçları. Available online: https://biruni.tuik.gov.tr/medas/?kn=95&locale=tr (accessed on 8 April 2020).
47. T.C. Fındıklı Kaymakamlığı. Fındıklı'nın Coğrafi Yapısı. Available online: http://www.findikli.gov.tr/cografi-konum-ve-temel-zellikler (accessed on 8 April 2020).
48. Beydere. Turkey Page. Available online: https://web.archive.org/web/20200218105717/http://www.fallingrain.com/world/TU/53/Beydere.html (accessed on 25 June 2021).
49. Fındıklı Kaymakamlığı. *Fındıklı*; Fındıklı Kaymakamlığı: Fındıklı, Rize, Turkey, 2017.
50. Korgavuş, B. Rize Merkez İlçgesi Kültürel Peyzaj Alanlarında Zamansal Değişimin Coğrafi Bilgi Sistemleri İle Belirlenmesi. *Artvin Çoruh Üniversitesi Orman Fakültesi Derg.* **2015**, *15*, 96. [CrossRef]
51. Aktürk, G. Remembering traditional craftsmanship: Conserving a heritage of woodworking in Rize, Turkey. *Int. J. Intang. Herit.* **2020**, *15*, 134–146.
52. Karadeniz Kultur Envanteri. "Sarıçam". Available online: https://karadeniz.gov.tr//saricam-7 (accessed on 16 June 2021).
53. Karadeniz Kultur Envanteri. Ihlamurlu. Available online: https://karadeniz.gov.tr//ihlamurlu (accessed on 16 June 2021).
54. Fidan, M.S.; Yaşar, Ş.Ş.; Mehmet Yaşar, E.A. Features Design and Traditional Architecture of the Turkish Eastern Black Sea House. *Mugla J. Sci. Technol.* **2016**, *2*, 24–29. [CrossRef]
55. Sümerkan, M.R. Building Characteristics of the Traditional Houses in Respect to Shaping Factors at Eastern Black Sea Region. Unpublished. Ph.D. Thesis, Karadeniz Technical University, Trabzon, Turkey, 1990.
56. Pınar, E. The Examination of Structural Details of Eastern Black Sea Region's Vernacular Architecture Int. *J. Soc. Sci.* **2018**, *6*, 61–74.
57. Dipasquale, L.; Mecca, S. Local Seismic Culture in Mediterrenean Region. In *Seismic Retrofitting: Learning from Vernacular Architecture*; Correia, M.R., Laurenço, P.B., Varum, H., Eds.; Taylor & Francis Group: London, UK, 2015.
58. Güler, K.; Bilge, A.C. Construction Techniques of the Vernacular Architecture of the Eastern Black Sea Region. In *International Conference on Vernacular Heritage and Earthen Architecture: Towards a Sustainable Future, Proceedings of the International Conference on Vernacular Heritage, Sustainability and Earthen Architecture Correia, Valencia, Spain, 11–13 September 2014*; Carlos, G.M., Rocha, S., Eds.; CRC Press: Valencia, Spain, 2014.
59. Eruzun, C. Ahşabın Kimlik Bulduğu Doğu Karadeniz Mimarisi. In Proceedings of the 5th Turkish Folklore Congress, Ankara, Turkey, 24–29 June 1996; pp. 175–182.
60. Şen, N. *Five Houses in Rize*; Fono Press: İstanbul, Turkey, 1967; Volume 53.
61. Eskiçırak, D. Doğu Karadeniz Bölgesi Geleneksel Konutlarının İyileştirilmesine Yönelik Yapım Sistemi Ve Malzeme Kullanımı Analizi—Örnek Konutların Mevcut Durum Değerlendirmesi. Master's Thesis, Istanbul Teknik Universites, Istanbul, Turkey, 2009.
62. Taha Toros Archives, Les maisons de la region orientale de la Mer Noire (In English: Architecture: Houses in the Eastern Black Sea Region of Anatolia). Available online: http://openaccess.marmara.edu.tr/handle/11424/169957 (accessed on 2 February 2022).
63. Özgüner, O. Village Architecture in the Eastern Black Sea Region. Ph.D. Thesis, Middle East Technical University, Ankara, Turkey, 1970.
64. Bayram, Ö.F. Doğu Karadeniz Bölgesinde Geçmişten Günümüze Vernaküler Mimari. Master's Thesis, Yıldız Teknik Üniversitesi, Istanbul, Turkey, 2014.
65. Čadež, T.; Hevia, M.H. Environmental Migration in Turkey: Challenges, Recognition and Implications for Policy. *Migr. Environ. Clim. Change Policy Brief Ser.* **2016**, *8*, 1–9.
66. Noy, C. Sampling Knowledge: The Hermeneutics of Snowball Sampling in Qualitative Research. *Int. J. Soc. Res. Methodol.* **2008**, *11*, 327–344. [CrossRef]
67. Matos, A.; Barraza, L.; Ruiz-Mallén, I. Linking Conservation, Community Knowledge, and Adaptation to Extreme Climatic Events: A Case Study in Gorongosa National Park, Mozambique. *Sustainability* **2021**, *13*, 6478. [CrossRef]
68. Urquhart, C. *Grounded Theory for Qualitative Research: A Practical Guide*; SAGE: Thousand Oaks, CA, USA, 2013.
69. Warrick, O. Local Voices, Local Choices? Vulnerability to Climate Change and Community-Based Adaptation in Rural Vanuatu. Ph.D. Thesis, University of Waikato, Waikato, New Zeeland, 2011.
70. Byrne, D. A worked example of Braun and Clarke's approach to reflexive thematic analysis. *Qual. Quant.* **2021**. [CrossRef]
71. Türkiye Cumhuriyeti Çevre ve Şehircilik Bakanlığı. *Karadeniz Bolgesi İklim Degisikligi Eylem Plani*; 2019; pp. 1–31. Available online: https://doi-org.tudelft.idm.oclc.org/10.1007/s11135-021-01182-y (accessed on 2 February 2022).
72. Krauß, W.; Bremer, S. The role of place-based narratives of change in climate risk governance. *Clim. Risk Manag.* **2020**, *28*, 100221. [CrossRef]
73. Jigyasu, R. Managing Cultural Heritage in the Face of Climate Change. *J. Int. Aff.* **2019**, *73*, 87–100. [CrossRef]
74. Fernandes, J.; Mateus, R.; Bragança, L. *The Potential of Vernacular Materials to the Sustainable Building Design*; Taylor and Francis: London, UK, 2013; pp. 623–629.
75. Elena, S.; Alexandre, S.G.; Chiara, B.; John, H. Adapting Cultural Heritage to Climate Change Risks: Perspectives of Cultural Heritage Experts in Europe. *Geosciences* **2018**, *8*, 305. [CrossRef]

land

Article

Scenario Analysis for Resilient Urban Green Infrastructure

Bo Li [1], Yue Wang [1], Tong Wang [2,*], Xiaoman He [1] and Jan K. Kazak [3]

1. School of Architecture and Art, Central South University, Changsha 410083, China
2. Faculty of Architecture and the Built Environment, Delft University of Technology, 2628 BL Delft, The Netherlands
3. Institute of Spatial Management, Wrocław University of Environmental and Life Sciences, 50-357 Wrocław, Poland

* Correspondence: t.wang-12@tudelft.nl

Abstract: With the advancement of urbanization, the stress on the green infrastructure around the urban agglomeration has intensified, which causes severe ecological problems. The uncertainty of urban growth makes it difficult to achieve effective protection only by setting protection red lines and other rigid measures. It is of practical significance to optimize the resilience of the stressed green infrastructure. To this end, we explore a scenario simulation analysis method for the resilience management of green infrastructure under stress. This research applies artificial neural network cellular automata to simulate the impacts of the Chang-Zhu-Tan urban agglomeration expansion on the green infrastructure in 2030 in three scenarios: no planning control, urban planning control, and ecological protection planning control. Based on the analysis, we identify four green infrastructure areas under stress and formulate resilience management measures, respectively. The results show that: (1) The distribution pattern of green infrastructure under stress is different in three scenarios. Even in the scenario of ecological protection planning and control, urban growth can easily break through the ecological protection boundary; (2) Residential, industrial, and traffic facility land are the main types of urban land causing green infrastructure stress, while forest, shrub, and wetland are the main types of the stressed green infrastructure; (3) Efficient protection of green infrastructure and the management of the urban growth boundary should be promoted by resilient management measures such as urban planning adjustment, regulatory detailed planning, development strength control and setting up the ecological protection facilities for the stressed green infrastructure areas of the planning scenarios and the no-planning control scenarios, for the areas to be occupied by urban land, and for the important ecological corridors. The results of this study provide an empirical foundation for formulating policies and the methods of this study can be applied to urban ecological planning and green infrastructure management practice in other areas as well.

Keywords: green infrastructure; resilience management; biodiversity; scenario analysis; cellular automata model

Citation: Li, B.; Wang, Y.; Wang, T.; He, X.; Kazak, J.K. Scenario Analysis for Resilient Urban Green Infrastructure. *Land* **2022**, *11*, 1481. https://doi.org/10.3390/land11091481

Academic Editor: Francisco J. Escobedo

Received: 29 July 2022
Accepted: 1 September 2022
Published: 4 September 2022

1. Introduction

According to the data from the seventh census of China and the national population development plan, the urbanization rate of Chinese permanent residents has reached 63.89% in 2020 and will rise to 70% in 2030, while the figure was only 36.09% in 2000. Each 1% increase in China's urbanization rate requires 3459 km^2 of construction land [1]. Urban agglomerations, as the key receiving areas for the urbanized population, are expanding faster than independent cities in general, and therefore, the green infrastructure around them will be under more serious stress. Green infrastructure refers to the natural ecosystem with important ecological value, including green space such as forest and grassland, and blue space such as wetland and water systems, designed and managed to deliver different kinds of ecosystem services [2–4]. The green infrastructure of urban agglomerations not only has various functions such as maintaining regional biodiversity, regulating climate, providing recreational services, and guaranteeing water supply, but also plays an important role in promoting urban transformation and enhancing urban development dynamics [5–8]. Because of the need for construction land in the urbanization process, the green infrastructure of urban agglomerations is often more vulnerable to stress [9,10]. The green infrastructure of three major urban agglomerations in China—Yangtze River Delta, Pearl River Delta, and Beijing-Tianjin-Hebei—decreased by 951 km^2, 97 km^2, and 212 km^2, respectively, from 1992 to 2010 [11]. In this regard, China's territorial spatial planning system puts forward the spatial control requirements of "three zones and three lines", which sets the urban development boundary, permanent basic agricultural land, and ecological protection red lines. However, such boundaries do not specify rigid space and flexible space. Therefore, it is of great practical significance to understand the dynamics of green infrastructure of urban agglomerations, and to protect the original function of green infrastructure while meeting the development of urban agglomerations with empirical analysis.

Green infrastructure management can be divided into two categories. One is focused on meeting urban development, mainly through the method of cellular automata simulation to delineate the urban-growth boundary (UGB), and the construction of green infrastructure outside the boundary is prohibited, as represented by Beijing, Huizhou, and other cities [12,13]. The other category takes green infrastructure protection as the starting point and delineates the green infrastructure protection boundary mainly through the evaluation results of ecological suitability, with Wuhan, Chengdu, and other cities as representatives [14].

Although the two management methods have different focuses, they both implement static rigid control of green infrastructure. While both cities and green infrastructure are complex network systems that are constantly changing dynamically, traditional static and rigid control measures can hardly solve the complex contradictory problems between urban development and green infrastructure protection [15,16]. This has led to the breakthrough of the red line of green infrastructure control. The core green infrastructure is difficult to hold, and different types of ecological protection red lines have been broken one after another.

A resilient urban green infrastructure can help to solve the problem of urban growth stressing green infrastructure, because of its adaption to a constantly changing environment dynamically. Resilience as a term was used by physicist Thomas Young in 1807 to describe elastic deformation in the context of materials science. While the traditional concept of resilience in ecology was used to describe the persistence of natural systems in response to changes in external elements and human factors, the concept of resilience has evolved to focus on "transformational capacity", i.e., the ability of ecosystems to change, adapt, and change in response to pressures and constraints. Ecological resilience, on the other hand, emphasizes the amount of disturbance that an ecosystem can withstand without changing its self-organizing processes and structure, and refers to the degree to which an ecosystem can adapt to change in the face of external pressures and before reforming a stable structural system [17]. The main purpose of optimizing the resilience of green infrastructure is to enhance the ability of green infrastructure to resist disturbances, cope with changes and adapt to changes through relevant control measures and strategies to achieve the harmonious development of urban-ecological systems [18].

Research on ecological resilience has focused on ecosystem resilience evaluation and construction [19–21], ecological resilience optimization [22,23], and the role of ecological resilience [24]. In China, the research on ecological resilience is still in its infancy, and the relevant studies mainly focus on the application [25,26] and the theory of ecological resilience [27] in the ecological design of resilience, and there is a relative lack of research on the optimization of ecosystem resilience. Ecological resilience has an important impact on improving the self-organization capacity of urban ecosystems [28] and plays an important role in promoting the coordination of multiple objectives between the complex systems of urban development and green infrastructure protection [15]. While studies on green infrastructure optimization do exist, they mainly focus on ecological spatial structure [29,30], governance control measures [31,32], and spatial networks [33,34]. There is a relative lack of research on optimization and guiding control strategies for rapidly developing urbanized areas, especially in the study of ecological spatial toughness control and optimization of urban agglomerations [35].

Moreover, the development process between urban growth and green infrastructure protection is less considered, and the relevant optimization measures are still for the improvement of static rigid control measures. Green infrastructure protection and urban growth are the results of the joint action of self-organization and other organizations [36], and the traditional research methods are unable to identify the dynamic diffusion process of the two types of space. Regarding this issue, the minimum cumulative resistance model (MCR) can well reflect the intrinsic linkage process of ecological processes [37], and the self-organizing kernel embedded in the cellular automata model (CA model) can simulate the spatio-temporal evolution process of urban expansion [38]. Additionally, both of these methods are widely applied in this field.

To summarize, the current rapid urban development in China has imposed great stress on green infrastructure which is essential for biodiversity, and the existing controlling measures have problems coping with the dynamic urban space and green infrastructure change process. Resilient control measures are desired to protect green infrastructure under uncertainty. Therefore, this research aims to formulate resilient strategies for the green infrastructure stressed by uncertain urban growth.

More specifically, to achieve this, the following research questions will be answered:

1. How can the dynamic urban growth stressing green infrastructure under uncertainty be presented?
2. What are the stressed areas of the green infrastructure and their characteristics accordingly?
3. What are the resilient strategies for the identified stressed green infrastructure based on their characteristics?

2. Context and Data

2.1. Context

Chang-Zhu-Tan urban agglomeration is located in the central-eastern part of Hunan Province, including Changsha, Zhuzhou, and Xiangtan cities, which are distributed in the shape of a triangle along Xiangjiang River, with less than 20 km between them. With the development of the urban agglomeration, the green infrastructure of Chang-Zhu-Tan urban agglomeration is seriously reduced [39,40], among which the area of important green infrastructure is reduced by construction land in Muyun town from 2004 to 2010 by 427.67 hm^2 [41]. At this stage, Changsha-Zhuzhou-Tan urban agglomeration is facing the development of urban integration. The pressure of green infrastructure around the urban agglomeration is further highlighted. Based on a collection of remote sensing images data, this study covers the main administrative areas of Changsha, Zhuzhou, and Xiangtan cities with a total area of 22,104.4 km^2, including most of the districts and counties under the jurisdiction of Changsha, Zhuzhou, and Xiangtan cities, but excluding a few areas such as Ningxiang City, Shaoshan City, the western part of Xiangxiang City, and Yanling County (Figure 1).

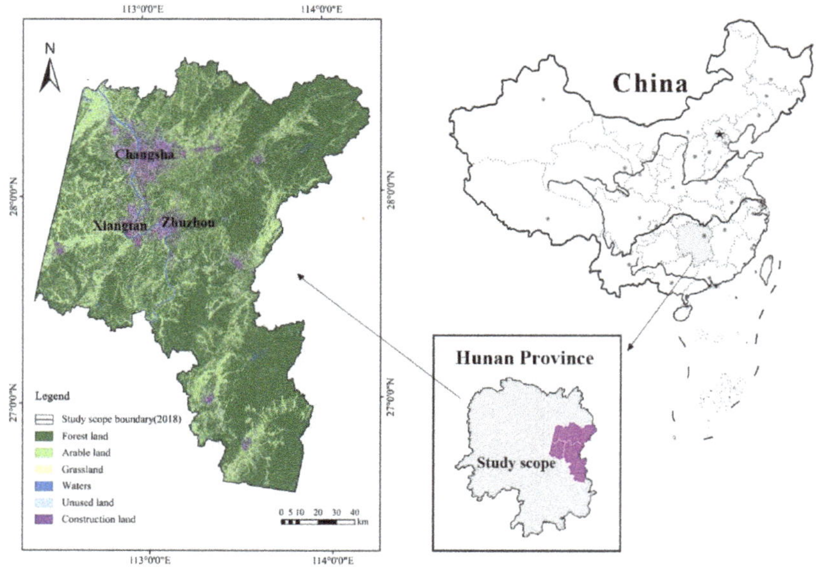

Figure 1. Location of the study area and its land-use map in 2018.

2.2. Data Sources

The data used include the following: 30 m resolution satellite image data, DEM data, and vegetation cover data of Chang-Zhu-Tan Urban agglomeration in 1999, 2008, and 2018 (source: Geospatial Data Cloud, http://www.gscloud.cn/ (accessed on 8 July 2020)), the Chang-Zhu-Tan Urban agglomeration Regional Plan (2008–2020) (adjusted in 2014) [42], the Master Plan of Ecological Green Heart Area of Chang-Zhu-Tan Urban agglomeration (2010–2030) (revised in 2018) [43], the Changsha City Urban Master Plan (2003–2020) (revised in 2014) [44], the Xiangtan City Urban Master Plan (2010–2020) (revised in 2017) [45], and the Zhuzhou City Urban Master Plan (2006–2020) (revised in 2017) [46]. Among them, the satellite image data of Chang-Zhu-Tan urban agglomeration in 1999, 2008, and 2018 were classified into arable land, forest land, grassland, water area, construction land, and unused land by using Erdas Imagine supervised interpretation function [47–49]. Additionally, the color difference of the surface coverage patches shown in different areas is combined with the Erdas Imagine classification to determine the main landmark elements of the geographical areas. The decoded data from the three years are then used to compare with satellite images using visual corrections to obtain the land-use data of the Chang-Zhu-Tan urban agglomeration. To reduce the number of operations, the final relevant land-use data are obtained by resampling in 100 m units with the GIS nearest neighbor assignment.

3. Methods

This research follows the following steps: Firstly, the green infrastructure of Chang-Zhu-Tan urban agglomeration is identified by using the principles of landscape ecology. Additionally, then the Cellular Automata (CA) model is used to simulate the green infrastructure in different control measure scenarios. Finally, the characteristics of the green infrastructure under stress are analyzed and the relevant resilience control and development strategies are formulated. The relevant methodological flow is shown in Figure 2.

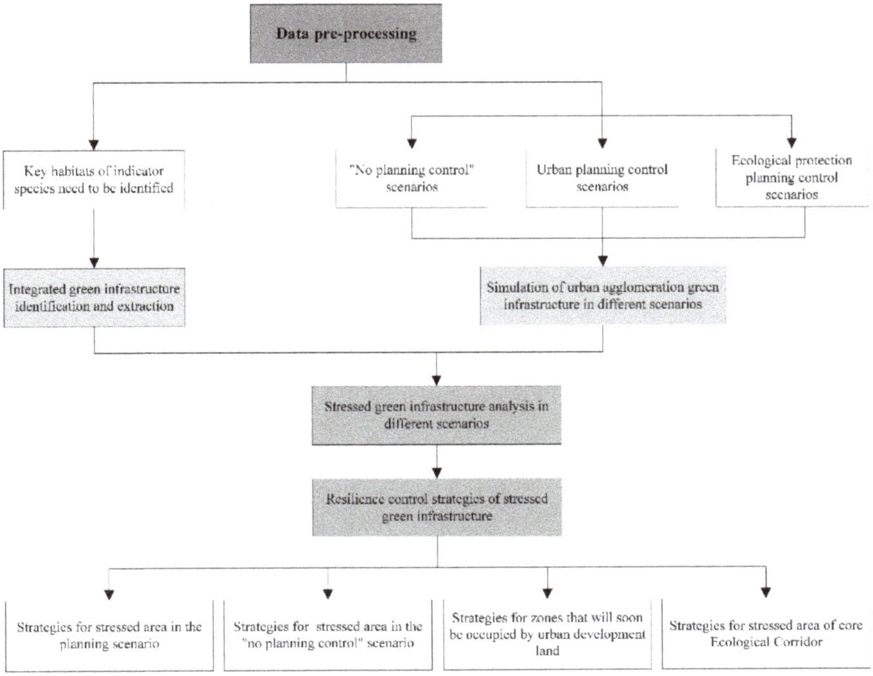

Figure 2. Research methodology.

3.1. Integrated Green Infrastructure of Urban Agglomeration Identification and Extraction

The green infrastructure range identification of the Chang-Zhu-Tan urban agglomeration should be integrated with the principle of landscape ecological security patterns [39,50]. Firstly, three representative indicator species representing the main environment characteristics of the study area are selected, i.e., Chinese pool heron, red-billed leiothrix, and Siberian weasel, representing medium pheasants living in the wetland ecosystem, small pheasants living in the forest ecosystem, and mammals inhabiting in multiple ecosystems, respectively. They are used to identify the green infrastructure and species migration ecological corridors that have important values for species survival and reproduction [51]. Secondly, key habitats of indicator species need to be identified as green infrastructure of significant value. Thirdly, the minimal cumulative resistance (MCR) model is used to simulate the process of the indicator species overcoming spatial resistance to spread out from the habitat, and the core habitat space range of different species is used as the "source" to identify the biological migration corridor between different ecological patches [52–54]. Fourthly, the migration corridors of different species are superimposed. Additionally, the multi-species migration corridors are treated as primary corridors with a width of 200 m, considering the conservation of biodiversity [55]. Finally, the whole green infrastructure range can be obtained by integrating the habitat and ecological corridor. As a result, the ecological spatial context of the Chang-Zhu-Tan urban agglomeration is obtained comprehensively.

The formula used is:

$$MCR = f\sum(D_{ij} \times R_i) \tag{1}$$

where *MCR* is the minimum cumulative resistance value, which reflects the minimum cost of indicator species in the process of moving from source to destination; f is a positive correlation function characterizing the relative accessibility of a path from a source patch to a point in space; D_{ij} is the spatial distance of the landscape basal plane i traversed by a species from source patch j to a point in space; and R_i is the relative resistance coefficient of indicator species (Table 1).

Table 1. Relative resistance coefficients of indicator species (0–100).

Resistance Coefficients/Land Use Type	Chinese Pool Heron (*Ardeolabacchus*) [56–59]	Red-Billed Leiothrix (*Leiothrix lutea*) [60–63]	Siberian Weasel (*Mustela sibirica*) [64–67]
Tillable field	15	5	10
Woodland	5	1	5
Grassland	5	5	5
Waters	5	10	35
Construction land	50	50	40
Unused land	20	10	5

3.2. Simulation of Urban Growth in Different Scenarios

The impacts of the urban growth of the Chang-Zhu-Tan urban agglomeration on the green infrastructure are simulated with the artificial neural network cellular automata model in the open source software "Geographic Simulation and Optimization System" (GeoSOS) [68]. This method is used to simulate the self-organized growth process of urban construction land expansion in a bottom-up manner and visualize the impact of urban growth on green infrastructure [69]. Additionally, this approach's biggest advantage is that the model information can be obtained by training the neural network, which is especially suitable for research in the field of complex systems [70], and the used detailed formula is as follows:

$$P_{d,ij}^{t} = \left(1 + (-1n\gamma)^{\alpha}\right) \times \sum\nolimits_{j} w_{j,i} \frac{1}{1 + e^{-net_j(k,t)}} \tag{2}$$

where $P_{d,ij}^{t}$ is the probability of a cell being developed, γ is a random number between 0 and 1, the value of α ranges from 1 to 10, *neti(k,t)* is the signal received by the *j*th neuron of the hidden layer, and *wj,l* is the weight between the hidden layer and the output layer [71]. The convertibility between different land uses is shown in Table 2, where 1 represents convertibility and 0 represents non-convertibility.

Table 2. Convertibility between different land-use types.

Land-Use Type	Grassland	Tillable Field	Woodland	Unused Land	Water	Construction Land
Grassland	1	1	1	1	0	1
Tillable field	1	1	1	1	0	1
Woodland	1	1	1	1	0	1
Unused land	0	0	0	0	0	1
Waters	0	0	0	0	1	0
Construction land	0	0	0	0	0	1

The spatial variables of the urban growth model are set with the 30 m resolution data of 2018 [72–77] (Table 3). All the influencing factors are normalized; this is the input into the artificial neural network CA model in GeoSOS as the parameter affecting the probability of land conversion, to simulate the urban growth in the year 2018 based on the urban growth law from the year 1999 to the year 2008. Additionally, after calibration, the simulated results are compared with the actual maps in 2018. In the simulation setting, α equals 5 and the conversion threshold is taken as 0.8 [78]. The validation results show that the overall accuracy of the model is 90.87% and the Kappa coefficient is 0.82, which is higher than the normally applied 0.8 [77]. Therefore, this calibrated model is used as the basis to simulate the expansion of Chang-Zhu-Tan urban agglomeration in the year 2030 with different control measures to understand the stress imposed on green infrastructure.

Table 3. Spatial variables of the urban growth model.

Type	Variables	Abbreviation	Contents
Terrain	Digital elevation model	DEM	Evaluate the impact of topography on scenario simulation.
Location	Distance to the central city	DisCentral city	Evaluate the impact of distance from the administrative center on scenario simulation.
Transportation	Distance to the road Distance to the railway Distance to the highway Distance to the national highway	DisRoad DisRail DisHighway DisNational highway	Evaluate the impact of surface road distance on scenario simulation.

The spatial expansion of urban growth is a self-organizing process, but other organizational factors such as different development policies and control strategies can have a significant impact on the development of urban agglomerations [79]. In terms of the driving factors of other organizations affecting urban growth mainly urban planning and ecological protection planning have been included, we have analyzed the main influencing factors on the development of urban agglomerations in China and determined three scenarios of green infrastructure stress in Chang-Zhu-Tan urban agglomeration with different control measures, namely, "no planning control", under control of urban planning, and ecological protection planning control. The "no planning control" scenario measure does not equal to the absence of control measures, but the simulation of urban growth with the original development pattern of urban agglomerations without new planning control measures. The second scenario is based on ongoing urban planning to predict future urban spatial development. According to urban planning, different conversion coefficients are assigned to land use, including urban comprehensive function area (1), urban new town group, high-tech group of science and education (3), industrial park (5), other non-urban land function areas (7), the ecological green heart protection area of urban agglomeration (9), and normalized processing is carried out [42]. The third scenario is to take the existing green infrastructure as the protection area of ecological planning and simulate its future situation under the stress of urban construction land. The conversion coefficient of green infrastructure is assigned 3, and the non-green infrastructure is assigned 7 [80].

3.3. Stressed Green Infrastructure Identification

By superposition analysis of the spatial changes of Chang-Zhu-Tan urban agglomeration in 2030 in three scenarios, the conflict areas between urban growth and green infrastructure in different scenarios are obtained, namely the stressed areas of the green infrastructure of urban agglomeration. This serves as the area for further analysis.

3.4. Green Infrastructure Stress Characteristics and Resilience Measures

By superimposing and analyzing the green infrastructure under stress in different areas, we have classified the areas into three categories based on degrees of stress on green infrastructure: first, the part of green infrastructure that will not be affected by urban growth in the three scenarios can be divided into the safety zone of green infrastructure; second, the green infrastructure affected by urban growth in two or one of the scenarios is under the stress of uncertainty, which requires the adoption of compatible and flexible management and control measures; third, in the three scenarios, the part of the green infrastructure to be occupied by urban growth will face the greatest pressure from urbanization. The development control measures, taking into account both ecological functions and construction land functions, should be considered. Therefore, the second and third parts of green infrastructure are threatened by urban growth and need to be included in key management zones.

In terms of the main driving force of urban expansion, self-organization growth, and other-organization planning, an ecological corridor plays an important role in maintaining regional green infrastructure [81–83]. The key management zones of green infrastructure are divided into the following categories (Table 4): (1) external factors that cause changed green infrastructure with urban planning control and the ecological protection planning control scenario; these spaces are further divided into stress zones with the urban planning control alone, with ecological protection planning control alone and joint stress zones of both; (2) the no planning control self-organizational model developed stressed green infrastructure, and the area affected by it alone are together classified as the stress zones in the no planning control, excluding the areas of the next type; (3) in all the three scenarios, the part of green infrastructure under the stress of urban growth are the zones that will soon be occupied by urban development land; (4) the biological migration corridor affected by urban growth is classified as the stress zones of the core ecological corridor.

Table 4. Stressed green infrastructure types and strategies.

Stressed Green Infrastructure Types	Stressful Situation	Strategies
Zones stressed by urban growth in one scenario	(1) Stressed areas in the urban planning control scenario	Adjustment of planning
	(2) Stressed areas in the ecological protection planning control scenario	Strengthening protection management
	(3) Stressed areas in the "no planning control" scenario	Formulation ecological protection planning
Zones stressed by urban growth in two scenarios	(1) Stressed areas in the urban planning control scenario and the ecological protection planning control scenario	Adjustment of planning and strengthening protection management
	(2) Stressed areas in the urban planning control scenario and "no planning control" scenario	Adjustment of urban planning and formulation ecological protection planning
	(3) Stressed areas in the ecological protection planning control scenario and "no planning control" scenario	Replacement of urban development land and control of development intensity
Zones stressed by urban growths in all three scenarios	Zones that will be occupied by urban development land	Development compatible landscape ecological protection measures

The stress zones of green infrastructure may be invaded by urban growth, and there is great uncertainty in future development and construction, so resilience optimization measures need to be formulated. To do so, we firstly analyze the causes of stress in different areas and then the planning and control measures accordingly, such as adjustment of planning, replacement of land, and control of development intensity are advised, in order to enhance the ability of green infrastructure to adapt to the change [18,84,85]. On the other hand, the zones that will soon be occupied by urban development land are the key areas for future construction activities; control and guidance strategies can be formulated according to the compatibility of different development land types to different habitats [25,86–88] (Table 4).

4. Results

4.1. Green Infrastructure and Corridor in the Chang-Zhu-Tan Urban Agglomeration

The green infrastructure of the Chang-Zhu-Tan urban agglomeration is obtained by identifying the main habitats of different species and superimposing them together (Figure 3). The green infrastructure of the Chang-Zhu-Tan urban agglomeration has a total area of 15,690.19 km^2, which is mainly concentrated in the western and eastern parts of the Changsha built-up area, the northwestern and southern parts of the Xiangtan built-up area, and the eastern and northern parts of the Zhuzhou built-up area. The length of primary corridors is 363.19 km long and areas are 143.56 km^2, and they are mainly concentrated in the peripheral area of the built-up area of the urban agglomeration and the southern part of Changsha city, the northern part of Xiangtan city, and the southern part of Zhuzhou city.

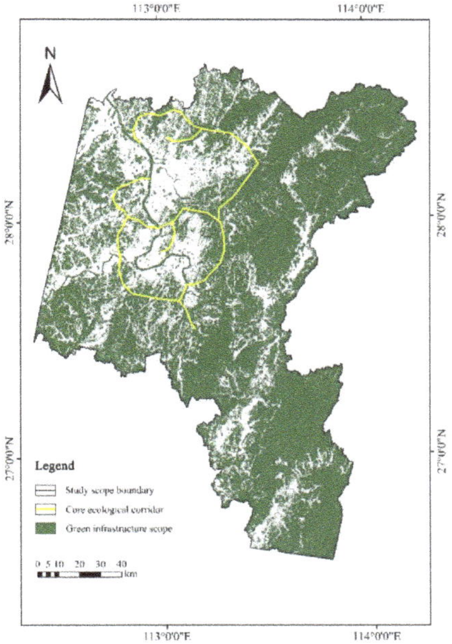

Figure 3. Green infrastructure and corridors in 2018.

4.2. Stressed Green Infrastructure Analysis in Different Scenarios

4.2.1. Stressed Green Infrastructure Characteristics

The simulation results show the patterns of the stressed green infrastructure in the "no planning control" scenario (Figure 4), the stressed green infrastructure in the urban planning control scenario (Figure 5), and the stressed green infrastructure in the ecological protection planning control scenario (Figure 6). The details of the stressed area and their characteristics are shown in Table 5.

The analysis of green infrastructure under stress in the three scenarios shows that the stressed zone of green infrastructure is the smallest in the ecological protection planning control scenario, the second largest in the "no planning control" scenario, and the largest in the urban planning control scenario. This indicates that when the construction is carried out according to urban planning, the surrounding green infrastructure is under greater pressure and the green infrastructure is more affected. Meanwhile, by analyzing the characteristics of the stressed areas, we find that the green infrastructure in the southwest of Changsha City, the north and east of Xiangtan City, and the north and west of Zhuzhou City are under greater pressure in the three scenarios, which also coincides with the development direction of the city integration and urban agglomeration development strategy of the Changsha-Zhuzhou-Tan urban agglomeration. The western area of Changsha City, as the key construction area of Xiangjiang New District, has stressed green infrastructure in all three scenarios. However, the stressed area in the northern part of Changsha City in the urban planning control scenario is significantly larger than that in the other two scenarios, which is also related to the fact that the northern part of Changsha City is considered the key development area in the urban general planning. Notably, this analysis proves that the planning and control as "other-organized" measures can only affect urban growth to a certain extent, but cannot fundamentally change the urban "self-organized" growth process. Secondly, the three scenarios have different impacts on the local green infrastructure, indicating that there are areas of the green infrastructure under uncertain urbanization stress, which are the areas for resilient green infrastructure control.

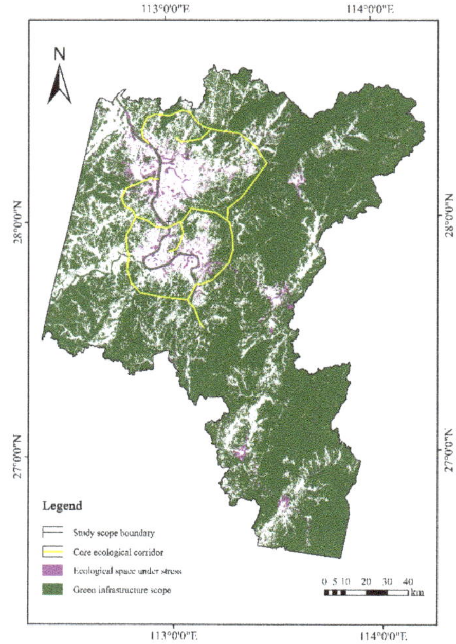

Figure 4. Stressed green infrastructure in the "no planning control" scenario in 2030.

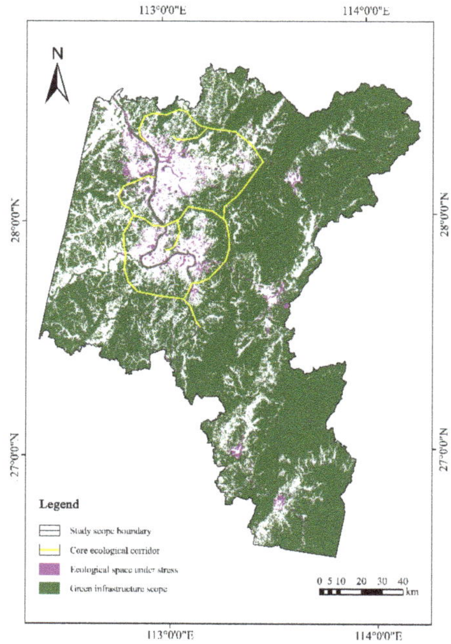

Figure 5. Stressed green infrastructure in the "urban planning control" scenario in 2030.

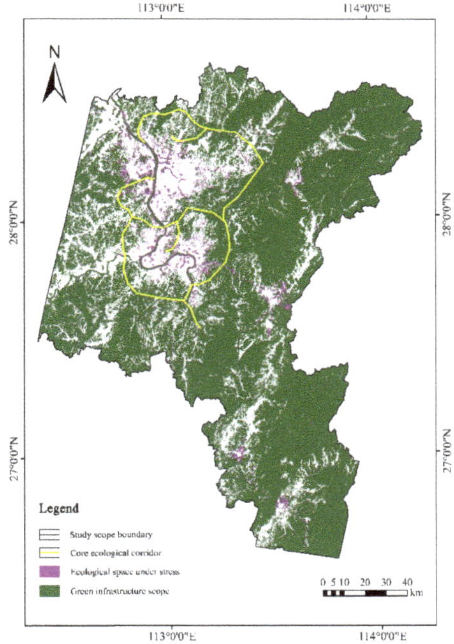

Figure 6. Stressed green infrastructure in the "ecological protection planning control" scenario in 2030.

Table 5. Characteristics of stressed green infrastructure in different scenarios.

Scenarios	Areas	Ranges	Characteristics
"No planning control"	202.80 km^2	The western and southern parts of the built-up area of Changsha, the northeastern part of the built-up area of Xiangtan, and the eastern and western parts of the built-up area of Zhuzhou.	The urban agglomeration is in a "spread-out" expansion mode. The western and southern parts of Changsha, the northeastern part of Xiangtan, and the eastern and western parts of Zhuzhou, as areas of high ecological value, have relatively more stressed green infrastructure.
Urban planning control	210.35 km^2	The western and northern parts of the built-up area of Changsha, the northern part of the built-up area of Xiangtan, and the southeastern part of the built-up area of Zhuzhou.	The expansion of the urban agglomeration to the north and west is evident, and the relevant areas are planned as key areas in the urban agglomeration plan, with urban functional areas such as new riverfront areas, ecological new towns, and industrial parks planned for the long term, exacerbating the state of green infrastructure under stress.
Ecological protection planning control	182.64 km^2	The western and southern parts of the built-up area of Changsha, the eastern and northern parts of the built-up area of Xiangtan, and the southeastern part of the built-up area of Zhuzhou.	The extent of the green infrastructure under stress corresponds to the key development areas. The western and southern parts of Changsha, as the key areas for the construction of the Xiangjiang New District, have been developing rapidly in recent years, while the eastern and northern parts of Xiangtan, as the direction of the development of the integrated city of Chang-Zhu-Tan Urban agglomeration, have been under greater pressure of green infrastructure.

4.2.2. Identification of the Control Area of the Green Infrastructure

A comprehensive analysis of the stressed green infrastructure reveals that in the three scenarios, the safety zone of green infrastructure, which is not affected by urban growth, accounts for 98% of the total green infrastructure, while the total stressed zone of green infrastructure is only 269.86 km^2 (Table 6 and Figure 7).

Table 6. Different control areas of the green infrastructure.

Types	Characteristics	Areas	Locations
The safety zone of green infrastructure	Green infrastructure without stress	15,420.33 km^2	Western and eastern Changsha, northwestern and southern Xiangtan, eastern and northern Zhuzhou
Green infrastructure with alternative future in multi-scenarios	Stressed green infrastructure	133.64 km^2	Areas to the north and south of Changsha, northeast of Xiangtan, and south of Zhuzhou
Zones that will soon be occupied by urban development land		136.22 km^2	The western part of the built-up area of Changsha, the northern part of the built-up area of Xiangtan, and the southeastern part of the built-up area of Zhuzhou

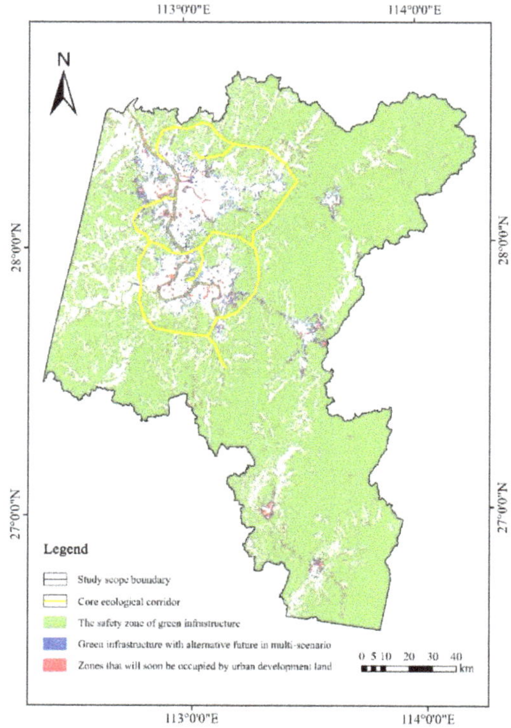

Figure 7. Different control areas of the green infrastructure.

4.3. Resilient Strategies for the Identified Key Green Infrastructure Control Areas

4.3.1. Identification of Key Green Infrastructure Control Area

By overlaying the stressed green infrastructure patterns and urban land-use planning maps in different scenarios in GIS, four detailed types of key ecological control spaces are identified. The detailed information is shown in Table 7.

Table 7. Identification of key green infrastructure control areas.

Types	Area	Location	Stressing Urban Development Land	Stressed Ecosystem	Disturbed Species
Stressed area in the planning scenario	67.12 km^2	North and west of the built-up area of Changsha, north of the built-up area of Xiangtan, and south of the built-up area of Zhuzhou	Residential land, industrial land, and public service land	Forest, shrub, and wetland	Small pheasant species, small mammal species, and wetland medium-sized pheasant species
Stressed area in the "no planning control" scenario	23.00 km^2	The western and southern parts of the built-up area of Changsha, the northeastern part of Xiangtan, the western and southeastern part of Zhuzhou	Residential land, public green land, and industrial land	Forest, shrub, and meadow	Small pheasant species and small mammal species
Zones that will soon be occupied by urban development land	43.52 km^2	North and west of the built-up area of Changsha, north of the built-up area of Xiangtan, northeast, and south of the built-up area of Zhuzhou	Residential land, public service land, and land for roads and transportation facilities	Forest, shrub, and wetland	Small pheasant species and wetland medium-sized pheasant species
Stressed area of core Ecological Corridor	19.54 km^2	Biomigratory corridors in the north of Changsha and the east of Xiangtan	Residential land, public green land, and land for roads and transportation facilities	Forest, shrub, and wetland	Small pheasant species, small mammal species, and wetland medium-sized pheasant species

4.3.2. Resilience Management Strategies of the Stressed Green Infrastructure

The protection measures in Table 4 are applied to the stressed green infrastructure in the empirical case, and resilience management strategies are formulated according to the stress characteristics of the four key control areas of green infrastructure (Table 8).

Table 8. Measures to improve the resilience of the stressed green infrastructure.

Types	Resilience Management Strategies
Stressed area in the planning scenario	**Residential land:** The planning of urban residential land should be adjusted, the important forest and wetland ecosystems should be retained, the scale and intensity of residential area construction should be strictly restricted, and the interference with the existing green infrastructure should be reduced in combination with the construction of community parks and residential green space. **Industrial land:** Adjust the planning of urban industrial land, change the industrial land within the region to non-construction land, or change it into urban park green space. **Public service land:** Adjust urban public service land planning to preserve important ecosystems; restrict the type of land used for low-density education, culture, and sports facilities, and strictly control their scales.
Stressed area in the "no planning control" scenario	**Residential land:** It should be included in the scope of urban ecological protection planning and control of detailed planning to avoid development as residential land. **Public green land:** Special planning for urban public green space has been formulated; the original forest and wetland ecosystems should be retained, and local plants should be used to construct parks. The construction intensity of urban hard squares should be strictly controlled. **Industrial land:** The key monitoring areas included in the urban ecological protection planning shall be strictly monitored and managed, and the industrial construction projects shall be guided to be replaced with industrial parks in non-stress areas.
Zones that will soon be occupied by urban development land	**Residential land:** Priority should be given to the formulation of detailed control planning, strict restrictions should be given to building density and floor area ratio, and important woodland patches and wetland patches should be reserved as residential park green space. **Public service land:** Priority should be given to the preparation of a detailed control plan, strictly limiting the building density and plot ratio, and retaining important ecological patches as public green space inside the park. **Land for roads and transportation facilities:** Priority should be given to the compilation of detailed control planning, and the construction of road traffic attached green space should be combined with the requirements of urban ecological protection planning to avoid aggravating habitat fragmentation of protected species.
Stressed area of core Ecological Corridor	**Residential land:** Priority should be given to the preparation of a detailed control plan and the corridor area should be designated as a residential green space to avoid the layout of residential buildings interfering with biological migration activities. **Public green land:** Priority should be given to the formulation of detailed control planning, and local plants should be strictly used in the construction of green space in the park while ensuring the width of the biological migration corridor. **Land for roads and transportation facilities:** Priority should be given to the preparation of detailed control plans and the construction of underpass culverts or ecological bridges for protected species to cross, to avoid roads interrupting the migration process.

5. Discussions

The research on urban growth and ecological resilience focuses on the representation of their feature of coupling relations [25,28,32,35] but there is a lack of comprehensive analysis in the context of urban growth uncertainty. Both cities and green infrastructure are complex network systems that are constantly changing, and traditional static and rigid control measures can hardly solve the complex contradictory problems between urban development and green infrastructure protection [15,16]. Uncertainty exists in organizational drivers, such as planning policies and their implementation, leading to different future patterns of urban growth. Therefore, this study proposes a scenario-analysis approach by combining the minimum cumulative resistance model (MCR) with cellular automata (CA) to reflect the intrinsic dynamic linkage process of ecological processes under uncertainties [37,38]. Based on our simulation results, urban land expansion in urban agglomeration is a self-organizing growth process under the influence of organizational planning measures which are also consistent with other studies [17,28,36]. The distribution pattern of green infrastructure under stress is different in three scenarios. Even in the scenario of ecological protection planning and control, urban growth can easily break through the ecological protection boundary. The applied artificial neural network CA model can be used to predict the uncertain future of urban growth scenarios, and it could be more conducive to solving the practical problems that may be faced by the stressed green infrastructure [89,90].

Furthermore, most of the existing studies on the ecosystem resilience assessment and management are limited to the evaluation and improvement of indicators [20,22–24] and ignore the difference in the diversification pattern caused by the spatial game of the future urban growth. The validation and calibration process and results confirmed our hypotheses that the process of urban growth stressing green infrastructure in various scenarios could be simulated and the stressed green infrastructure could be identified with different characteristics. Residential, industrial, and traffic facility land are the main types of urban land causing green infrastructure stress, while forest, shrub, and wetland are the main types of the stressed green infrastructure in our case. The green infrastructure stressed by urban growth is mainly concentrated around the urban built-up area, but the distribution pattern of the stressed green infrastructure varies with different scenarios. This is because the driving factors of urban self-organization determine the main trend of urban growth. For instance, the urban population of Chang-Zhu-Tan urban agglomeration is about 10.2 million in 2020, and it will rise to 14 million in 2030 as compared. Even in the scenario of ecological protection planning control, urban land use is easy to break through the protection boundary. The identification of stressed green infrastructure in various scenarios and the detailed analysis of their characteristics help identify the diversities and form resilience optimization measures for the stressed green infrastructure afterwards.

The importance of resilience for the green infrastructure has been discussed in the case study of Detroit in America [91] and in the review on relative research of America and Europe [92] but still in its infancy in China. The analysis results have proved clearly the hypothesis that the current rigid measures cannot cope with the dynamic, complex, changing processes of urban land use and green infrastructure, which, on the other hand, confirms that resilient measures are needed to improve the green infrastructure resilience for urban agglomerations in China. In contrast with the more frequent design proposals for green infrastructure protection, which almost exclusively gives static and rigid conservation boundaries, our approach provides alternatives for changing the landscape of urban agglomeration with more resilience. The in-depth analysis proposes that the resilience of green infrastructure could be achieved by resilient management measures, such as urban planning adjustment, regulatory detailed planning, development strength control, and setting up the ecological protection facilities for the stressed green infrastructure.

To summarize, this study has successfully formulated an effective methodology that provides a scenario analysis approach to identify the stressed ecological spatial patterns and form resilient measures for urban agglomerations in China. We comprehensively analyze the dynamic game process of the urban growth stress green infrastructure and identify the different types of stressed green infrastructure. This approach helps formulate efficient green infrastructure resilience control strategies according to the uncertainty of urban growth, ensuring that stressed ecosystems can resist disturbance, respond, and adapt to changes in different scenarios.

Nevertheless, this study has some limitations. The application of the resilience concept in green infrastructure is still in the exploration stage, and the future of urban growth under the influence of multiple driving forces is more complex than the scenario hypothesis in this research. The research on the stress impacts and resilience response of different urban land-use types on different green infrastructure needs to be further studied.

6. Conclusions

The resilience optimization of green infrastructure in rapidly developing urbanized areas, as a key yet difficult point in the ecological construction of urban agglomerations, has not received sufficient attention in relevant studies. The uncertainty of urban growth leads to the uncertainty of urban agglomeration green infrastructure. As it is difficult to adapt to such uncertainty, the previous planning measures for green infrastructure protection based on the demarcation of rigid protection red lines are easy to fail. To solve this problem, we have proposed a scenario simulation analysis method to identify the types of stressed green infrastructure and conduct resilience management based on the characteristics of self-organization growth of urban land and other-organization growth of urban planning. In this method, the cellular automata model of urban growth has been constructed by using an artificial neural network, and four types of green infrastructure under urban growth stress have been identified in three scenarios. Resilience management strategies have been proposed to adapt to the alternative futures of urban growth in detail based on land-use types.

Currently, China is still in a period of rapid urbanization, and the research on the optimization of the resilience of the stressed green infrastructure not only helps to solve the conflicts between urban expansion and ecological protection but also contributes to the construction of an ecological city. Through the study on the optimization of the resilience of the stressed green infrastructure of Chang-Zhu-Tan urban agglomeration, it is found that the strong urban self-organizing driving force dominates the future urban growth process, while the other organizing driving force of planning affects the future urban growth pattern to a certain extent. Although the protection planning of green infrastructure has a certain constraint effect on urban growth, its protection boundary is easily breached by urban land use. The identification of four types of stressed spaces in Chang-Zhu-Tan urban agglomeration and their resilience management is beneficial to improving the conservation efficiency of green infrastructure and the management efficiency of the urban-growth boundary by targeting the most suitable measures for the spaces with resilient potentials.

Author Contributions: Conceptualization, B.L. and T.W.; methodology, X.H.; formal analysis, X.H.; resources, T.W. and B.L.; writing—original draft preparation, Y.W., B.L., T.W. and J.K.K.; writing—review and editing, B.L., Y.W., T.W. and J.K.K.; visualization, X.H. and Y.W.; supervision, B.L. and T.W.; funding acquisition, T.W. and J.K.K. All authors have read and agreed to the published version of the manuscript.

Funding: This research was supported by the Hunan Province Philosophy and Social Science Foundation of China (09yba158). And the article processing costs are waived.

Institutional Review Board Statement: Not applicable.

Informed Consent Statement: Not applicable.

Data Availability Statement: Not applicable.

Acknowledgments: The authors would like to thank the reviewers for their valuable comments and suggestions.

Conflicts of Interest: The authors declare no conflict of interest.

References

1. National Bureau of Statistics of China. *Degree of Urbanization in China from 1980 to 2021*; National Bureau of Statistics: Beijing, China, 2022.
2. Brzoska, P.; Spage, A. From City- to Site-Dimension: Assessing the Urban Ecosystem Services of Different Types of Green Infrastructure. *Land* **2020**, *9*, 150. [CrossRef]
3. Chatzimentor, A.; Apostolopoulou, E.; Mazaris, A.D. A review of green infrastructure research in Europe: Challenges and opportunities. *Landsc. Urban Plan.* **2020**, *198*, 103775. [CrossRef]
4. Monteiro, R.; Ferreira, J.C.; Antunes, P. Green Infrastructure Planning Principles: An Integrated Literature Review. *Land* **2020**, *9*, 525. [CrossRef]
5. Chu, M.R.; Lu, J.Y.; Sun, D.Q. Influence of Urban Agglomeration Expansion on Fragmentation of Green Space: A Case Study of Beijing-Tianjin-Hebei Urban Agglomeration. *Land* **2022**, *11*, 275. [CrossRef]
6. Shen, Z.; Wu, W.; Chen, M.; Tian, S.Q.; Wang, J. Linking Greenspace Ecological Networks Optimization into Urban Expansion Planning: Insights from China's Total Built Land Control Policy. *Land* **2021**, *10*, 1046. [CrossRef]
7. Yang, Y.Y.; Bao, W.K.; Li, Y.H.; Wang, Y.S.; Chen, Z.F. Land Use Transition and Its Eco-Environmental Effects in the Beijing-Tianjin-Hebei Urban Agglomeration: A Production-Living-Ecological Perspective. *Land* **2020**, *9*, 285. [CrossRef]

8. Wang, Y.C.; Shen, J.K.; Xiang, W.N. Ecosystem service of green infrastructure for adaptation to urban growth: Function and configuration. *Ecosyst. Health Sustain.* **2018**, *4*, 132–143. [CrossRef]
9. Colding, J.; Gren, A.; Barthel, S. The Incremental Demise of Urban Green Spaces. *Land* **2020**, *9*, 162. [CrossRef]
10. Sarkar, A.; Chouhan, P. Dynamic simulation of urban expansion based on Cellular Automata and Markov Chain Model: A case study in Siliguri Metropolitan Area, West Bengal. *Modeling Earth Syst. Environ.* **2019**, *5*, 1723–1732. [CrossRef]
11. Wang, C.; Ding, L. Regional characterization of environmental problems in urban agglomerations in the context of urban expansion. *Acta Ecol. Sin.* **2017**, *37*, 8058–8066. [CrossRef]
12. Tang, Y.; Zhan, L. Delineation of Urban Development Boundary Based on Coupled Neural Network and Cellular Automata-Taking Huizhou City as an Example. *Planners* **2018**, *34*, 101–106.
13. Liao, H.; Zhang, J.; Zhao, Z.; Cui, J. Tongzhou District, Beijing: A Study on the Delimitation Method of "Two Lines in One" of Urban Development Boundary and Ecological Red Line. *Beijing Plan. Constr.* **2019**, *2*, 88–91.
14. He, M.; Wang, Y. Research on the Construction and Protection Countermeasures of Urban Ecological System in Wuhan. *Planners* **2009**, *25*, 30–34. [CrossRef]
15. Bajc, K.; Stokman, A. Design for Resilience: Re-Connecting Communities and Environments. *Landsc. Archit. Front.* **2018**, *6*, 14–31. [CrossRef]
16. Xiao, X.; Li, M.; Yuan, L. Sustainable Green Space System in High-density Cities-Based on the GI Theory. In Proceedings of the AASRI International Conference on Industrial Electronics and Applications (IEA), London, UK, 27–28 June 2015; pp. 575–579.
17. Gunderson, L.H. Ecological resilience—In theory and application. *Annu. Rev. Ecol. Syst.* **2000**, *31*, 425–439. [CrossRef]
18. Hong, W.Y.; Jiang, R.R.; Yang, C.Y.; Zhang, F.F.; Su, M.; Liao, Q. Establishing an ecological vulnerability assessment indicator system for spatial recognition and management of ecologically vulnerable areas in highly urbanized regions: A case study of Shenzhen, China. *Ecol. Indic.* **2016**, *69*, 540–547. [CrossRef]
19. Ingrisch, J.; Bahn, M. Towards a Comparable Quantification of Resilience. *Trends Ecol. Evol.* **2018**, *33*, 251–259. [CrossRef]
20. Sterk, M.; Gort, G.; Klimkowska, A.; van Ruijven, J.; van Teeffelen, A.J.A.; Wamelink, G.W.W. Assess ecosystem resilience: Linking response and effect traits to environmental variability. *Ecol. Indic.* **2013**, *30*, 21–27. [CrossRef]
21. De Meulder, B.; Shannon, K. Designing Ecologies for Resilient Urbanisms. *Landsc. Archit. Front.* **2018**, *6*, 12–33. [CrossRef]
22. Barnett, L.A.K.; Baskett, M.L. Marine reserves can enhance ecological resilience. *Ecol. Lett.* **2015**, *18*, 1301–1310. [CrossRef]
23. Schippers, P.; van der Heide, C.M.; Koelewijn, H.P.; Schouten, M.A.H.; Smulders, R.; Cobben, M.M.P.; Sterk, M.; Vos, C.C.; Verboom, J. Landscape diversity enhances the resilience of populations, ecosystems and local economy in rural areas. *Landsc. Ecol.* **2015**, *30*, 193–202. [CrossRef]
24. Mumby, P.J.; Chollett, I.; Bozec, Y.M.; Wolff, N.H. Ecological resilience, robustness and vulnerability: How do these concepts benefit ecosystem management? *Curr. Opin. Environ. Sustain.* **2014**, *7*, 22–27. [CrossRef]
25. Wu, X.; Zhang, J.J.; Geng, X.L.; Wang, T.; Wang, K.; Liu, S.D. Increasing green infrastructure-based ecological resilience in urban systems: A perspective from locating ecological and disturbance sources in a resource-based city. *Sustain. Cities Soc.* **2020**, *61*, 102354. [CrossRef]
26. Zhang, H.; Liang, X.Y.; Chen, H.; Shi, Q.Q. Spatio-temporal evolution of the social-ecological landscape resilience and management zoning in the loess hill and gully region of China. *Environ. Dev.* **2021**, *39*, 100616. [CrossRef]
27. Liu, J.G.; Cui, W.H.; Tian, Z.; Jia, J.L. Theory of stepwise ecological restoration. *Chin. Sci. Bull. Chin.* **2021**, *66*, 1014–1025. [CrossRef]
28. Pickett, S.T.A.; McGrath, B.; Cadenasso, M.L.; Felson, A.J. Ecological resilience and resilient cities. *Build. Res. Inf.* **2014**, *42*, 143–157. [CrossRef]
29. Wei, S.M.; Pan, J.H.; Liu, X. Landscape ecological safety assessment and landscape pattern optimization in arid inland river basin: Take Ganzhou District as an example. *Hum. Ecol. Risk Assess.* **2020**, *26*, 782–806. [CrossRef]
30. Fu, Y.J.; Shi, X.Y.; He, J.; Yuan, Y.; Qu, L.L. Identification and optimization strategy of county ecological security pattern: A case study in the Loess Plateau, China. *Ecol. Indic.* **2020**, *112*, 106030. [CrossRef]
31. Li, S.N.; Zhao, X.Q.; Pu, J.W.; Miao, P.P.; Wang, Q.; Tan, K. Optimize and control territorial spatial functional areas to improve the ecological stability and total environment in karst areas of Southwest China. *Land Use Policy* **2021**, *100*, 104940. [CrossRef]
32. Sun, C.Z.; Xu, S.; Qi, W.; Chen, C.; Deng, Y.; Pei, N.C.; Konig, H.J. Biodiversity constraint indicator establishment and its optimization for urban growth: Framework and application. *Environ. Res. Lett.* **2019**, *14*, 125006. [CrossRef]
33. Cui, L.; Wang, J.; Sun, L.; Lv, C.D. Construction and optimization of green space ecological networks in urban fringe areas: A case study with the urban fringe area of Tongzhou district in Beijing. *J. Clean. Prod.* **2020**, *276*, 124266. [CrossRef]
34. Shen, Q.-w.; Lin, M.-l.; Mo, H.-p.; Huang, Y.-b.; Hu, X.-y.; Wei, L.-w.; Zheng, Y.-s.; Lu, D.-f. Ecological network construction and optimization in Foshan City, China. *Yingyong Shengtai Xuebao* **2021**, *32*, 3288–3298. [CrossRef] [PubMed]
35. Wang, S.J.; Cui, Z.T.; Lin, J.J.; Xie, J.Y.; Su, K. The coupling relationship between urbanization and ecological resilience in the Pearl River Delta. *J. Geogr. Sci.* **2022**, *32*, 44–64. [CrossRef]
36. Cheng, K.; Chen, Y. Progress and review of urban self-organization studies at home and abroad. *Urban Probl.* **2006**, *25*, 21–27. [CrossRef]
37. Yu, Q.; Yue, D.P.; Wang, J.P.; Zhang, Q.B.; Li, Y.T.; Yu, Y.; Chen, J.X.; Li, N. The optimization of urban ecological infrastructure network based on the changes of county landscape patterns: A typical case study of ecological fragile zone located at Deng Kou (Inner Mongolia). *J. Clean. Prod.* **2017**, *163*, S54–S67. [CrossRef]

38. Lei, Z.K.; Feng, Y.J.; Tong, X.H.; Liu, S.; Gao, C.; Chen, S.R. A spatial error-based cellular automata approach to reproducing and projecting dynamic urban expansion. *Geocarto Int.* **2022**, *37*, 560–580. [CrossRef]
39. Lu, X.; Li, N.; Li, Z. Research on the protection and utilization of urban and rural green infrastructures in urban cluster regions: An example of Changzhutan ecological green heart region. *Urban Dev. Stud.* **2013**, *20*, 82–87. [CrossRef]
40. Tang, Y.; Zou, B.; Xu, S.; Chen, J.; Li, M. Analysis of anthropogenic disturbance processes in the ecological landscape of Changzhutan in the past 40 years. *Geomat. Spat. Inf. Technol.* **2017**, *40*, 74–79. [CrossRef]
41. Tang, C.; Fan, J.; Chen, X. Research on the coordination of land use based on regional functions—Taking the ecological green heart of Changzhutan Muyun Town as an example. *J. Nat. Resour.* **2012**, *27*, 1645–1655. [CrossRef]
42. Zhang, P.; Zhu, Y.; Shi, Y.; Hu, Y. Regional Planning of Chang-Zhu-Tan City Cluster (2008–2020) (Adjusted in 2014). Available online: http://fgw.hunan.gov.cn/xxgk_70899/ghjh/201605/t20160517_3058399.html (accessed on 8 July 2020).
43. Hunan Provincial Department of Natural Resources. Master Plan of Ecological Green Heart Area of Changzhutan City Cluster (2010–2030) (Revised in 2018). Available online: http://zrzyt.hunan.gov.cn/zrzyt/xxgk/tzgg/201909/t20190926_10466709.html (accessed on 8 July 2020).
44. Changsha Planning & Design Survey Research Institute. Changsha City Master Plan (2003–2020) (Revised in 2014). Available online: http://www.csgky.net/product/30.html (accessed on 8 July 2020).
45. Xiangtan City Natural Resources and Planning Bureau. Xiangtan City Urban Master Plan (2010–2020) (Revised in 2017). Available online: http://zygh.xiangtan.gov.cn/10604/10606/17643/content_828376.html (accessed on 8 July 2020).
46. Zhuzhou Municipal Commission of Development and Reform. Zhuzhou City Urban Master Plan (2006–2020) (Revised in 2017). Available online: http://fgw.zhuzhou.gov.cn/c14814/20190926/i984185.html (accessed on 8 July 2020).
47. Dorgan, S. Image Processing and Data Analysis with ERDAS IMAGINE. *Photogramm. Eng. Remote Sens.* **2020**, *86*, 597–598. [CrossRef]
48. Long, W.; Sriharan, S.; IEEE. Land cover classification of SSC image: Unsupervised and supervised classification using ERDAS imagine. In Proceedings of the IEEE International Geoscience and Remote Sensing Symposium, Anchorage, AK, USA, 20–24 September 2004; pp. 2707–2712.
49. Lei, N.; Hu, X.Y.; Zhou, X.H. Remote Image's Geometry Revision Based on ERDAS IMAGINE: A Case Study in Jiaozhou Bay. In Proceedings of the AASRI Conference on Environmental Management and Engineering (AASRI-EME 2011), Wuhan, China, 26–27 November 2011; pp. 56–60.
50. He, Y.; Tang, C.; Zhou, G.; He, S.; Qiu, Y.; Shi, L.; Zhang, H. Spatial Conflict Measurement in Rapidly Urbanizing Regions Based on Geography Perspective: An Example of Chang-Zhu-Tan City Cluster Region. *J. Nat. Resour.* **2014**, *29*, 1660–1674. [CrossRef]
51. Yu, K. Security patterns and surface model in landscape ecological planning. *Landsc. Urban Plan.* **1996**, *36*, 1–17. [CrossRef]
52. Huang, X.X.; Wang, H.J.; Shan, L.Y.; Xiao, F.T. Constructing and optimizing urban ecological network in the context of rapid urbanization for improving landscape connectivity. *Ecol. Indic.* **2021**, *132*, 108319. [CrossRef]
53. Tang, F.; Zhang, P.-t.; Zhang, G.-j.; Zhao, L.; Zheng, Y.; Wei, M.-h.; Jian, Q. Construction of ecological corridors in Changli County based on ecological sensitivity and ecosystem service values. *Yingyong Shengtai Xuebao* **2018**, *29*, 2675–2684. [CrossRef] [PubMed]
54. Li, Y.Y.; Zhang, Y.Z.; Jiang, Z.Y.; Guo, C.X.; Zhao, M.Y.; Yang, Z.G.; Guo, M.Y.; Wu, B.Y.; Chen, Q.L. Integrating morphological spatial pattern analysis and the minimal cumulative resistance model to optimize urban ecological networks: A case study in Shenzhen City, China. *Ecol. Processes* **2021**, *10*, 63. [CrossRef]
55. Zhu, Q.; Yu, K.; Li, D. Ecological corridor width in landscape planning. *Acta Ecol. Sin.* **2005**, *25*, 2406–2412. [CrossRef]
56. Liang, W.; Wong, L.C.; Wong, J.Y.P. Ardeid nesting colony survey in Hainan, China. *Waterbirds* **2006**, *29*, 69–75. [CrossRef]
57. Hwang, J.-W.; Lee, J.-Y.; Choi, Y.-S.; Nam, H.-K.; Park, J.-Y.; Hur, W.-H.; Jeong_Chil, Y. Distribution and breeding of the Chinese Pond Heron (Ardeola bacchus) in Korea. *Korean J. Ornithol.* **2021**, *28*, 59–65. [CrossRef]
58. Fasola, M.; Galeotti, P.; Dai, N.H.; Dong, Y.H.; Zhang, Y.M. Large numbers of breeding egrets and Herons in China. *Waterbirds* **2004**, *27*, 126–128. [CrossRef]
59. Li, J.-Y.; Li, S.-P.; Sun, Y.-F.; Wu, Y.-F.; Wu, M.-L. Population dynamics and breeding space niche of four heron species in Tanghai wetlands. *Zool. Res.* **2006**, *27*, 351–356. [CrossRef]
60. Ma, Q.; Xiao, W.-F.; Su, H.-L. The Preliminary Study on Breeding Habit of Leiothrix lutea at Xingshan County, Hubei. *Chin. J. Zool.* **2010**, *45*, 46–51. [CrossRef]
61. Zhang, Z.Q.; Hou, D.H.; Xun, Y.; Zuo, X.W.; Yang, D.D.; Zhang, Z.W. Nest-site microhabitat association of red-billed leiothrix in subtropical fragmented forest in central China: Evidence for a reverse edge effect on nest predation risk? *J. Nat. Hist.* **2016**, *50*, 1483–1501. [CrossRef]
62. Herrando, S.; Llimona, F.; Brotons, L.; Quesada, J. A new exotic bird in Europe: Recent spread and potential range of Red-billed Leiothrix Leiothrix lutea in Catalonia (northeast Iberian Peninsula). *Bird Study* **2010**, *57*, 226–235. [CrossRef]
63. Tojo, H.; Nakamura, S. The breeding ecology of the introduced red-billed leiothrix on Mt. Tsukuba, central Japan. *J. Ornithol.* **2006**, *147*, 263. [CrossRef]
64. Yadav, K.; Ghimirey, Y.; Ghimire, S.K.; Prajapati, A.; Acharya, R. Observations of Siberian Weasel Mustela sibirica in Api-Nampa Conservation Area, Darchula district and Humla district, Nepal. *Small Carniv. Conserv.* **2019**, *57*, 14–19.
65. Hua, Y.; Zhang, W.; Xu, Y. Seasonal variation of pelage characteristics in Siberian weasel (Mustela sibirica) of Xiaoxing' anling area, Heilongjiang, China. *Acta Theriol. Sin.* **2010**, *30*, 110–114. [CrossRef]

66. Zhigalin, A. Siberian weasel Mustela sibirica Pallas, 1773 predatism on bats during winter period. *Subterr. Biol.* **2019**, *32*, 111–117. [CrossRef]
67. Sasaki, H.; Ohta, K.; Aoi, T.; Watanabe, S.; Hosoda, T.; Suzuki, H.; Abe, M.; Koyasu, K.; Kobayashi, S.; Oda, S. Factors affecting the distribution of the Japanese weasel Mustela itatsi and the Siberian weasel M-sibirica in Japan. *Mammal Study* **2014**, *39*, 133–139. [CrossRef]
68. Li, X.; Chen, Y.M.; Liu, X.P.; Li, D.; He, J.Q. Concepts, methodologies, and tools of an integrated geographical simulation and optimization system. *Int. J. Geogr. Inf. Sci.* **2011**, *25*, 633–655. [CrossRef]
69. Gantumur, B.; Wu, F.L.; Vandansambuu, B.; Tsegmid, B.; Dalaibaatar, E.; Zhao, Y. Spatiotemporal dynamics of urban expansion and its simulation using CA-ANN model in Ulaanbaatar, Mongolia. *Geocarto Int.* **2022**, *37*, 494–509. [CrossRef]
70. Li, X.; Yeh, A.G.O. Calibration of cellular automata by using neural networks for the simulation of complex urban systems. *Environ. Plan. A Econ. Space* **2001**, *33*, 1445–1462. [CrossRef]
71. Chen, Y.; Li, X.; Liu, X.; Li, S. Early warning of farmland protection zones based on GeoSOS, a coupled geosimulation optimization system. *Acta Geogr. Sin.* **2010**, *65*, 1137–1145. [CrossRef]
72. Zhang, Y.H.; Li, X.; Liu, X.P.; Qiao, J.G. Self-modifying CA model using dual ensemble Kalman filter for simulating urban land-use changes. *Int. J. Geogr. Inf. Sci.* **2015**, *29*, 1612–1631. [CrossRef]
73. Li, X.; Yeh, A.G. Neural-network-based cellular automata for simulating multiple land use changes using GIS. *Int. J. Geogr. Inf. Sci.* **2002**, *16*, 323–343. [CrossRef]
74. Li, X.; Yeh, A.G.O. Data mining of cellular automata's transition rules. *Int. J. Geogr. Inf. Sci.* **2004**, *18*, 723–744. [CrossRef]
75. Li, X.; Chen, G.Z.; Liu, X.P.; Liang, X.; Wang, S.J.; Chen, Y.M.; Pei, F.S.; Xu, X.C. A New Global Land-Use and Land-Cover Change Product at a 1-km Resolution for 2010 to 2100 Based on Human-Environment Interactions. *Ann. Am. Assoc. Geogr.* **2017**, *107*, 1040–1059. [CrossRef]
76. Li, X.; Yeh, A.G.O. Knowledge discovery for geographical cellular automata. *Sci. China Ser. D Earth Sci.* **2005**, *48*, 1758–1767. [CrossRef]
77. Ren, J.; Zhou, W.; Liu, X.L.; Zhou, L.; Guo, J.; Wang, Y.H.; Guan, Y.J.; Mao, J.T.; Huang, Y.H.; Ma, R.R. Urban Expansion and Growth Boundaries in an Oasis City in an Arid Region: A Case Study of Jiayuguan City, China. *Sustainability* **2020**, *12*, 210. [CrossRef]
78. Xia, L. GeoSOS Tutorial. Available online: https://www.geosimulation.cn/GeoSOS/index.html (accessed on 8 July 2022).
79. You, H.Y.; Yang, X.F. Urban expansion in 30 megacities of China: Categorizing the driving force profiles to inform the urbanization policy. *Land Use Policy* **2017**, *68*, 531–551. [CrossRef]
80. Li, B.; Zhang, Y.; Gan, T. Analysis of spatial and temporal changes in the landscape pattern of ChangZhuTan city cluster. *Chin. Foreign Archit.* **2018**, *11*, 57–61.
81. Huang, L.; Wang, D.R.; He, C.L. Ecological security assessment and ecological pattern optimization for Lhasa city (Tibet) based on the minimum cumulative resistance model. *Environ. Sci. Pollut. Res.* **2022**, 29. [CrossRef]
82. Han, Y.; Yu, C.Y.; Feng, Z.; Du, H.C.; Huang, C.S.; Wu, K.N. Construction and Optimization of Ecological Security Pattern Based on Spatial Syntax Classification-Taking Ningbo, China, as an Example. *Land* **2021**, *10*, 380. [CrossRef]
83. Li, H.; Zhang, T.; Cao, X.S.; Zhang, Q.Q. Establishing and Optimizing the Ecological Security Pattern in Shaanxi Province (China) for Ecological Restoration of Land Space. *Forests* **2022**, *13*, 766. [CrossRef]
84. Li, W.J.; Xie, S.Y.; Wang, Y.; Huang, J.; Cheng, X. Effects of urban expansion on ecosystem health in Southwest China from a multi-perspective analysis. *J. Clean. Prod.* **2021**, *294*, 126341. [CrossRef]
85. Tajuddin, N.; Dabrowski, M. Enabling Socio-Ecological Resilience in the Global South: Insights from Chennai, India. *Sustainability* **2021**, *13*, 10522. [CrossRef]
86. Hou, W.; Zhou, W.; Li, J.Y.; Li, C. Simulation of the potential impact of urban expansion on regional ecological corridors: A case study of Taiyuan, China. *Sustain. Cities Soc.* **2022**, *83*, 103933. [CrossRef]
87. Meneguetti, K.S.; de Oliveira, F.L. Reconsidering green belts, green wedges and greenways. *Acta Sci. Technol.* **2021**, *43*, e55196. [CrossRef]
88. Zheng, H.W.; Shen, G.Q.; Wang, H.; Hong, J.K. Simulating land use change in urban renewal areas: A case study in Hong Kong. *Habitat Int.* **2015**, *46*, 23–34. [CrossRef]
89. Mitsova, D.; Shuster, W.; Wang, X.H. A cellular automata model of land cover change to integrate urban growth with open space conservation. *Landsc. Urban Plan.* **2011**, *99*, 141–153. [CrossRef]
90. Yang, J.; Shi, F.; Sun, Y.Z.; Zhu, J. A Cellular Automata Model Constrained by Spatiotemporal Heterogeneity of the Urban Development Strategy for Simulating Land-use Change: A Case Study in Nanjing City, China. *Sustainability* **2019**, *11*, 4012. [CrossRef]
91. Meerow, S.; Newell, J.P. Spatial planning for multifunctional green infrastructure: Growing resilience in Detroit. *Landsc. Urban Plan.* **2017**, *159*, 62–75. [CrossRef]
92. Schiappacasse, P.; Müller, B. Planning Green Infrastructure as a Source of Urban and Regional Resilience—Towards Institutional Challenges. *Urbani Izziv* **2015**, *26*, S13–S24. [CrossRef]

land

MDPI

Article

Urban Sensory Gardens with Aromatic Herbs in the Light of Climate Change: Therapeutic Potential and Memory-Dependent Smell Impact on Human Wellbeing

Izabela Krzeptowska-Moszkowicz, Łukasz Moszkowicz and Karolina Porada *

Faculty of Architecture, Cracow University of Technology, 31-155 Krakow, Poland; ikrzepto@pk.edu.pl (I.K.-M.); lmoszkowicz@pk.edu.pl (Ł.M.)
* Correspondence: kporada@pk.edu.pl

Abstract: The aim of this study was to analyze urban sensory gardens containing aromatic herbs in terms of the plants used in them. The analysis considered the impact of climate change, particularly of higher temperatures, which may affect the character of contemporary urban gardens. The study was planned primarily in the context of the gardens' therapeutic significance to their users. An important part of the work was to analyze how particular aromatic plants are perceived and received by the inhabitants, using the example of one of Poland's largest cities, Kraków, to assess whether they can have an impact on the inhabitants' positive memories and thus improve their well-being. Initially, the plant composition of gardens located in Poland that feature aromatic herbs was analyzed. This was followed by a survey and an analysis of therapeutic gardens using the Trojanowska method as modified by Krzeptowska-Moszkowicz et al. The plant composition analysis of sensory gardens featuring herbs demonstrated that vulnerable plants in the Central European climate are being introduced to urban sensory gardens. In terms of major aromatic plants, it was found that almost every respondent reported the existence of scents that had some form of essential significance associated with personal memories. Considering the important sensory impact of water elements in therapeutic gardens, as well as problems related to the acquisition of drinking water or water used in agriculture or horticulture, the paper also addresses this topic. It was found that the city dwellers who filled in the questionnaire strongly preferred the introduction of more ecological solutions in the gardens related to water use—to collect and use rainwater, e.g., for watering, instead of piped water.

Keywords: sensory gardens; climate change; urban green spaces; aromatic herbs; plant smell memory; stress; human wellbeing in urban areas; restorative environments

Citation: Krzeptowska-Moszkowicz, I.; Moszkowicz, Ł.; Porada, K. Urban Sensory Gardens with Aromatic Herbs in the Light of Climate Change: Therapeutic Potential and Memory-Dependent Smell Impact on Human Wellbeing. *Land* **2022**, *11*, 760. https://doi.org/10.3390/land11050760

Academic Editors: Jan K. Kazak, Magdalena Wilkosz-Mamcarczyk and Alexandru-Ionuţ Petrişor

Received: 28 March 2022
Accepted: 18 May 2022
Published: 23 May 2022

1. Introduction

1.1. Sensory Gardens with Herbs as Therapeutic Places in the Urban Tissue

Publicly accessible sensory gardens are a part of urban public greenery [1]. Similarly other well-designed green areas in cities, they can act primarily as restorative gardens if they guide the user to achieving an internal bodily balance (homeostasis) [2], which results in an improvement in one's emotional and physical fitness, even if by merely reducing stress. A sensory garden is a special type of garden, listed among therapeutic gardens, and as such has significant use in therapy [3]. Such gardens should be designed so that humans are able to experience them from up close [4]. Although the primary function of a sensory garden is to affect its users' senses, it is defined in various ways in the literature, with listings of commonly used definitions provided by Husein [5], Krzeptowska-Moszkowicz et al. [6], Wajchman-Świtalska et al. [7]. To us, the sense of a publicly accessible urban sensory garden is expressed most precisely in the definition of the British Sensory Trust, which states that a garden is: "a self-contained area that concentrates a wide range of sensory experiences [...], such an area, if designed well, provides a valuable resource for a wide range of uses, from education to recreation" [8]. In an urban setting, sensory gardens can be established for various types of users and are thus intended for different forms of activity [6].

Sensory gardens should be designed with special care so that they can fulfil their assigned therapeutic tasks [3,4,9,10]. Plants are a crucial component of these gardens, with aromatic herbs being particularly of value [3]. Herbs are defined as "any plant with leaves, seeds or flowers used for flavoring, food, medicine, fragrance production, etc." [11]. In ages past, herbs also held spiritual significance—they had a religious, ritual, or symbolic value [12]. This highlights their important and wide ranging effects on people. Many herbs that appeared in Central and Eastern Europe by the Middle Ages and arrived through convents and monastic gardens, were originally prevalent in warmer areas of Europe and Asia [12–14]. They made their way from monastic gardens to lay ones, and to the collective conscious as part of medicinal preparations, seasoning for meals, or in various types of ceremonies, which contributed to their widespread recognition [12]. Accounting for the origins of many such plants, it was not always possible to cultivate them in the soil in the climate of Poland. However, the currently observed climate change could visibly alter the appearance and extend the significance of gardens that feature herbs.

At present, herbs are desirable element of contemporary gardens and play a crucial role in selected therapeutic programs [3]. This is due to the influence of the scent of aromatic herbs on the sense of smell that does not limit itself to the flowers of such plants, as it also includes their vegetative elements, which extends their application scope. A study by Zajadacz et al. [15] noted that scents play a greater role in the spatial orientation of blind people in sensory gardens than they do outside of them. Arslan et al. [16] also discussed the general matters of the application of aromatic herbs in therapeutic gardens. Aromas, including herbal ones, are also crucial for another reason, as the key significance of scents in stimulating memory-dependent bodily functions was observed, allowing us to "transfer the past into the present" [17,18] and to build certain new relations based on previous experiences. This is used in practice in, among other places, sensory gardens near special education facilities—in children's therapy, where a familiar plant scent, especially one associated with home and family, contributes to better performance in spatial orientation tasks and building self-confidence and autonomy [4].

1.2. Climate Change, Greenery, and Human Health

Global climate change results in local phenomena that affect specific regions and countries [19]. Rising temperatures affect people who live in cities in an especially negative way, with high summer temperatures having a profoundly negative impact. This problem is further compounded by the so-called urban heat island effect that occurs in urbanized areas [20]. Heat waves, describing long periods of elevated temperature, are another compounding effect of climate change, and were found to have a particularly detrimental effect on human health and wellbeing, especially in seniors [21,22]. Such phenomena have been observed in Kraków, one of Poland's largest cities, among others [23].

The presence of urban greenery can alleviate the negative consequences of high temperatures, for instance by plant transpiration or through the visibly lower heat accumulation of plant-occupied surfaces in comparison to paved surfaces or open areas [24,25]. A green environment can also positively affect the health and wellbeing of urban residents, even via its passive observation [26]. Numerous studies found that humans desire contact with nature and display a preference for a green setting instead of a typically urbanized one when choosing a place to regenerate after mental exhaustion. Being present in a greenery-rich environment was found to facilitate psychological renewal [27,28]. As opposed to stress and the rapid pace of urban life, a garden offers a slow rhythm and an absence of disruptiveness via its changing plants, which is noted by its users, and brings calmness and peace [29]. Studies performed in Great Britain found that, in a big-city setting, visiting urban green areas led to different types of wellbeing benefits than strolling outside the city. It was found that it reduces anxiety [30]. The positive impact of green spaces on various health-related human problems was investigated in numerous studies, which were collected and briefly discussed by Chiabai et al. [31]. Papers that describe its impact on the mental wellbeing of adults were summarized by Houlden et al. [32]. Urban greenery positively affects the residents of urban areas and can remediate the negative consequences of climate change, yet it is also impacted by such consequences. Positive phenomena in this respect include an extension of the vegetation period both in Poland and other European countries [33], which contributes to gardens appearing attractive for longer periods, and in the case of therapeutic gardens, extends the season during which they can affect their users. In the light of climate change, a simulated temperature increase of 1 °C relative to the final three decades of the twentieth century was performed and showed a significant decrease in the temperate cool region's reach, and an increase in the area of the temperate warm region, along with the appearance of a warm region [19], which can contribute to changes in the composition of species introduced into green areas in individual territories, and can lead to the spread of thermophilic flora, especially in cities [34,35]. The estimated temperature increase is expected to lead to an increased water deficit [21], which is a climate-change consequence that negatively affects plants and gardens.

1.3. Goal of the Study

The purpose of the survey study among the residents of a large city was to investigate the way respondents perceived the smells of various herbs and to reach deeper to determine whether specific herbs had positive memory-dependent significance to respondents. The practical applications of the answers to these questions may aid in better understanding the restorative and therapeutic impact of sensory gardens featuring herbs on city residents. Therefore, the study also featured an analysis of the therapeutic potential of existing Polish gardens with sensory features which primarily targeted the sense of smell. The outcome of this analysis was to indicate how their therapeutic scope can be extended. Problems associated with climate change were also investigated. Our work also takes into account the problems associated with climate change, particularly that of rising temperatures in cities, which may affect existing urban gardens.

2. Materials and Methods

2.1. Research Sites

The study investigated six publicly accessible urban gardens located in Poland (Table 1). The gardens were selected due to aromatic herbs being their major component, in addition to being intentionally designed as fragrant gardens. The selected gardens were relatively well-known domestically, as they had either already appeared in studies on therapeutic gardens, were featured in garden trails, or were located in areas frequently visited by people from various areas of Poland and abroad. Some of the gardens directly referenced the Middle Ages or the Renaissance, periods when the sense of smell was held in high regard. All of the analyzed gardens were contemporary, although some of them were located in areas previously occupied by historical gardens.

Publicly accessible gardens are defined as gardens that are accessible to all visitors and can be visited either during specific hours or are always open. The following urban gardens with sensory features and located in Poland were selected for the study:

1. Frombork: Herbal garden near a historical medieval hospital building belonging to the Holy Spirit Hospital—currently a museum;
2. Kamień Śląski: The S. Kneipp herbal garden—near a Kneipp Institute therapeutic facility [10];
3. Kraków: The herbal garden of the Czapski Pavilion—a museum;
4. Sandomierz: The Garden of Marcin of Urzędów—a garden referencing a medicinal Renaissance garden [36];
5. Kraków: Zapachowo fragrance garden and sensory path at the S. Lem Science Garden [7,37];
6. Solec Zdrój: Educational Path—Aromatherapeutic Avenue—a fragrant garden built by the Solec Zdrój Municipality.

Table 1. Overview of the gardens under study.

	GARDENS FEATURING HERBS				FRAGRANT GARDENS	
	1	**2**	**3**	**4**	**5**	**6**
Purpose of establishment	Educational, collecting plants used in medicine	Garden used in Kneipp therapy, intended for aromatherapy strolls	Container garden intended to affect the senses and support pollinating insects living in the city	Educational, used to present selected medicinal plants that used to grow in M. of Urzędów's Renaissance medicinal garden	Intended to stimulate the sense of smell, located in a sensory education park that references the thoughts and deeds of M. Kukelhaus	Intended to stimulate the sense of smell, located in a sensory education park that references the thoughts and deeds of M. Kukelhaus
Primary target users	Museum visitors	Kneipp Institute patients undergoing therapy	Museum visitors and city residents	Museum visitors	Primarily children and youth	City residents, patients of the nearby spa treatment center
Garden composition	Freeform layout, herbs are grown in various places around the garden in marked beds, the path has a freeform outline	Geometric layout, the path runs between square-shaped beds with individual herb species, a pergola covered with creepers runs perpendicular	Geometric layout, composed of a square-shaped lawn and a path along one of its sides, with plant containers alongside it	Geometric composition, modeled after a Renaissance garden, a circular bed occupies a central position	Freeform composition, the path meanders between clusters of aromatic herbs and bushes	Generally geometric composition, large beds divided by slanted paths into quadrangles of varying size

Poland is in Central Europe, in a temperate climate zone. Poland is noted to have a transitional climate influenced by Western Europe's marine climate and Eastern Europe's continental climate. Additionally, due to the influx of a diverse range of air masses, the Polish climate is characterized by a degree of variance. Poland's territory can be divided into subregions that differ in temperature, the course of the seasons and amount of rainfall that occurs [19,38]. Temperatures, especially temperature extremes, and particularly those in winter, have a significant impact on the plant types cultivated in gardens. Therefore, so-called frost-resistance zones are delineated, with three such zones in Poland, which are taken into consideration during species-composition formulation for gardens. See Figure 1.

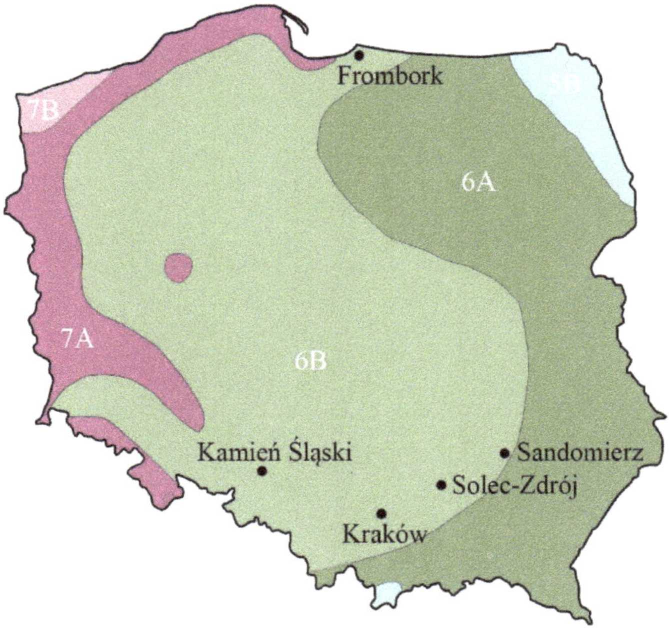

Figure 1. This illustration shows a map of Poland with the location of the herb gardens analyzed in this work. Colors and numbers with letters: 5B, 6A, 6B, 7A, 7B indicate frost zones of plants in Poland.

2.2. Methods

At the start of the study, the selected plant material present in the gardens was analyzed, with a focus on species that provide olfactory stimulation. This was achieved using the available species lists and with surveys performed in all the gardens. A list of aromatic herbs used in the gardens was compiled. In addition, the species that were listed in the literature as struggling to survive winters in the Polish climate or unable to survive them while planted in the soil were enumerated. The frequency of using temperature-sensitive plants in the sensory gardens under study was investigated, as well as the ways the most vulnerable plants are able to survive winter in urban conditions in the current Polish climate.

A survey study was used to investigate how the individual herb fragrances present in such gardens were perceived by the residents of a large city—Kraków. The survey sample consisted of 73 people. The sample group was small and therefore cannot be considered representative, but in our opinion it is sufficient for the preliminary study described. In future research, the target group of respondents will be expanded.

Seeing as sensory gardens are often dedicated to specific groups of city-dwelling users [6] with distinctive needs, the respondents were divided into the following three age groups: group A—young people between 18 and 29 years of age, typically students or engaging in first-time employment; group B—people between 30 and 59 years of age, either employed or active in their respective families, group C—seniors aged 60 and above. The respondents were asked six questions, with three concerning aromatic herbs and the remaining three concerning water in herbal gardens. Those aromatic herbs that appeared the most often in Polish gardens with sensory features associated with the sense of smell as well as species sensitive to low temperatures were selected for this study. Concerning water in herbal gardens, the questions in the survey asked whether respondents found it to be significant in herbal gardens, its significance in those gardens and whether the respondents found where said water was sourced from to be significant.

In the first stage, an analysis of the gardens' therapeutic potential was performed using the Trojanowska method [39,40], as adapted to gardens with sensory features by Krzeptowska-Moszkowicz et al. [6]. The method is based on an assessment of the gardens' attributes, whereby an attribute is understood as "a feature of space or the presence of features" [39]. In this method, as the number of attributes increases, so does an area's therapeutic potential. It is a method that has, upon suitable adaptation, been successfully used to analyze therapeutic areas of various sizes, such as sensory gardens [4], parks [39,40] and large areas such as a seacoast fragment with natural and cultural value [41].

The attributes were grouped into design stages as follows: (1) functional program, (2) functional–spatial structure, (3) the design of internal spaces and architectural form, (4) placemaking, (5) accounting for sustainability requirements [39]. The last group of attributes, namely sustainability requirements, is essential for two reasons. First, it brings to mind the problems that appear in green areas, including in response to climate change, and the proper management of natural resources by humans. The second crucial aspect concerns the associated properties of a sensory garden that determine its perception by users.

3. Results

3.1. Aromatic Herbs in the Gardens with Sensory Features

Table 2 shows the species used in the gardens under study and how often they appeared. The listing includes plants from warmer areas of the world, which are vulnerable during winter, and thus are interesting from a climate-change standpoint. Furthermore, two species of aromatic herbs that appeared less often in these areas while being well-known in Poland were of note (Levisticum officinale W.D.J. Koch and Melissa officinalis L.). The nine species identified were later featured in the survey intended for later stages of the study.

True lavender appeared in all the gardens, in addition to *Salvia officinalis* L. or other species of salvia. Both *Nepeta* × *Faasenii Bergmans ex Stearn* and *Origanum vulgare* L. were not as prevalent, but still common, as they were present in five cases. Mint, *Mentha x piperita* L., was also used relatively frequently, but mostly peppermint, in addition to *Hyssopus officinalis* L., which was noted in four of the six analyzed gardens. *Melissa officinalis* L., lemon balm, as well as a number of herbs used as dried seasoning in cooking, were rarer, present in only a half of the gardens, which is why they were not included in the survey. Our study showed that other aromatic herbs were only present in gardens in singular cases, and as such cannot be said to be popular as plants used in publicly accessible urban herbal and fragrant gardens.

The group of aromatic herbs that were previously considered not to winter in the soil or only partially winter this way includes as many as four species found in the gardens under study. Interestingly, plants from the following three genera: *Lavandula*, *Salvia* and *Hyssopus*, are very popular in these gardens and were found in either all or nearly all the gardens. Even rosemary, *Rosmarinus officinalis*, which is highly sensitive to low temperatures and was seen as a plant that does not winter in Poland, was present in two gardens. Therefore, it can be stated that sensitive plants were boldly introduced into contemporary urban gardens.

Table 2. Presence of aromatic herbs in the gardens under study. The plants typically used in these gardens, as well as more vulnerable plants originating from warmer areas of the world. V is a stamp which should be read as "yes", in this case a confirmation that the plant is in the assigned garden.

PLANT	RESILIENCE/WINTERING IN POLAND	GARDENS FEATURING HERBS						NUMBER OF GARDENS
		1	2	3	4	5	6	
PERENNIALS THAT WINTER IN POLAND								
Nepeta × *Faasenii Bergmans ex Stearn*	perennial	v	v	v	-	v	v	5/6
Origanum vulgare L.	perennial	v	v	v	v	v	-	5/6
Mentha × *piperita* L.	perennial	v	v	v	-	v	-	4/6
Other species and varieties: *Mentha* sp.		-	v	v	-	v	-	
Melissa officinalis L.	perennial	v	v	-	-	v	-	3/6
Levisticum officinale W. D. J. Koch	perennial	v	-	v	-	-	-	2/6
VULNERABLE SPECIES FROM WARMER REGIONS								
Lavandula angustifolia Mill.	prostrate shrub/partially winters	v	v	v	v	v	-	5/6
Salvia officinalis L.	subshrub/partially winters	v	v	v	-	v	-	6/6
Other species: *Salvia* sp.		-	-	-	v	-	v	
Hyssopus officinalis L.	subshrub/partially winters	v	v	-	v	v	-	4/6
Rosmarinus officinalis L.	wintergreen prostrate shrub/winters poorly	v	-	v	-	-	-	2/6

3.2. Kraków's Residents' Associations with the Smell of Herbs

The results of the survey performed as a part of this study showed that the vast majority of aromatic herbs and their smells were positively perceived by the respondents—residents of Kraków (Figure 2). Although, the smell of catnip *Nepeta* sp. and lovage *Levisticum* sp. were indicated as liked by a smaller group. A number of respondents identified the smell of lovage *Levisticum* sp. as one they did not like, with a similar sentiment expressed for lavender *Lavandula* sp. and salvia *Salvia* sp. An altogether different result in comparison to the previous herbs was returned for the smell of hyssop *Hyssopus* sp., with the majority of respondents reporting this smell to be either completely unknown to them or that it was neither pleasant nor unpleasant. This can mean that although it is a generally well-known plant, as a plant mentioned in the Bible, it was not known to the respondents from direct experience, due to its infrequent use in private or public gardens in Poland.

Almost all of the respondents reported having a favorite aromatic plant, whose smell they saw as particularly pleasant and bringing to mind positive associations. The most important plants to which the respondents ascribed such significance were herbs such as lavender *Lavandula* sp. and mint *Mentha* sp. (Figure 3). There was a slight difference according to the respondents' age. Among young persons between 18 and 29 years of age, the highest number of respondents pointed to lavender *Lavandula* sp., followed by mint *Mentha* sp., which was in turn followed by lemon balm *Melissa* sp., oregano *Origanum* sp., and rosemary *Rosmarinus* sp., which were picked much less often. Among older persons, aged 30 and above, the greatest number of respondents listed their most liked and positively associated herbs as mint *Mentha* sp., followed by lavender *Lavandula* sp., rosemary *Rosmarinus* sp., and lovage *Levisticum* sp., respectively.

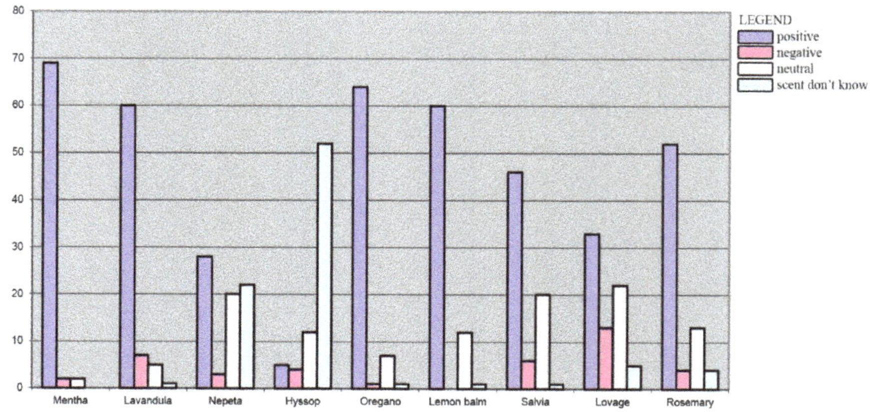

Figure 2. Results showing the perception of the smell of each herb as reported by respondents.

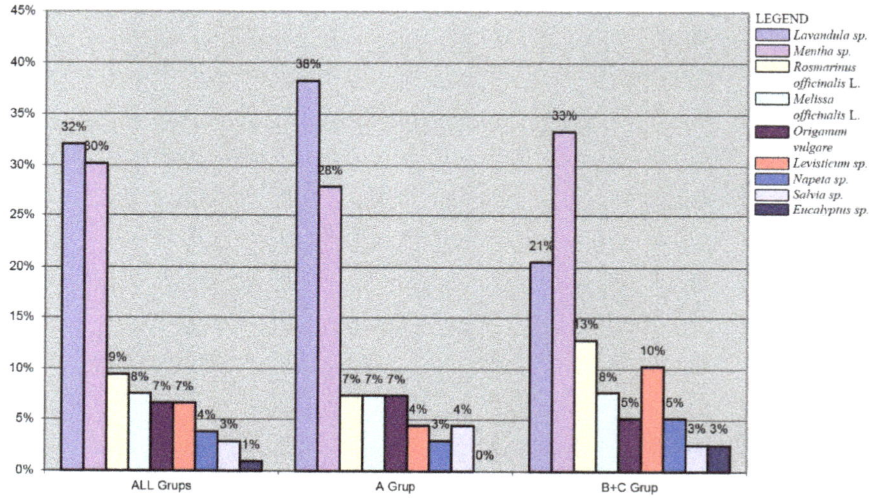

Figure 3. Results presenting associations with a smell that held particular significance to respondents divided into two age groups—young people up to 29 years of age (group A) and aged 30 and above (groups B and C).

The respondents reported that the smells in question brought clearly positive associations, primarily linked with home and family, often simultaneously with childhood and youth, the vacation season, as well as memories of exceptional travels, mostly during summer (Table 3). In every age group, positively perceived smells carried at least one of the above meanings. However, a difference in the percentage shares in their ascription to each meaning was observed. In the group with the youngest respondents (group A), the greatest significance was found for associating the smell of one's favorite herbs with home and family (70%), the middle-aged group saw smells associated with home, travel, and summer as equally important, while seniors reported positive associations mostly with home and family (60%).

Table 3. Specific associations with past events listed by respondents as tied with preferred aromatic herbs.

	ALL GROUPS	AGE GROUP A	AGE GROUP B	AGE GROUP C
Home, family (%)	60	70	37	60
Exceptional travels—often during vacation season, the vacation season (%)	26	21	43	10
Other—typically described as a personal preference or non-descript (%)	14	10	20	30

In terms of the specific significances that respondents ascribed to the smell of individual herbs (Table 4), most respondents across all age groups reported that it was the smell that was the most important, yet a large group of respondents tied it with the addition of a given herb to a specific beverage or meal. The significance of the application of a given herb as a home remedy and the association of its smell with a preparation intended to aid in minor health disorders was reported to be smaller, yet still substantial.

Table 4. The significance of favorite herbs and their smell.

	ALL GROUPS	AGE GROUP A	AGE GROUP B	AGE GROUP C
The smell of the plant itself (%)	56	61	50	40
The smell of herbs added to beverages or meals (%)	30	27	32	40
The smell of a plant as a homemade medicine (%)	9	9	7	20
The smell of a plant used as an anti-insect agent (%)	3	1	7	0
Symbolic plant with spiritual significance (%)	2	1	4	0

3.3. Therapeutic Potential of Gardens Featuring Aromatic Herbs

The results of the analysis of the therapeutic potential of the gardens under study are presented in Tables 5 and 6 and are graphically illustrated in Figure 4. Most gardens with aromatic herbs were rated to have a potential of around 50%, with the range being 46–59%. Only one garden, near the J. Czapski pavilion, was found to have a slightly higher therapeutic potential of 66%.

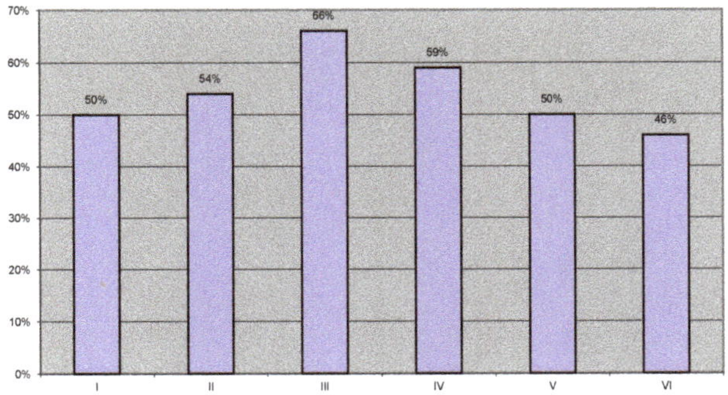

Figure 4. Overall results of the therapeutic potential analysis of the gardens under study.

Table 5. Analysis of the therapeutic properties of herbal gardens with sensory features, part 1—functional program.

	ATTRIBUTES	HERBAL GARDENS				FRAGRANT GARDENS	
		1	2	3	4	5	6
FUNCTIONAL PROGRAM							
A. Enabling physical and psychological re-generation	A-1. Places for rest that facilitate experiencing the surroundings from up close	YES (1) Multiple seats	YES/NO (0.5) None in the garden; behind the pergola there is a meadow with sunbeds for patients	YES/NO (0.5) Seating only on the platform, no other seating	YES/NO (0.5) Near the path adjacent to the garden	YES/NO (0.5) A few benches, some deep inside the garden	YES (1) Numerous seats
	A-2. Isolation from the urban environment, noise, smells, and the pressure of time and fast living	YES/NO (0.5) Partially, there is a street just behind a fence	YES (1) The entire garden is walled, the herbal garden is in its corner	YES (1) The garden is in an internal courtyard	YES/NO (0.5) The museum grounds are fenced, the garden is near a fence that separates it from a street	YES (1) The fragrance garden is at the edge of the science garden, located at the side of an extensive park	YES/NO (0.5) Fenced from one side, close to a street from another
	A-3. Ability to easily observe animals or people	YES/NO (0.5) Observable insects and birds, herbal plants are up close, but there are taller perennials, bushes, and trees	YES/NO (0.5) Observable insects—the herbs are short, but in large clusters, there are also creepers	YES (1) Observable insects—the plants are in rather tall pots	YES/NO (0.5) Observable insects—the herbs are placed low, and grow in extensive clusters	YES/NO (0.5) Observable insects—many flowering plants are close to the soil, there are also bushes	YES/NO (0.5) The herbs are placed low and often grow in extensive surfaces, there are also creepers and bushes
B. Facilitating social contact	B-1. Ability to meet as a group	YES/NO (0.5) Only on the lawn	YES/NO (0.5) On the lawn near the garden, and for patients—on the lawn with sunbeds	YES (1) There is a dedicated meeting space—there are chairs and tables on the terrace, which can be moved	YES/NO (0.5) Only on the lawn near the outer edge of the garden	YES/NO (0.5) There is a small space with benches, but the main footpath crosses it	YES (1) There is a separate square-shaped interior with benches all around; There is another space under a pergola

Table 5. *Cont.*

ATTRIBUTES		HERBAL GARDENS				FRAGRANT GARDENS	
		1	2	3	4	5	6
C. Facilitating physical activity	C-1. Places for play and recreation	NO (0) The garden itself offers no such capacity; it serves other purposes	NO (0) The herbal garden is not intended for this, but there are other gardens nearby	YES/NO (0.5) There is a lawn, but it is not dedicated to physical exercise	NO (1) There is a lawn, but it is not intended for physical exercise due to the presence of the museum	YES (1) Numerous grassy spaces with devices to perform experiments nearby	NO (0) The garden is not intended for this purpose
	C-2. Place dedicated to gardening classes or hortitherapy	NO (0) Absent	NO (0) Absent	NO (0) Absent	NO (0) Absent	NO (0) Absent	NO (0) Absent
D. Meeting essential user needs	D-1. Safety in the garden space	YES (1) The area is adjacent to the museum and is fenced	YES (1) The garden is in a fenced area	YES (1) The garden is adjacent to the museum, in a courtyard	YES (1) The garden is adjacent to the museum, it is surrounded by a tall fence	YES (1) The garden is located within a fenced science garden	YES/NO (0.5) There is a street nearby shielded by a row of trees
	D-2. Safety in direct contact with plants	YES/NO (0.5) Many of the plants are safe, but some are not	YES/NO (0.5) Numerous safe plants: there are rose bushes	YES (1) Safe plans were used, primarily herbs	YES (1) Most plants are safe	YES/NO (0.5) Many of the plants are safe, there are rose bushes	YES (1) Most plants are safe
	D-3. Seating or shelter	YES (1) Shelter inside the museum	YES (1) Shelter under a pergola and in a gazebo near the garden	YES (1) Shelter inside a coffee shop with a terrace inside the garden	YES (1) Shelter in the museum	YES/NO (0.5) Shelter under tree canopies nearby	YES/NO (0.5) Shelter under a pergola, no other shelters
	D-4. Sunny and shaded places	YES (1) Both types of places are present	YES (1) There are sunny places and a shaded space under a pergola	YES (1) Both types of places are present	YES/NO (0.5) Both types of places are partially present	YES/NO (0.5) The benches are not shaded in any way	YES (1) Places in the sun and in the shade—under a pergola and tree canopies

Table 5. *Cont.*

ATTRIBUTES		HERBAL GARDENS				FRAGRANT GARDENS	
		1	2	3	4	5	6
	D-5. Amenities for the disabled	NO (0) No dedicated amenities for the disabled	YES/NO (0.5) Wide, smooth paths; no other amenities	YES/NO (0.5) Wide paths; no dedicated amenities	NO (0) Absent	NO (0) Aggregate footpath with narrow paved strips, not intended for wheelchair use	YES/NO (0.5) The paths are wide and smooth; no amenities for the visually impaired
	D-6. Elements that indirectly affect comfort of use: access to food and drink, toilets, and others	YES (1) Inside the museum	NO (0) None in the immediate vicinity	YES (1) There is a coffee show with a garden view	YES (1) Inside the museum	YES (1) Elements are present nearby, in the park	NO (0) Absent
E. Cognitive support	E-1. Features that facilitate education in the garden	YES/NO (0.5) Only plaques with plant names	YES (1) There are plaques with educational content and plant names	NO (0)	YES (1) There are plaques with educational content and plant names	YES (1) There are plaques with plant and genus names, as well as information on origin	YES (1) There are plaques with plant names and information about them; there is a garden plan
Score		7.5/13	7.5/13	9/13	8.5/13	8/13	7.5/13

Table 6. Analysis of the therapeutic properties of herbal gardens with sensory features, part 2—other attributes.

ATTRIBUTES		HERBAL GARDENS				FRAGRANT GARDENS	
		1	2	3	4	5	6
FUNCTIONAL STRUCTURE							
A. Functional–spatial structure	A-1. Isolation of the garden from its surroundings, creating a separate, intimate space	YES/NO (0.5) From one side, the fence is see-through, with a street alongside it. The fence is partially covered in vines	YES (1) The garden is surrounded by a tall wall	YES (1) The garden is in an internal courtyard surrounded by walls	YES (1) The garden is close to a fence entirely covered in vines; the street nearby is obscured	YES/NO (0.5) The garden is near the outer edge of a science garden, there are no tall insulation plants	YES/NO (0.5) There is a parking lot nearby, a street runs alongside it and is visible despite the presence of a row of trees

Table 6. *Cont.*

	ATTRIBUTES	HERBAL GARDENS			FRAGRANT GARDENS		
		1	2	3	4	5	6
	A-2. Siting in a place that retains fragrances and sounds inside the garden	YES/NO (0.5) Only in some places	YES (1) A tall wall shields from the wind and produces a quiet place	YES (1) The plants in pots receive adequate sunlight, the garden is surrounded by a solid fence and buildings	YES/NO (0.5) Fragrances are retained fully, but sounds only partially	YES (1) Behind the garden there is a small, elevated plateau that shields it from the north, there are also places with shrubs	YES/NO (0.5) Partially
Score		1/2	2/2	2/2	1.5/2	1.5/2	1/2
INTERNAL SPACE AND ARCHITECTURAL FORM DESIGN							
A. Internal space and architectural form design	A-1. Garden complexity, presence of various garden interiors, proper path system	YES/NO (0.5) Diverse interiors, the paths form loops; there are places for rest, there are no isolated interiors	YES/NO (0.5) Simple layout: there is a pergola with vines	YES/NO (0.5) Simple layout: a path runs around a lawn; plant pots only from one side	YES/NO (0.5) Intriguing garden layout that references a historical model, but the garden is small and has no distinct interiors	YES/NO (0.5) The path has an interesting course, but there are no distinct interiors that would allow for longer stays	YES (1) Interesting sensory garden layout, there are pergolas with resting places, there is a separate interior outside of the main area
	A-2. Legibility of composition	YES (1) The composition is legible	YES (1) The composition is legible	YES (1) The composition is geometric and legible	YES (1) The composition is geometric and legible	YES/NO (0.5) The composition is not fully legible, the path has dead ends	YES (1) The composition is geometric and legible
	A-3. Presence of water, especially water in motion	NO (0) None	YES/NO (0.5) There are none in the garden, but there is a stream nearby	NO (0) None	NO (0) None	NO (0) None	NO (0) None

Table 6. *Cont.*

	ATTRIBUTES	HERBAL GARDENS			FRAGRANT GARDENS		
		1	2	3	4	5	6
	A-4. Plant sensory impact on each of the senses	YES (1) Numerous senses are stimulated, but the plants cannot be tasted	YES (1) All five senses are stimulated	YES (1) The herbs provide diverse stimuli and can be tasted	YES (1) Numerous senses are stimulated, but the plants cannot be tasted	YES/NO (0.5) It is primarily a fragrance garden	YES (1) Numerous senses are stimulated, but the plants cannot be tasted
	A-5. Intensity of plant sensory impact (e.g., diversity of species, large spaces, elevated beds)	YES (1) The species are diverse	YES (1) Large herbal plant beds, there is a pergola with fragrant vines	YES (1) The plants are elevated and placed in large pots	YES (1) the herbs are placed in large groups, the path surface is lined with fragrant thyme	YES (1) Large surfaces covered with aromatic, closely placed plants, along almost the entire path	YES (1) A diverse range of sensory stimuli: there are fragrant herbs, creepers, and bushes
	A-6. Other sensory active elements (e.g., labyrinth, sensory path)	NO (0) None	NO (0) None	NO (0) None	YES/NO (0.5) There is a stone path inlaid with joints inlaid with thyme	YES (1) There is a sensory path and labyrinths close to the fragrant garden	NO (0) None
Score		3.5/6	4/6	3.5/6	4/6	3/6	4/6
PLACEMAKING							
A. Placemaking— personalization and animation of the space, art in the garden and special uses	A-1. Ability to personalize the space	NO (0) None	NO (0) None	NO (0) None	NO (0) None	NO (0) None	NO (0) None
	A-2. Ability to animate the space	YES/NO (0.5) A lecture on the history of medicine and herbs can be organized here	NO (0)	YES (1) Multimedia presentations, films on the museum's wall (seats on the grass)	YES/NO (0.5) A lecture on the plants from the work by Marcin of Urzędów can be organized	YES/NO (0.5) Open-air lessons for children, the youth and adults can be organized	NO (0)

Table 6. *Cont.*

ATTRIBUTES		HERBAL GARDENS				FRAGRANT GARDENS	
		1	2	3	4	5	6
	A-3. Artistic creations	NO (0) None	NO (0) None	YES (1) The museum wall features an artistic exhibition	YES/NO (0.5) There is an ornamental wall that references the garden's style	NO (0) None	NO (0) None
	A-4. Special indications for use	NO (0) None	NO (0) None	YES (1) The herbs can be tasted	NO (0) None	NO (0) None	NO (0) None
Score		0.5/4	0/4	3/4	1/4	0.5/4	0/4
SUSTAINABILITY							
A. Sustainability criteria	A-1. Biodiversity preservation: use of domestic plant species and plants attractive to various groups of animals, creating habitats for animals	YES (1) Some of the plants are attractive to insects; domestic plants; densely planted specimens of varying height offer room for habitats	YES/NO (0.5) Certain plants may be attractive to butterflies or other insects	YES/NO (0.5) Deliberate application of species attractive to hymenoptera	YES (1) Certain plants are attractive to insects; some of the plants are domestic species common in the Sandomierz area	YES/NO (0.5) Some plants are attractive to insects; domestic plants	YES/NO (0.5) Certain plants can be attractive to butterflies and other insects
	A-2. Sustainable water management, e.g., stormwater collection and use	NO (0) None	NO (0) None	NO (0) None	NO (0) None	NO (0) None	NO (0) None
	A-3. Natural energy sources	NOT APPLICABLE No electrical appliances	NOT APPLICABLE No electrical appliances	NO (0) None	NOT APPLICABLE No electrical appliances	NOT APPLICABLE No electrical appliances	NOT APPLICABLE No electrical appliances

Table 6. *Cont.*

	ATTRIBUTES	HERBAL GARDENS				FRAGRANT GARDENS	
		1	**2**	**3**	**4**	**5**	**6**
	A-4. Natural garden mainte-nance methods	No data available	YES (1) Yes, the plants can be harvested	YES (1)	No data available	No data available	No data available
Score		1/2	1.5/3	1.5/4	1/2	0.5/2	0.5/2
Total score		13.5/27	15/28	19/29	16/27	13.5/27	12.5/27

3.4. Water in Sensory Gardens Featuring Herbs

Due to global warming and the existence of the urban heat island effect, along with the occurrence of various associated problems concerning urban greenery and their impact on human health and wellbeing, we investigated the role of water in city gardens. This is why the survey performed as part of this study included questions on whether water and its sound can have a positive impact on the perception of gardens that include herbs and the significance it could have. Most respondents reported that it can have a positive impact to humans and city-dwelling animals—see Table 7.

Table 7. Problems associated with water in sensory gardens—survey responses.

	Respondent Responses
-Is the presence of water in a garden, e.g., as a fountain, important?	Yes—62 No—5 I have no opinion—6
-What does the presence of water mean to you?	Relaxation—26 Humidification—5 Relaxation and humidification—25 Water source for birds and insects—2 Others—3 It does not mean anything—8
-Where should the water used in a garden come from?	Harvested stormwater and rainwater—62 From the grid—2 It is irrelevant—9

In the light of the problems caused by water deficits and those of cities, the respondents were asked whether they found the source of water used in a garden to be significant. The majority of the respondents replied that they found it significant and that they believed that rainwater and stormwater should be the main source of water used in gardens.

4. Discussion

In gardens with sensory features, especially those that feature aromatic herbs, the impact of climate change, and especially global warming, has already become somewhat noticeable. Plants that were rarely seen in urban spaces in the twentieth century due to their possible freezing in the Polish climate [42] are currently more often encountered in cities. One such case is true lavender *Lavandula angustifolia* Mill., which in Poland was first introduced into private gardens, and in recent years it has increasingly been planted in public gardens, near parking lots and on church grounds. It was observed that this species was present in all of the herbal gardens investigated in this study. Rosemary *Rosmarinus officinalis* L. is another such example. It was seen as a pot plant in the Polish climate, originally present on the edges of the Mediterranean Sea, and is typically cultivated in European countries with mild winters [43]. This species has only recently been introduced in Polish public spaces, on a much smaller scale than lavender, as it is a more sensitive species. Rosemary was introduced into the garden near the J. Czapski Museum, which has sensory garden features. In Kraków's city center, the urban heat island effect has been observed, while along the area's edges, temperatures are not always higher than in the surrounding rural areas. This depends on a range of factors [23]. The aforementioned area near the Museum is located in the heart of the city, where temperatures are the highest. In terms of aromatic plants, high temperatures make the fragrances produced by them more intense. The area is also shielded from cross-ventilation by buildings, while from the north it is fenced by an insolated garden wall that is warmed by the sun. These are factors that contribute to the presence of favorable conditions for sensitive plants, including in winter. Rosemary is able to survive winter in such a setting without cold protection, which was observed for several consecutive years. As an evergreen plant, it greatly enlivens garden spaces with its vivid green color. Shanahan et al. argued that the visual features of a green environment and the biophysical parameters of greenery, stemming from the features of plant structure and physiological activity, both play a part in affecting human health [25].

Expanding the set of plants used in sensory gardens to include species that were previously seen as vulnerable, and which are familiar to and liked by city residents can enhance health benefits. A study by Bengtsson and Grahn [9] demonstrated that a significant role in rehabilitation and convalescence is played by high species diversity in therapeutic gardens, especially during the later stages of the process. These observations can also be applied to city parks, which is why apart from plants introduced by people, it is important to note factors that can contribute to the diversity of domestic species in park cover [44,45]. Many domestic plants or aromatic herbs are attractive to insects and observing animals in therapeutic gardens enhances the scope of sensory stimuli [46] and supports therapy [47]. A study by Cooper Marcus et al. [48] also showed just how important the inclusion of animals is in improving wellbeing. It presented how observing wildlife, including insects and birds, was identified as a major activity among hospital gardens.

The findings indicate that most city residents can be said to positively perceive the smells of various herbs. In addition, regardless of age, many residents expressed that they preferred herbal smells and reported that such smells elicited personal and highly positive associations. In our study, the respondents' favorite aromas were typically associated with the home, childhood, as well as the vacation season, representing a time of rest and relaxation. Winterbottom and Wagenfeld argued that bringing to mind pleasant past experiences can be significant to people who have had to leave their place of residence for various reasons such as illness, homelessness, as well as for people who find themselves in a completely alien environment [3]. The period of childhood, linked with nature or specific plants, remains strongly embedded in a person's memory, as observed by Lohr et al., who observed that childhood experiences can impact one's attitude towards wildlife and gardening in adulthood [49]. Other researchers also pointed to the significance of memories tied with garden settings, as memories of previously encountered gardens were observed in the elderly [17]. The results of our survey clearly showed that aromas can also bring to mind summertime relaxation and pleasant memories of vacation trips. The application of plants such as lavender or rosemary in Central European climate conditions, despite previously not being recommended, can stimulate these distinct memories, especially of travel to warmer areas of Europe.

Smells that bring to mind positive associations, when accounting for clearly pleasant memories, can improve the mood and sense of wellbeing of those who experience them. In our survey, a significant percentage of respondents, who were big-city residents, associated smells with good memories, which means that it can be argued that sensory gardens that stimulate the sense of smell of residents of big cities can improve the quality of life for individuals in a city and their overall wellbeing. Winterbottom and Wagenfeld [3] observed that the smell of lavender *Lavandula* sp., rose *Rosa* sp. and salvia *Salvia* sp. improved mood and eased stress, while the smell of mint *Mentha* sp. and citruses were perceived as invigorating or overwhelming. Lavender and mint were often identified in our study as a respondent's preferred plant, which brought to mind highly positive associations. However, salvia, despite reactions to its smell being reported as either positive or indifferent, was identified as personally significant or bringing to mind special memories by very few respondents. These results are important indicators for designers of therapeutic sensory gardens intended for construction in urban settings.

Based on the survey's results, we also noted that a significant portion of the respondents tied their favorite smells with beverages and food overall. This may indicate that adding such herbs to meals served in cafes operating alongside sensory gardens can enhance and extend the positive effect that appears as a result of a good memory and can thus result in greater therapeutic benefits and an elevated sense of wellbeing. Trojanowska argued that access to food and drink should be an attribute of a park with therapeutic features [39]. There are cases of Polish sensory gardens that feature a coffee shop, such as the garden near the J. Czapski Museum in Kraków.

Our analysis of Polish herbal gardens with sensory features using the Trojanowska method [39,40] as adapted by Krzeptowska-Moszkowicz et al. [6] demonstrated that such gardens have a medium therapeutic potential. It also revealed their limitations. One major problem in cities may be the existence of poor-quality greenspaces, which is a factor that contributes to human health, as it was observed that in urbanized areas such greenery, despite being present, may not impart as many health benefits as good-quality greenspace [50]. The higher adaptability of such gardens to disabled persons (attribute 1: D-5) is also notable, as sensory gardens can be visited by this demographic. The potential personalization of space (attribute 4: A-1) was also found to be overlooked and may prevent visitors from forming a deeper connection with a garden.

In the herbal gardens under analysis, the absence of water features was a major problem (attributes 3: A-3 and 5: A-2). Panel paintings depicted medieval European gardens as equipped with fountains, including in settings with aromatic plants [51]. Water is a significant element in therapeutic gardens and its application, particularly when it is in motion, greatly enhances their therapeutic potential [3,48]. Water is an excellent tool of expanding sensory stimuli, especially auditory, visual, and tactile ones [3]. Most respondents were of the opinion that in a garden with aromatic plants, sounds associated with water, e.g., produced by a small fountain, would be perceived positively. They also noted the positive impact of water on a garden's microclimate, e.g., by humidifying the air and lowering temperature. In our opinion, this highlights the existence of a need for flowing water among city residents and shows that water can positively affect people during periods of high temperature that occur in cities during summer.

The growing water deficit linked to climate change and progressively increasing global warming and water evaporation leads to the problem of supplying plants with water in agriculture and horticulture, and it is not limited to Poland [21]. It also concerns urban sensory gardens, wherein the urban heat island effect further escalates the situation [23,52]. Our survey revealed that the majority of the surveyed residents of a big city, in this case Kraków, found the source of water used to irrigate gardens to be a significant issue. Most respondents expressed the opinion that harvesting stormwater and rainwater for this purpose would be preferable to the use of grid-sourced water. This may indicate that city residents have an awareness of contemporary water-accessibility problems, the significance of water overall and pro-environmental solutions in this regard.

Author Contributions: Methodology, I.K.-M.; writing—original draft preparation, I.K.-M.; writing—review and editing, Ł.M. and K.P.; visualization, Ł.M. and K.P. All authors have read and agreed to the published version of the manuscript.

Funding: This research received no external funding.

Data Availability Statement: Not applicable.

Acknowledgments: We would like to thank the reviewers and editor for their accurate comments and assistance in improving the manuscript.

Conflicts of Interest: The authors declare no conflict of interest.

References

1. Harasimowicz, A. Green spaces as a part of the city structure. *Ekon. Sr.* **2018**, *2*, 45–62.
2. Westphal, J.M. Hype, Hyperbole, and Health: Therapeutic site design. In *Urban Lifestyles: Spaces, Places People*; Benson, J.F., Rowe, M.H., Eds.; Brookfield A. A. Balkema: Rotterdam, The Netherlands, 2000.
3. Winterbottom, D.; Wagenfeld, A. *Therapeutic Gardens: Design for Healing Spaces*; Timber Press: Portland, OR, USA, 2015.
4. Hussein, H. Sensory Garden in Special Schools: The issues, design and use. *J. Des. Built Environ.* **2009**, *5*, 77–95.
5. Hussein, H. Affordances of Sensory Garden towards Learning and Self Development of Special Schooled Children. *Int. J. Psychol. Stud.* **2012**, *4*, 135–149. [CrossRef]
6. Krzeptowska-Moszkowicz, I.; Moszkowicz, Ł.; Porada, K. Evolution of the Concept of Sensory Gardens in the Generally Accessible Space of a Large City: Analysis of Multiple Cases from Kraków (Poland) Using the Therapeutic Space Attribute Rating Method. *Sustainability* **2021**, *13*, 5904. [CrossRef]

7. Wajchman-Świtalska, S.; Zajadacz, A.; Lubarska, A. Recreation and Therapy in Urban Forests—The Potential Use of Sensory Garden Solutions. *Forests* **2021**, *12*, 1402. [CrossRef]
8. Sensory Trust. Available online: Sensorytrust.org.uk/information/factsheets/sensory-garden-1.html (accessed on 11 February 2022).
9. Bengtsson, A.; Grahn, P. Outdoor environments in healthcare settings: A quality evaluation tool for use in designing healthcare gardens. *Urban For. Urban Green.* **2014**, *13*, 878–891. [CrossRef]
10. Górska-Kłęk, L. ABC Zielonej Opieki. In *Seria „Biblioteka Nestora", t. VIII.*; Dolnośląski Ośrodek Polityki Społecznej: Wrocław, Poland, 2016; p. 151.
11. Lange, D. *Europe's Medicinal and Aromatic Plants: Their Use, Trade and Conservation*; TRAFFIC International: Cambridge, UK, 1998.
12. Rostafiński, J. Wpływ przeżyć chłopięcych Mickiewicza na obrazy ostatnich dwu ksiąg Pana Tadeusza oraz o święceniu ziół na Matkę Boską Zielną. *Pol. Akad. Umiejętności. Rozpr. Wydziału Filoz.* **1922**, *61*, 1–36.
13. Hodor, K. Cloister gardens and their role in shaping the urban tissue of Cracow/Ogrody klasztorne i ich rola w kształtowaniu tkanki urbanistycznej miasta Krakowa. *Czas. Tech.* **2012**, *z. 6-A*, 57–64.
14. Majdecki, L. *Historia Ogrodów*; Wydawnictwo Naukowe PWN SA: Warszawa, Poland, 2009.
15. Zajadacz, A.; Lubarska, A. Sensory gardens as places for outdoor recreation adapted to the needs of people with visual impairments. *Studia Perieget.* **2020**, *2*, 25–43. [CrossRef]
16. Arslan, M.; Kalaylioglu, Z.; Ekren, E. Use of medicinal and aromatic plants in therapeutic gardens. *Indian J. Pharm. Educ. Res.* **2018**, *52*, 151–154. [CrossRef]
17. Scott, T.L.; Masser, B.M.; Pachana, N.A. Positive aging benefits of home and community gardening activities: Older adults report enhanced self-esteem, productive endeavours, social engagement and exercise. *SAGE Open Med.* **2020**, *8*, 1–13. [CrossRef] [PubMed]
18. Van Tilburg, W.A.P.; Sedikides, C.; Wildschut, T.; Vingerhoets, A.J.J.M. How nostalgia infuses life with meaning; From social connectedness to self-continuity. *Eur. J. Soc. Psychol.* **2019**, *49*, 521–532. [CrossRef]
19. Ziernicka-Wojtaszek, A. Weryfikacja rolniczo-klimatyczna regionalizacji Polski w świetle współczesnych zmian klimatu. *Acta Agrophysica* **2009**, *13*, 803–812.
20. McCarthy, M.P.; Best, M.J. Climate change in cities due to global warming and urban effects. *Geophys. Res. Lett.* **2010**, *37*, L09705. [CrossRef]
21. Szwed, M.; Karg, G.; Pińskwar, I.; Radziejewski, M.; Graczyk, D.; Kędziora, A.; Kundzewicz, Z.W. Climate change and its effect on agriculture, water resources and human health sectors in Poland. *Nat. Hazards Earth Syst. Sci* **2010**, *10*, 1725–1737. [CrossRef]
22. Jagiełło, P.; Jefimow, M.K.; Strużewska, J. Zmiana narażenia na wysoką temperaturę w Polsce w horyzoncie do 2100 roku na podstawie projekcji klimatycznych EURO-CORDEX. In *Współczesne Problemy Klimatu Polski*; Chojnacka-Ożga, L., Lorenc, H., Eds.; Instytut Meteorologii i Gospodarki Wodnej, Państwowy Instytut Badawczy: Warszawa, Poland, 2019; pp. 121–131.
23. Bokwa, A.; Hajto, M.; Walawender, J.P.; Szymanowski, M. Influence of diversified relief on the urban heat island in the city of Kraków, Poland. *Theor. Appl. Clim.* **2015**, *122*, 365–382. [CrossRef]
24. Winbourne, J.B.; Jones, T.S.M.; Garvey, S.M.; Harrison, J.L.; Wang, L.; Li, D.; Templer, P.H.; Hutyra, L.R. Tree transpirations and urban temperatures: Current understanding, implications, future research directions. *BioScience* **2020**, *70*, 576–588. [CrossRef]
25. Shanahan, D.F.; Bush, R.; Gaston, K.J.; Lin, B.B.; Dean, J.; Barber, E.; Fuller, R.A. Health Benefits from nature experiences depend on dose. *Sci. Rep.* **2016**, *6*, 28551. [CrossRef]
26. Ulrich, R.S. Health benefits for gardens in hospitals. In Proceedings of the Plants for People International Exhibition Floriade, Haarlemmermeer, The Netherlands, 6 April–20 October 2002.
27. van den Berg, A.; Hartig, T.; Staats, H. Preference for Nature in Urbanized Societies: Stress, restoration, and the pursuit of sustainability, *J. Soc. Issues* **2007**, *63*, 79–96. [CrossRef]
28. Berman, M.G.; Jonides, J.; Kaplan, S. The cognitive benefits of Interacting with nature. *Psychol. Sci.* **2008**, *1*, 1207–1212. [CrossRef]
29. Adevi, A.A.; Mårtensson, F. Stress rehabilitation through garden therapy: The garden as a place in the recovery from stress. *Urban For. Urban Green.* **2013**, *12*, 230–237. [CrossRef]
30. Coldwell, D.F.; Evans, K.L. Visits to urban green-space and the countryside associate with different components of mental well-being and are better predictors than perceived or actual local urbanisation intensity. *Landsc. Urban Plan.* **2018**, *175*, 114–122. [CrossRef]
31. Chiabai, A.; Quiroga, S.; Martinez-Juarez, P.; Higginns, S.; Taylor, T. The nexus between climate change, ecosystem services and human health; towards a concepyual framework. *Sci. Total Environ.* **2018**, *635*, 1191–1204. [CrossRef] [PubMed]
32. Houlden, V.; Scott, W.; de Albuquerque, J.; Javris, S.; Rees, K. The relationship between greenspace and the mental wellbeing of adulds: A systematic review. *PLoS ONE* **2018**, *13*, e0203000. [CrossRef] [PubMed]
33. Kępińska-Kasprzak, M.; Struzik, P. Reakcja roślin dziko rosnących na obserwowane ocieplenie klimatu. In *Współczesne Problemy Klimatu Polski*; Chojnacka-Ożga, L., Lorenc, H., Eds.; Instytut Meteorologii i Gospodarki Wodnej, Państwowy Instytut Badawczy: Warszawa, Poland, 2019; pp. 233–244.
34. Krzeptowska-Moszkowicz, I.; Moszkowicz, Ł. Selected problems of *Ailanthus altissima* (Mill.) Swingle presence in urban spaces: The case of the city centre of Kraków. In Proceedings of the Plants Urban Areas Landscape International Scientific Conference, Nitra, Slovakia, 14–15 May 2014; pp. 56–62. [CrossRef]
35. Sudnik-Wojcikowska, B. The effect of temperature on the spatial diversity of urban flora. *Phytocoen. Suppl. Cartogr. Geobot.* **1998**, *10*, 97–105.

36. Suchecka, A.; Giergiel, T. Otwarcie Sandomierskiego Ogrodu Marcina z Urzędowa. *Wiadomości. Bot.* **2015**, *59*, 178–185.
37. Zajadacz, A.; Lubarska, A. *Ogrody Sensoryczne Jako Uniwersalne Miejsca Rekreacji Dostosowane Do Potrzeb Osób Niewidomych W Kontekście Relacji Człowiek-Środowisko*; Bogucki Wydawnictwo Naukowe: Poznań, Poland, 2020.
38. Medwecka-Kornaś, A. Czynniki naturalne, wpływające na rozmieszczenie geograficzne roślin w Polsce. In *Szata Roślinna Polski*; Szafer, W., Zarzycki, K., Eds.; Wydawnictwo Naukowe PWN: Warszawa, Poland, 1972; pp. 35–94.
39. Trojanowska, M. *Parki i Ogrody Terapeutyczne*; Wydawnictwo Naukowe PWN: Warszawa, Poland, 2017; p. 231.
40. Trojanowska, M. The Universal Pattern of Design for Therapeutic Parks. *Methods of Use. Int. J. Sci. Eng. Res.* **2018**, *9*, 1410–1413.
41. Trojanowska, M. Therapeutic qualities and sustainable approach to heritage of the city. The coastal strip in Gdańsk, Poland. *Sustainability* **2020**, *12*, 9243. [CrossRef]
42. Seneta, W.; Dolatowski, J. *Dendrologia*; Wydawnictwo Naukowe PWN: Warszawa, Poland, 2011; p. 544.
43. Muza SA. *Wielka Encyklopedia Przyrody. Rośliny Kwiatowe*; Muza SA: Warszawa, Poland, 1998.
44. Moszkowicz, Ł.; Krzeptowska-Moszkowicz, I. Impact of the public parks location in the city on the richness and diversity of herbaceous vascular plants on the example of Krakow Southern Poland. In Proceedings of the Plants Urban Areas Landscape International Scientific Conference, Nitra, Slovakia, 7–8 November 2019; pp. 98–103.
45. Moszkowicz, Ł.; Krzeptowska-Moszkowicz, I.; Porada, K. Relationship between Parameters of Public Parks and Their Surroundings and the Richness, Diversity and Species Composition of Vascular Herbaceous Plants on the Example of Krakow in Central Europe. *Landsc. Online* **2021**, *94*, 1–16. [CrossRef]
46. Krzeptowska-Moszkowicz, I.; Moszkowicz, Ł.; Porada, K. Znaczenie miejskich ogrodów sensorycznych o cechach przyjaznych organizmom rodzimym, na przykładzie dwóch przypadków z terenu dużych miast europejskich: Krakowa i Londynu. In *Integracja Sztuki i Techniki w Architekturze i Urbanistyce*; Wydawnictwa Uczelniane Uniwersytetu Technologiczno-Przyrodniczego: Bydgoszcz, Poland, 2020; pp. 61–69.
47. Nowacki, J.; Bunalski, M.; Sienkiewicz, P. Fauna jako istotny element hortiterapii. In *Hortiterapia Jako Element Wspomagający Leczenie Tradycyjne*; Krzymińska, A., Ed.; Rhytmos: Poznań, Poland, 2012; pp. 117–144.
48. Cooper Marcus, C.; Barnes, M. *Gardens in Healthcare Facilities: Uses, Therapeutic Benefits, and Design Recommendations*; The Center for Health Design, Inc.: Concord, CA, USA, 1995; p. 70.
49. Lohr, V.I.; Pearson-Mims, C.H. Children's active and passive interactions with plants influence their attitudes and actions toward trees and gardening as adults. *Hort Technol.* **2015**, *15*, 472–476. [CrossRef]
50. Mitchell, R.; Popham, F. Greenspace, urbanity and health: Relationships in England. *J. Epidemiol. Community Health* **2007**, *61*, 681–683. [CrossRef] [PubMed]
51. Kulesza, P. *Plants in 15th-Century Netherlandish Painting: A Botanical Analysis of Selected Paintings*; Wydawnictwo KUL: Lublin, Poland, 2014.
52. Kalbarczyk, E.; Kalbarczyk, R. Typology of climate change adaptation measure in Polish cites up to 2030. *Land* **2020**, *9*, 351. [CrossRef]

land

MDPI

Article

Gardens of Historic Mental Health Hospitals and Their Potential Use for Green Therapy Purposes

Anna Staniewska

Chair of Landscape Architecture, Faculty of Architecture, Cracow University of Technology, Warszawska 24, 31-155 Kraków, Poland; astaniewska@pk.edu.pl

Abstract: Gardens of historic psychiatric institutions represent a special type of heritage garden that possess both aesthetic and therapeutic purposes. Their existence and current state are affected by changes in the organisation of mental treatment. The article focuses on the possible use of these gardens as places of modern green therapies carried out in, and connected with, nature. Taking into account the state of the art on the beneficial influence of nature on human health and well-being, the paper provides an overview of historic and modern nature-based activities considered therapeutic. Subsequently, three case studies of contemporary psychiatric facilities operating in historic mental hospital sites are examined. Many activities linked to nature exercised historically in those gardens bear similarities to a contemporary spectrum of ecotherapies. An analysis of historic and contemporary plans of the sites and gardens and a description of the therapeutic activities carried out in nature are provided. Results prove that their potential is promising, but not yet used to its full extent because of organisation and financing within the context of health care systems. Using those gardens for the spectrum of green therapies may bring benefits for patients and the historic substance alike.

Keywords: historic psychiatric hospitals; therapeutic landscape; green therapy; horticulture therapy

Citation: Staniewska, A. Gardens of Historic Mental Health Hospitals and Their Potential Use for Green Therapy Purposes. *Land* **2022**, *11*, 1618. https://doi.org/10.3390/land11101618

Academic Editors: Jan K. Kazak, Katarzyna Hodor and Magdalena Wilkosz-Mamcarczyk

Received: 18 August 2022
Accepted: 12 September 2022
Published: 21 September 2022

1. Introduction

1.1. Historic Psychiatric Hospitals and Their Gardens—From Lunatic Asylum to Contemporary Mental Health Centres

Historic psychiatric hospitals were built in the early 19th century as specific places of treatment of manifold disorders described usually as 'lunacy', during the era of psychiatry's beginnings as a distinct medical specialization. Lunatic asylums—as they were called—represented special types of buildings and landscapes, which were designed according to the theoretical works of medical doctors and took into account their experiences gathered while managing those institutions and collected during study travels undertaken to visit model institutions. With time, these institutions evolved just like psychiatry developed. At the beginning of the 19th century, a significant advance was brought by the writings and practice of Pinel in France and Chiarugi in Italy [1,2], and the introduction of moral treatment practised in England in a private Quaker institution called The Retreat and led by William Tuke, who handed it to his grandson Samuel, who contributed greatly to publicizing his achievements [3]. The humane attitude implemented at this small establishment was widely propagated and treated as a model for many following decades. It was based on the assumptions that the asylum's interior and surroundings should be pleasant to distract the troubled mind, and that appropriate daily routine, diet, and occupation contributed to healing.

The structure of the asylum and its landscape was similar to an English country house estate [4], but it had a clear therapeutic function. Although asylums evolved, some principles remained the same. Gardens of historic psychiatric institutions represent a special type of historic garden that served not only aesthetic but also therapeutic purposes [5,6]. At the turn of the 19th and 20th centuries, psychiatric hospitals were aspiring to be more than just institutions of isolation, aiming at providing a truly restorative setting with countryside locations, and accommodating patients in detached pavilions instead of large corridor buildings. In addition, occupational therapy was widely introduced along with agricultural and horticultural activities on hospital farms. Various types of physical activities, such as '*tennis, bowls, badminton, football and amusements*' (such as the three-legged race) were also considered beneficial and advertised in theoretical writings [7] (pp. 100–114).

Interestingly, radial plans of several mental institutions from the beginning of the 20th century in Europe visually resemble a diagram of garden cities as presented in the writings of Ebenezer Howard (Figure 1). While no direct inspiration for the design and building process is usually found in historic sources [8], this type of circular plan and similarities to garden cities contributed to these institutions being described as 'garden cities for the insane' [9].

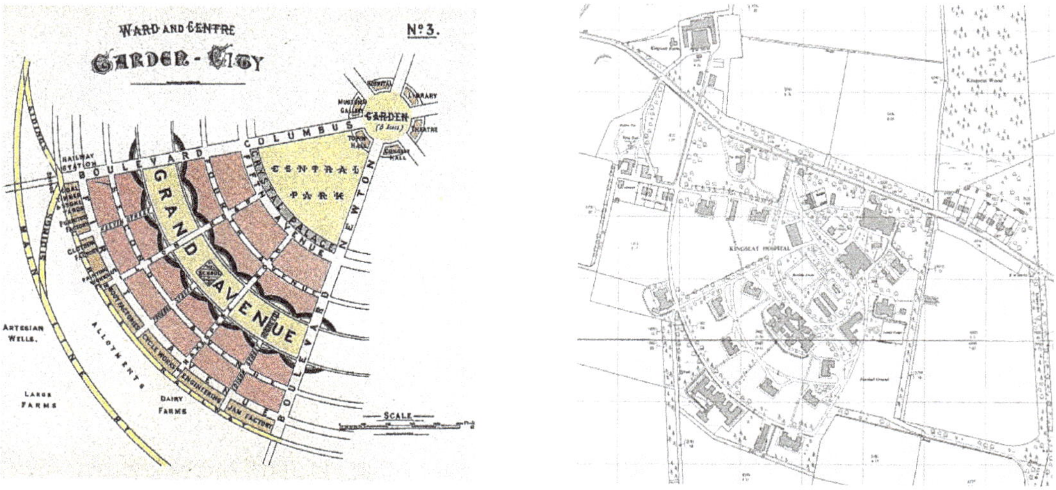

(a) (b)

Figure 1. Similarities in plans between the garden city diagram and the plan of a mental asylum of the early 20th century: (**a**) Diagram No. 3 by Ebenezer Howard showing the ward and centre of the garden city (from Howard, Ebenezer, *To-morrow: A Peaceful Path to Real Reform*, London: Swan Sonnenschein & Co., Ltd., 1898; source: [10] public domain, (**b**) The plan of the Kingseat Asylum in Scotland, built from 1900 to 1904, as pictured on the Ordnance Survey Map from 1944; source: the collection of the National Library of Scotland, CC-BY (NLS).

Despite all these efforts and the therapeutic regime implemented by early psychiatrists working there, asylums were heavily criticized by the advocates of clinical treatment, which was considered more effective than alienists' work. Finally, large institutions were defeated by numbers, because in an institution that sometimes holds as many as 2,000 patients, any therapy was condemned to failure [11]. The first three decades of the 20th century brought an almost double increase in the number of psychiatric patients in Britain [12]. Despite the introduction of brain surgery and electric shock treatment, psychiatric patients demonstrating socially unacceptable behaviour were administered sedatives, and the era of confinement in large psychiatric institutions lasted until the 1950s. A significant breakthrough in the treatment of mental disorders was achieved with the humanitarian effect of antipsychotic drugs. This, combined with a strong antipsychiatry movement [13], contributed to the process of change and closure of the asylums.

In the UK, the decision to close Victorian asylums was strongly linked with the famous 'Water Tower speech' delivered by the minister of health, Enoch Powell, at the annual conference of the National Association for Mental Health in March 1961. Referring to specific elements of asylum sites, he described these institutions as *'isolated, majestic, imperious, brooded over by the gigantic water-tower and chimney combined, rising unmistakable and daunting out of the countryside'* [14]. Services provided by psychiatric hospitals were to be delivered by special acute-care psychiatric wards of general hospitals and community care. This resulted in the dissolution of the asylums and plans to reduce the number of mental health beds. This policy, in terms of patient comfort and quality of mental healthcare services, has brought mixed results [15]. While, generally, large-scale institutions were heavily criticized for their overcrowding and scandalous cases of violent treatment of patients, and community care had its evident benefits, the fate of the elderly and patients requiring long-term treatment, who were permanent residents of old institutions, became difficult [16]. This wave of asylums' closures was called deinstitutionalization, and led, in many cases, to the complete demolition or redevelopment of numerous Victorian asylum sites. Many of them stayed vacant for a long time, because it was difficult to obtain planning permission for redevelopment, which significantly increased the value of the area [17,18]. Many were irreversibly destroyed and demolished, while numerous sites were turned into up-market housing developments where all traces of the old asylum were often deliberately obliterated [19,20]

Deinstitutionalization of psychiatric hospitals affected not only UK institutions, but also historic asylums located in Italy, the United States, Australia, and New Zealand [21,22].

Redevelopment strategies for former asylum sites' conversion usually focus on their attractive locations (on the edge of the city, with pleasant views, and greenery in abundance) and they often lead to the process of strategic forgetting of the past, or to selectively remembering it to repackage the asylum as housing [23].

Yet, some contemporary mental health centres still operate in historic psychiatric hospitals in several European countries, such as France [24], Switzerland [25], Germany [26], and Poland [27]. They offer the most up-to-date treatment in the renovated buildings surrounded by historic green spaces and parks.

The existence and current state of historic mental hospital gardens is inevitably bound to changes in the organisation of mental therapies and treatment. What is more, like many other historic gardens, they are nowadays facing numerous challenges linked with the need for proper maintenance and issues connected to climate change, along with the ageing of the original planting. The potential use of their gardens for the spectrum of green therapies seems to bring benefits for patients and the historic substance alike, although it is not fully used.

1.2. Aim of the Study

The article presents the therapeutic potential of historical gardens and parks of former psychiatric hospitals, indicating the possibilities of using these gardens to conduct green therapies. Furthermore, it explores some problems related to the maintenance of historical vegetation, focusing primarily on the main benefits of nature-based interventions.

Links between nature and health have been explored since ancient times dating back to Hippocrates, who in *On Airs, Waters, and Places* referred to *'the body [which] was inseparable from inquiry into places and directions, seasons and winds [...] human being was being embedded in a world'* [28] (p.661). Early research was based on several theories that refer to medical geography. The most important of them focus on the concept of therapeutical landscapes [29,30] and evolutionary perspectives that encompass the prospect-refuge theory [31], the functions-evolutionary model [32,33], and the psychoevolutionary model of R. Ulrich [34–36].

Recently, numerous studies have explored not only the historical and modern paradigms (of the various disciplines) which determine the discourse of nature concerning human health and well-being research [37], but have referred to a multitude of particular aspects connected with these issues. While gardens have always been an appreciated element [38] of historic cities, providing public green leisure areas such as commons, pleasure grounds, and public parks [39], they have nowadays become a vital part of green infrastructure

(GI) [40] which may provide, among other services, therapeutic functions. Historic gardens, along with other types of garden spaces [41,42], may be treated as resources in sustainable urban development in a time of accelerating densification and climate change. Relationships between green infrastructure and ecosystem, and human health, construction, evaluation and management of green infrastructure, and analysis of a special aspect of green infrastructure are research topics in the field of landscape planning and urban studies that are of great importance recently [43]. Many studies explore the benefits of nature [44–46] and sustainable urban living spaces that are vital in an urban context, as the population of urban dwellers grows on a global scale and is confronted with the climate crisis [47,48]

Apart from the studies focusing on physical health [49], there is a growing body of research that focuses on a positive association between urban green space and attention, mood, and physical activity [50]. Other research aims at measuring the impact on mental health of relations to, and activities carried out in, natural environments [51]. They often examine the link between visiting green spaces and well-being [52], indicating that more time spent in green spaces is associated with better mood [53] and higher scores on mental health and vitality scales, independent of cultural and climatic contexts [54]. An important role is especially attributed to physical activity during leisure in green space [55], and interactions with nature, such as walking [56], which are likely to bring numerous benefits, especially for people with mental issues such as depression [53].

Research on various nature-based interventions (NBIs) is of vital importance, as mental disorders are increasingly diagnosed in recent years. For this reason, therapeutic measures connected with nature are gaining importance. The WHO states that in 2019, one in every eight people, which equals nearly 970 million people around the world, were living with a mental disorder [57], and estimates that the COVID-19 pandemic contributed to an increase in anxiety and major depressive disorders in only one year [58]. In the United States, 19.86% of adults are experiencing a mental illness, which is equivalent to nearly 50 million Americans [59].

Because natural healing powers are particularly relevant in the case of stress reduction, a wide spectrum of NBIs seem to present promising opportunities to support not only treatment, but also the prevention of several mental disorders. Various stress-related disorders [60] may be caused by psychosocial stress, such as fatigue, burnout, exhaustion, depression, anxiety, or adjustment disorder [61].

While most of the research referring to horticultural therapy was carried out to examine the impact of gardening on vulnerable groups, such as the elderly [62], children and youth [63], patients with mental illness [64] or particular problems such as depression and related issues [65], PTSD [66–69], and dementia [70,71], only some attention was paid to the setting in which the therapy was conducted [72]. Of course, the features that define a therapeutic garden [73–75] and park [76] are formulated, but they are applied mostly to newly established environments.

There are no studies on therapeutic activities supporting psychiatric treatment carried out in the parks and gardens of historic psychiatric hospitals, and relations between their heritage values and composition. Therefore, this study aims to describe relations between the landscape and garden composition of heritage mental hospital sites and a spectrum of green therapies in the past and nowadays. Green-based interventions in those particular places will be examined with reference to historic sources, the contemporary literature, and research. In addition, three selected case studies of contemporary mental health centres operating on historic psychiatric hospital heritage sites will be examined.

2. Materials and Methods

To explore the links between the historic mental hospital landscapes and contemporary green therapies supporting mental health and well-being, the study employs a mixed-methods approach and is based on secondary desk research and structured interviews, which together contribute to case study analyses and provide data for discussion.

Firstly, the connections to nature offered in the past, in former asylums from the beginning of the 20th century, are explored. This part of the paper is based on historic materials and included primary sources, such as writings of medical doctors and asylum superintendents summarised previously in secondary sources, including the previous body of research on the landscape composition and arrangement of historic asylums and their gardens in Europe [6,77].

Subsequently, the contemporary spectrum of green therapies widely described in recent reviews is analysed with reference to the historic use of green spaces and activities carried out in historic psychiatric hospitals, and possible implementation within the historic model.

The last stage is an exploratory case study based on structured interviews carried out in three mental health facilities which operate in historic buildings located in a park setting. It describes the location and brief history of each particular hospital and provides information on the contemporary spectrum of green therapies conducted in those selected sites. Structured in-depth interviews were used to collect relevant source data to obtain comparable materials and qualitative information on the types of green therapies conducted and their relation to the heritage park site. It was particularly important if nowadays historic landscape features were intentionally used. Interviews are an acknowledged method of data collection in many research fields [78–80] including heritage science [81] and medicine alike [82]. As an interactive method, an interview enables mutual learning and allows the researcher to focus on the issues that the interviewer might not have previously considered important [83].

In the discussion, reference is made to other relevant examples of historic hospital settings and their therapeutic use, although these may not exactly match the case study criteria.

Case Study Choice and Location

Although numerous contemporary institutions still operate in historic buildings and sites of old psychiatric hospitals across Europe, three were chosen for this study. The case studies were selected to present the most developed places where all possible elements of the functional programme were implemented at the stage of the design. For this reason, mental hospital sites which were built in 1900 or later were chosen. In addition to this, selected hospitals represent institutions built in pavilion style on a circumferential plan resembling the garden city diagram—an example of harmonious development which resembles a small community—and which still operate nowadays. In addition, they are of comparable size in terms of surface. The original plans of those establishments were well preserved and accessible. All three institutions declare that they appreciate their historic setting and try to make contemporary use of historic green spaces, and that they adapt heritage buildings to current treatment and patients' needs. The following mental health institutions (Table 1) were selected for investigation of the contemporary green therapies actions in a historic setting:

1. The Babiński Specialized Hospital in the southern suburb of Kraków, former second National Institute for the Nervous and Mentally Ill, Kraków-Kobierzyn (Małopolska Voivodeship, Poland);
2. The Voivodeship Mental Hospital Lubiąż, former Provincial Hospital for the Nervously and Mentally Ill (so-called 'New Institution', to differentiate it from the old hospital situated in a former Cistercian abbey), Lubiąż (Lower Silesia Voivideship, Poland);
3. Klinikum am Weissenhof Centre for Psychiatry Weinsberg, former Königliche Heilanstalt Weinsberg, Staatliche Irrenanstalt, Weissenhof-Weinsberg (Baden-Würtemberg, Germany).

Table 1. An overview of the historic mental hospitals and their gardens under study.

	Kraków-Kobierzyn	Lubiąż	Weissenhof-Weinsberg
Historic name	The second National Institute for the Nervous and Mentally Ill	Provincial treatment and care facility in the town of Lubiąż	Königliche Heilanstalt Weinsberg, Staatliche Irrenanstalt
Contemporary surface in ha	48 ha	21 ha	45 ha
Construction dates	1909-1919	1902-1910	1900-1903
Architects involved	Władysław Klimczak	Eduard Blümner	Carl Hees
Landscape architects	Wiktor Żochowski	nn	Albrecht Lilienfein und Sohn

3. Results

Activities in nature have been used as means of therapy in the past, and they were taken into account as important features to be considered when establishing an asylum from the turn of the 19th and 20th centuries.

This section provides an overview of links to nature-based activities in the past and a broad spectrum of contemporary opportunities for green therapies, and places them in the context of historic mental hospital landscapes. Subsequently, three case studies are described with particular attention to their landscape composition and layout, and green therapies conducted nowadays.

3.1. An Overview of Connections to Nature and Green Therapies Historically Used in Mental Hospitals from the Turn of the 19th and 20th Centuries

Firstly, the setting of the institution was important: on the outskirts of the city, in the country, and preferably on a hill [84] with beautiful views to ease the troubled mind and isolate the mentally ill from harmful industrialised and overpopulated cities [85]. Gardens often used borrowed views from elevated mounds to secure the pleasant, picturesque views for patients without being seen [86] (pp. 191–192). To restrict the visibility of curious onlookers, the institutions were surrounded by fences and walls often constructed as ha-has [87] (p. 185). Water was also important, and was presented as a fountain or a well in the middle of the rectangular inner court or garden [88] (pp. 114–115).

A range of activities was encouraged and carried out on the hospital grounds. Walking took place in airing courts; however, it was restricted to the wards for agitated and criminal patients. In Brislington House, which was located in the vicinity of the Avon River, therapy included walks in nature [89], and some institutions even organised excursions for patients (such as Hanwell Asylum [84], or Illenau, near Achern, in Germany [88]).

Physical exercise was encouraged in the airing courts and on the hospital grounds (battledore, shuttlecock, football, and cricket). In many places there were special amenities designed within the parks, such as a bowling green and football pitch, and even tennis courts for upper-class patients (as reported by the director of the Kobierzyn institution in the 1920s [90]).

More sophisticated activities were often aimed at and exercised by educated, wealthy, and noble patients. An interesting collection of drawings and paintings by Charles Altamont Conan-Doyle (father of Sir Arthur Conan-Doyle), who spent several years in Montrose Asylum in Scotland, is well known [91].

John Conolly, who introduced occupational therapy while acting as superintendent in Hanwell Asylum, admitted that gardening was one of the meaningful occupations for patients, who took care in turning airing courts into gardens and made representative entrance areas beautiful. He wrote:

The cultivation of the gardens, and of the ground called the farm, as well as of the extensive ornamental ground in front of the asylum, is entirely effected by the labour of numerous male patients, superintended by gardeners, or by steady workmen. The cheerfulness with which their work is performed, and the satisfaction with which, at stated hours, they assemble for their allowance of beer, sufficiently attest that calming and remedial influences are thus exercised [84] (p. 51).

Many other asylums employed some of the patients in agricultural work. Obviously, nowadays a strong tension can be observed between the primary therapeutic aim of these activities and the economic necessity to provide the institution with its fruit, vegetables, and dairy products. The numbers of patients involved in the work on hospital farmland or in so-called 'agricultural colonies' varied. As Sarah Rutherford discovered, it was on average up to 30%. But there was an intention that regular work in the open air was not only to keep patients busy with the purposeful work, but also to provide them with visible satisfaction from the results of their work, physical exercise, and fresh air. Interestingly, hospital records cited by Rutherford show that the productivity of hospital farms was higher in the Broadmoor institution, which was a criminal asylum, than in less corrective asylums such as Hanwell [4] (pp. 226–228). While some doctors advising on the theory of insane asylum construction primarily considered an asylum farm as an important element of the institution's functioning, some indicated that animal keeping could also evoke positive feelings and cheerfulness, and mentioned that some animals could be kept for pleasure and entertainment including, among others, '*rabbits, sea-gulls, hawks and poultry*' [3] (p. 63).

All those activities were conducted in separated male and female groups, taking into account social norms and cultural habits attributed to some of them, especially to men or women. Women were employed in all types of housework in the kitchen and vegetable kitchen garden, in addition to lighter work in the fields, gardening, washing, and repairing laundry. Other types of occupational therapy included wickerwork and linen production. Men were involved in heavier agricultural, building, and construction work, often helping to complete the asylum complex, which saved large sums of money, as in the case of Brassens in France [92] (p. 62).

In institutions which also admitted tuberculosis patients, there were often verandas and special sunny terraces or so-called 'Liegehallen' [93] that facilitated passive rest with a view of the garden for those suffering from pulmonary problems.

3.2. Nature and Green Therapies Supporting Mental Health and Well-Being Nowadays

Therapeutic activities which use nature and natural materials or natural settings, also referred to as nature-based interventions (NBIs), were often traditionally employed as calming and relaxing activities in historic asylums. In the last 50 years, they fall under two most frequently used umbrella terms: green therapies and ecotherapy. While green therapies and NBIs describe a range of outdoor activities which use plant material, planted garden settings, or natural environments, ecotherapy results from the deeper philosophical concept of exploring particularly strong bonds and mutual relations between humans and nature [94]. Ecotherapy is sometimes referred to as 'a forgotten ecosystem service' [95] and, as such, is considered as offering a range of therapeutic programmes based on NBIs [96], which support conventional treatment by reducing the number of pharmaceuticals [97] and preventing mental health problems.

Ecotherapy is often defined as therapeutic treatment consisting of regular structured activity, led by trained professionals, taking place in a green environment, focused on performing an activity, and relating to exploration and appreciation of nature in its various forms and aspects. It often involves social contacts generated by spending this time with other people, although the interactions are never forced. It includes a few main groups of programmes (see Figure 2), such as:

- Social and therapeutic horticulture (passive: simply spending time and admiring gardens and plants; or active: focused on gardening, tending the food-growing plants. May also take place indoors, in greenhouses);

- Green exercise therapy (doing exercise in green spaces: yoga, walking, running, or cycling);
- Care farming (the therapeutic use of agricultural landscape and farming practices such as growing crops, looking after farm animals, or helping to manage woodland);
- Animal-assisted interventions (spending relaxed time in contact with animals in spaces like farms, especially introduced in groups of young patients);
- Animal-assisted therapy (meant as building a therapeutic relationship with animals, especially dogs and horses);
- Environmental conservation (activities focused on protecting and caring for natural spaces, often combining physical exercise with conservation tasks);
- Nature arts and crafts (creating art in green spaces, or with nature and natural materials, such as painting, sculpture, and creating land-art, and also referring to use of the environment as inspiration);
- Adventure therapy (focused on adventurous physical group activities like rafting or rock climbing);
- Wilderness therapy (spending time in the wild and in remote locations, performing activities together in a group, such as making shelters and hiking) [98,99].

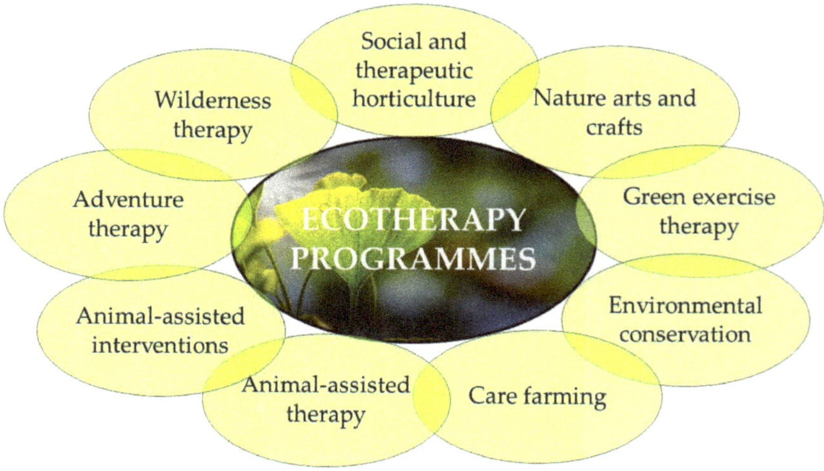

Figure 2. A variety of ecotherapy programmes, own elaboration based on a paper by MIND [98,99].

3.3. A spectrum of Green Therapies Conducted in Selected Contemporary Facilities Operating within the Walls of Historic Psychiatric Hospitals

3.3.1. Kobierzyn, Krakow

The Babiński Specialized Hospital in Kobierzyn was established as the second National Institute for the Nervous and Mentally Ill in the western part of Galicia, a province of the Austro-Hungarian monarchy at the beginning of the 20th century. In 1903, a provincial diet decided that a suitable plot for building a 500-bed institution for the nervously and mentally ill should be found, and initial plans, along with cost estimates, should be prepared within a year. For several reasons, the process was delayed and a final resolution on building the hospital was passed by the parliament in the of autumn 1907. Doctor Jan Mazurkiewicz, a recognized specialist, was appointed as the director. The plan for the Kobierzyn hospital was initially prepared in Lviv under the supervision of Władysław Klimczak, a professor of the Lviv Polytechnic University, appointed as the construction manager. In April 1910, however, the entire hospital planning office was moved to the site at Kobierzyn, and construction of the pavilions began. The setting on the southern outskirts of Kraków offers distant views to the south, in the direction of Beskidy mountains, and to the north, on the valley of the Vistula, the city, and the woods of the so-called western green-wedge, with the hills of Sikornik topped with Kościuszko Memorial Hill. The hospital buildings and their gardens occupy more than half of the whole hospital area which covers about 48 ha. The hospital was constructed in the pavilion style and all units were immersed in the vast park. The hospital also had its farm and arable land (on the southern side), and a cemetery in the vicinity (apart from the main hospital area to the north). The hospital was built to accommodate 550 patients, with a possible extension to handle up to 800. Because it was built within the firing range of the southern front of the Krakow Fortress, it was in danger of being demolished in 1914, and it was only in early 1918 that the first patients were admitted [100].

Regarding landscape design, the hospital possesses the plan of the gardens and park from 1909, signed by Wiktor Żochowski, a co-owner of a gardening company operating in Kraków (Figure 3). This design was only partially implemented in a much-simplified form. While the main alleys, very simplified roundabout parterres, and geometric ward gardens are visible on an aerial picture taken in 1917, the rest of the landscaping was done after World War I, as mentioned in reports of the next director in 1925. Most of the trees that form the historic garden layout date from this time. During World War II, the hospital was under national socialist occupation and many patients were exterminated. The facility reopened after the war, and the number of patients increased significantly. The institution experienced some interventions in its main composition and structure, but still operates as a psychiatric hospital. Although agricultural work and land cultivation were carried out for a long time, they were finally abandoned as unnecessary for the continuation of the medical mission. Since 1988, the ensemble has been listed as a registered monument. Over the last 25 years, restoration of buildings has been intensified as necessary to improve conditions and the comfort of patients, and numerous treatment programmes reduced the number of patients requiring long-term stays on the wards [101].

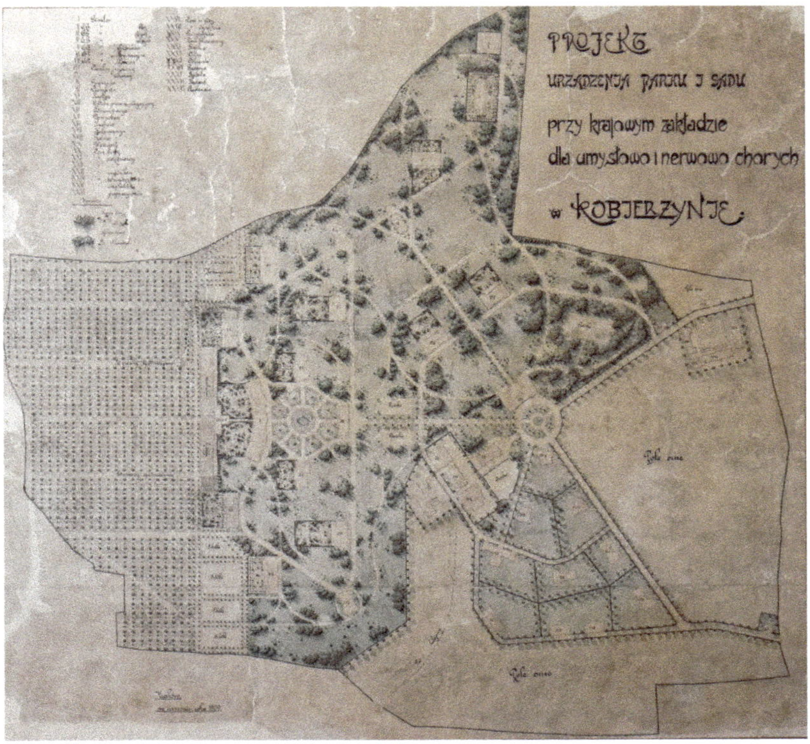

Figure 3. A plan of the establishment of a park and orchard around the National Institute for the Nervous and Mentally Ill in Kobierzyn, Kraków, 1909, drawing by Wiktor Żochowski; source: the archive of the Babiński Clinical Hospital.

Therapeutic gardening sessions in Kobierzyn hospital are carried out regularly once or twice a week in each ward, and they usually last one hour. The number of patients taking part in therapeutic gardening varies from 3 to 26 people, depending on the condition of patients in a given ward. Monthly, from 200 to 570 people participate. Since 2014, the number of patients involved in therapeutic gardening grew systematically from 11.2% to 29.5% in 2018 [102].

Within the framework of the garden therapies, two main types of sessions are offered. The first 'active' type is conducted in the ward gardens if the weather allows it. It covers the full range of activities performed to set up and maintain the gardens. If the gardens require revitalization or re-establishment (e.g., after a general renovation of the building), the ground for decorative flower beds is prepared by the patients acting together, and covered with garden cloth. Ornamental plants are planted and bark is laid down to prevent weeds.

Patients also regularly maintain the flowerbeds, undertaking care consisting of systematic weeding, pruning, fertilizing, and preparing beds for winter (covering plants to protect them from frost). Therapeutic gardening is also conducted in a greenhouse, which belongs to the Social Cooperative 'Kobierzyn' (social enterprise). In the greenhouse, patients learn to produce seedlings of vegetables and ornamental plants, and to make cuttings from ornamental shrubs. Garden therapy also includes dendrological walks, which aim to show patients the species of trees and shrubs that grow in the historic hospital park. Patients also collect and dry flower petals, which they use later during creative workshops in winter to create artwork and pictures. On the other hand, in winter and when it rains, hortitherapy is conducted in a 'passive way', via multimedia content on topics connected to the history of gardens and gardening issues. Patients become acquainted with the examples and pictures taken in the most beautiful gardens in Poland and abroad. Presentations also cover the issues of the health properties of various types of fruits and vegetables, along with the wealth of ornamental plant species, as well as birds and animals visiting the gardens. In addition, in winter, the seeds collected in autumn are cleaned and sorted to be used in the next year to create flowerbeds in the gardens of the wards. In addition, the garden therapist runs workshops for patients on designing decorative flower beds and herbal gardens.

Patients appreciate the historic park while taking walks and admiring nature, which helps them to relieve stress and escape from trauma. In the gardens of the wards, they organize barbecues during the warm time of the year. This is also a place to sit during visits by family and friends, in the case of a closed ward patient. Depending on the weather, patients spend several hours a day in these gardens. Most of them are on the southern side of unit buildings and are equipped with benches and tables, while a few have pavilions protecting against sun and rain [103].

The therapists are particularly fond of the gardens adjoining the wards as daily accessible green spaces and are happy to consult the experienced gardener and provide guidance on professional tools and equipment.

The sports ground is used occasionally for the annual sports celebration (meeting). In 2019, open-air gyms with some equipment were installed, and a chess table, which can be used by all visitors to the hospital grounds (two locations).

During the COVID-19 pandemic, some spaces were used. The unit gardens were treated as an extension of the ward and patients used them. During the first lockdown phase (from March to May 2020), no visitors were allowed to the park and patients did not go for any walks. In addition, the therapeutic gardening groups were suspended, but the therapists looked after the gardens, which they appreciated as a beneficial activity helping them to regenerate after work.

3.3.2. Lubiąż

The construction of the Provinzial heil- und Pflege-Anstalt Städtel Leubus (provincial treatment and care facility in the town of Lubiąż) was approved by the Parliament of the Province of Silesia in the early spring of 1901. It was supposed to be a new facility (often later called *'das Neue Anstalt'*) located approximately two kilometres from the facility established in 1830, and operating in a secularized post-Cistercian Baroque monastery complex. The selected plot was located on a low moraine hill on the outskirts of the small town of Lubiąż, near the old crossing on the Odra River, with good access to the main road leading to the town and the monastery. There was, however, one disadvantage of the location: the nearest railway stations were 8, 9, and 15 km away. The author of the project was Eduard Blümner, an architect and building construction counsellor (Baurat) from Wrocław specializing in public facilities (and hospital and care buildings). About 30 buildings were located in the initial area of 151.6 ha, which could serve nearly 1000 lower-class patients as well as employed doctors and staff. Construction began in 1902, the first patients were admitted in 1906, and the institution was completed in 1910 [104]. Now, the surface of the facility is significantly smaller (ca. 21 ha), since arable land was excluded.

The hospital buildings were arranged in a geometric, central-axial layout, and the pavilions for patients built on the peripheral alley were oriented in such a way that the patients' rooms and day rooms had windows opened to the southwest, and the verandas and terraces extended towards the pavilion gardens. The main axis of symmetry ran from the gatehouse at the entrance, then between the doctors' house and the admissions pavilion and the clerks' house, and up to the round lawn with a flower bed, in the centre of which stood the *Festsaalgebäude*—a building that served as a theatre, playroom, and concert hall. Behind the theatre, there was a kitchen and an ice cellar, and the axis was closed by the contagious pavilion, preceded by the symmetrical bipartite garden. The axis also divided the hospital complex into sections for women and men. On the female side, from the southeast, apart from the main part of the complex, there was a boiler room, a laundry room, and another house of doctors, while on the northwest side there was a farm outside with the inspector's house and livestock buildings (stables, pigsties, and cowsheds). Outside the layout, on the eastern side there was a water intake with a reservoir and a cemetery with a chapel (Figure 4).

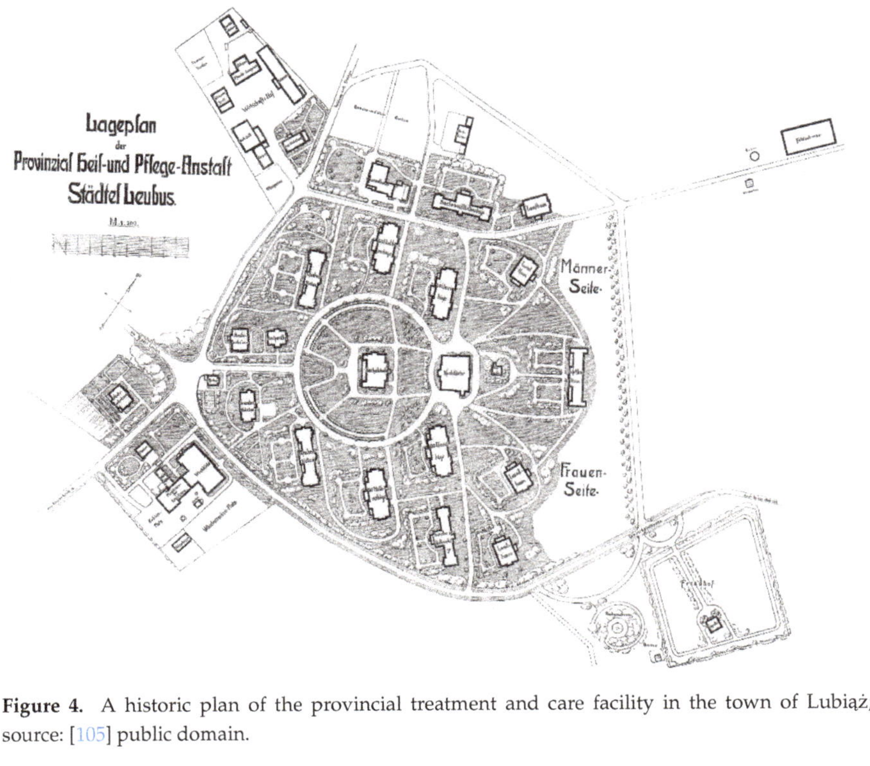

Figure 4. A historic plan of the provincial treatment and care facility in the town of Lubiąż; source: [105] public domain.

During World War II, the institution was under Nazi occupation and psychiatric patients of Jewish nationality were partially relocated there and killed during Aktion T4 in unexplained circumstances [106]. From 1942 to 1945, the site was used as a sanatorium for the troops returning from the front. After WWII, it was used for a short time as a field hospital for the Soviet army and later, buildings devastated by war activities were handed over to an organization dealing with agricultural education. The facility was turned into an agricultural mechanization training centre. It was not until 1957 that the institution again became a state-owned psychiatric treatment hospital. Nowadays, it is the most important stationary psychiatric institution for the Lower Silesia region, with seven wards (general and forensic wards and enhanced security wards for adults and adolescents) and a treatment and care facility for the elderly [107].

Nowadays, the time patients spend outside depends on the season and the weather. Most often, it is about an hour a day. There are usually organized group walks accompanied by therapists. During a pandemic, patients are not allowed to leave their unit building unaccompanied. In addition to two general departments, the hospital has two wards with enhanced security (juvenile and adults), two forensic departments, and a long-term care and treatment facility.

Closed wards have their own separate gardens. These areas are separated from each other, and walks are organized according to a fixed schedule. The other gardens are open, without any fences, and not regularly maintained. However, there are only a few benches in the park, and no gazebos or tables. Despite this fact, patients benefit from walks and report that these relax them, calm them down, and improve their mood.

The hospital does not conduct hortitherapy, but organizes artistic therapeutic classes, during which plant elements are used. During the walks, the patients collect flowers and leaves, with which they later make bouquets and paste into compositions on paper. The three-dimensional works are made of cones from the trees that grow in the park. Natural materials work well in occupational therapy because they offer more possibilities and sensory experiences than expensive plastic and stationery materials. Personal creativity provides patients with a great amount of satisfaction, allowing them to focus on something positive and celebrate their achievements when their works are displayed.

The green areas around the pavilions and the layout of the hospital are definitely an advantage, allowing for relaxing walks and enjoyment of the benefits of nature. Despite the transformations and the history of detrimental military use during World War II and the period directly after, a part of the oldest tree stand (plane trees, chestnut trees, and spruces) has survived and is inhabited by many birds. Both patients and staff appreciate it [108].

3.3.3. Weissenhof Weinsberg Klinikum

Klinikum am Weissenhof (nowadays: Zentrum für Psychiatrie Weinsberg) was built from 1900 and opened in 1903 as the fifth state-owned psychiatric institution in the province of Württemberg, initially aimed at providing treatment for 500 patients. To accommodate some separate wards and auxiliary buildings arranged on an organic plan with a circumferential road (Figure 5), the public domain of Weißenhof, a century-old estate, was chosen [105]. The plot was situated in a country area and offered magnificent views of the woody landscape to the southwest, with the romantic Burgruine Weibertreu on the hill. The setting was rural; however, the nearest railway station in Weinsberg was only two kilometres away and the historic city of Heilbronn was just six kilometres afar.

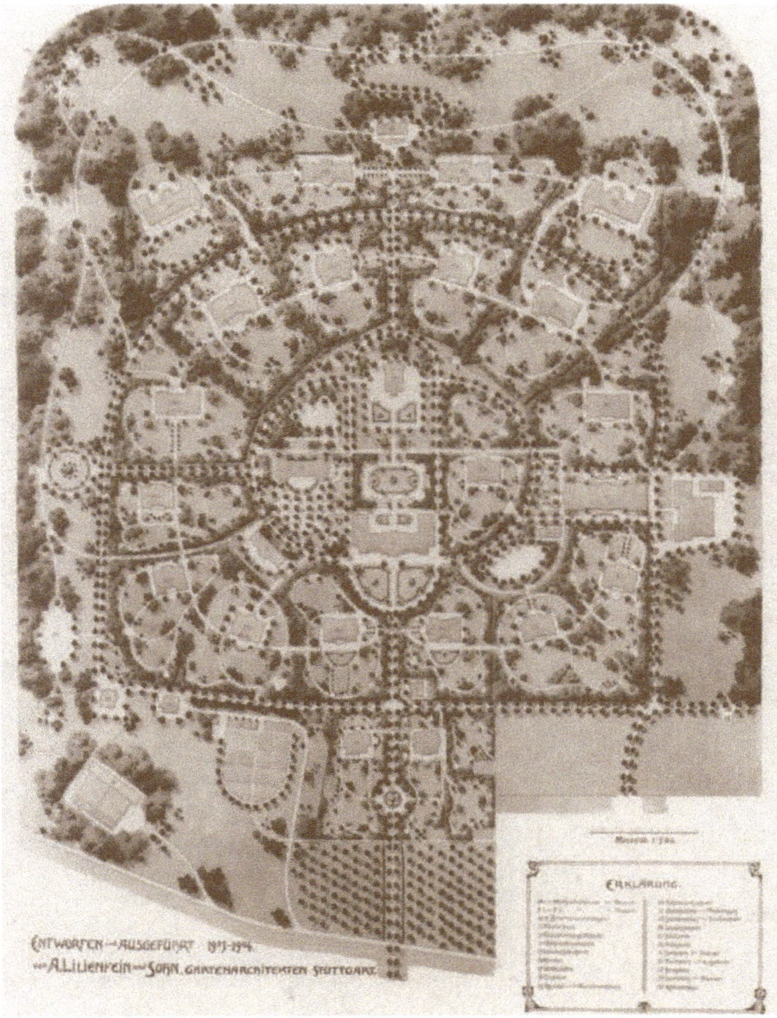

Figure 5. A plan of the park and gardens surrounding the institution in Weissenhof-Weinsberg, planned and implemented by A. Lilienfein and son, garden architects in Stuttgart from 1903–1904; source: courtesy of the archive of the Zentrum für Psychiatrie Weinsberg.

The last building of the pavilion system hospital was the institutional church, completed between 1913 and 1915. During the World War II, like many psychiatric hospitals under the national socialist regime in Germany, Aktion T4—a campaign to exterminate psychiatric patients and the mentally ill—took place here. Altogether, around 2000 patients lost their lives, transported to euthanasia centres from Weissenhof. After the war, the institution at Weissenhof-Weinsberg resumed its medical activity and was in part rebuilt. In 2003, a new forensic psychiatry department was built [109].

However, the overall composition remains still clearly visible, including the main longitudinal axis formerly dividing the ensemble into men's and women's sections. Nowadays, the clinic at the Weissenhof also offers treatment for patients other than psychiatric patients, and has 522 stationary beds, which is close to its historic capacity, while only limited cases need long-term stays. Several departments are located in a total of 97 buildings in the 43 ha park. At the time of construction, the park was laid out by a landscape architect, and apart from the common parts, farm, arable land, a cemetery, and ornamental grounds, each unit had its own garden enclosed by a fence and hedge.

The subjects of pride and care are 3800 trees, some of which are ageing and particularly vulnerable to weather and climate conditions, requiring supplementation, in recent years, by new and resilient species [110].

The historic park is used therapeutically in very different ways with the patients. Many activities within occupational therapy are carried out in the park. On weekdays, the care of the park is taken care of by the patients acting together. These tasks can range from raking leaves to cutting hedges and trees, and are always performed under the guidance and supervision of specialist staff. A small group of patients is also assigned to the animal enclosure within the former hospital farm. On average, 25 patients (divided into several small groups) are present on the site. The patients are usually employed in the park, either in the morning or in the afternoon, for 3 h each time, and always with a break.

A major advantage of occupational therapy in the park is that the result can usually be seen and experienced directly, and this gives patients a direct sense of achievement. In addition, the patients very often receive praise and recognition from passers-by. Since the clinic area has opened, there are many visitors and walkers around, in addition to employees of the clinic, who stop briefly when they pass, and happily let the patients know how beautifully the park has become thanks to the patients' work and involvement.

During the work, the age of the clinic and its park, as well as its history and its change over time, are regularly discussed in conversations. Both the changes in the flora and the park architecture, as well as the political past, are often topics of conversation and arouse the interest of the patients [111]. The dominant feature of the historic park landscape is the division into 'left' and 'right side', because at the time of construction all female patients were treated on one side and all male patients on the other side. It was, therefore, possible to draw a vertical line through the former park as a boundary between women and men, which is not valid anymore, but still can be seen in the overall layout of the site.

There are also many additional therapeutic activities carried out regularly in the park. These include exercise therapy, therapeutic walks (accompanied by professionals), and therapeutic farming. However, it is difficult to assess exactly how many patients are involved altogether in these practices, because several smaller care units which enjoy independence within the institution as a whole operate on the site, and no overall records concerning all therapies related to nature are kept.

4. Discussion

According to the American Horticultural Therapy Association (AHTA), horticultural therapy (HT) is a specialized type of therapeutic gardening programme. If one compares actions described in the case studies with the AHTA position paper on horticultural therapy [112], cases from Kobierzyn and Weinsberg are close to the definition. They offer voluntary regular activities focused on gardening tasks with a therapeutic aim. Weissenhof also offers additional programmes in the range of ecotherapy, such as exercise therapy, therapeutic walks (accompanied by professionals), and animal based-interventions within the hospital farm. Kobierzyn also attempts to introduce bee-therapy (apitherapy). In the case of Lubiąż, activities from the spectrum of nature arts and crafts are offered, and staff complains about the insufficient employment of qualified nature therapists.

There is, however, little known about the therapeutic reflection on those particular programmes or a systematic assessment of the well-being of the participants. Moreover, AHTA has its own system of certification and training, which is not a common training scheme for European countries that only, to some extent, have their own associations of horticultural therapists, such as those operating in Germany (Internationale Gesellschaft Gartentherapie e.V.) [113] and Switzerland (Schweizerische Gesellschaft Gartentherapie und Gartenagogik, SGGTA) [114].

One of the main advantages of the use of gardens in historic mental hospitals is that their landscape is perceived in a wider context of the whole institution. It is designed in a manner such that the whole area resembles an independent settlement, often referred to as a 'garden city', and a world in itself. Therefore, patients can work during their therapeutic gardening sessions in smaller enclosures where the results of their efforts are easier to notice and can be observed daily, which can contribute to their satisfaction by showing their impact on the reality of the garden. Successful shaping of a direct environment can contribute to the improvement of their mood. All the interviewees stressed that the lush greenery of the historic hospital site is an asset of the institution. They also indicated the challenges which result from the age of the trees and the necessary respect for the historic substance of the buildings and its network of paths.

Current research is very much engaged in creating guidelines for designing therapeutic green areas and gardens. Grahn's triangle of supporting environments [115] refers to aspects of both passive and active engagement with nature. This model can be applied in the context of the gardens in historic mental hospitals, indicating that patients with a subjective experience of low well-being are likely to manage inward-directed engagement and take part in more contemplative activities in hospital gardens. This can be a good guideline for the prevalent type of activities in the ward gardens. Consequently, those gardens would be the closest area enabling relaxation and contemplation.

Trojanowska offers a framework to be applied to public park design as an important element of public open spaces (POS) which can promote health and well-being [116]. While all these methodologies are established for the purpose of contemporary landscape and garden design, they can be to some extent useful when assessing therapeutic features of heritage gardens in historic mental hospitals. It is necessary, however, to be aware that those gardens were sometimes created more than 100 years ago and require a balanced approach respecting their cultural value. In addition, they were the product of an outdated understanding of mental disorders and they resulted from certain social circumstances. Nonetheless, they offer a good frame for the whole range of green interventions supporting the mental health and well-being of patients and several sustainable practices to be performed. What is more, some solutions connected with enhancing biodiversity or reducing intensive maintenance of undergrowth to secure bird habitats in historic parks can be problematic from the point of view of patients' subjective security perception. One of the challenges is protecting long-distance vistas, a feature indicated as vital in both historic and contemporary studies. Keeping them clear is one of the most difficult aspects to control, especially in those places where urban sprawl reached mental health hospitals once located on the outskirts of the city, because it does not depend on the institution administration only, but requires sensible urban planning regulations.

An important issue seems to be the need for specialist horticulture skills intended for historic garden maintenance, if it is to be treated as an occupational therapeutical gardening therapy programme. Social enterprises, which employ people suffering from mental disorders, attempt to fill that gap and already operate on some sites of historic mental hospitals, such as in Trieste (La Cooperativa Agricola Monte San Pantaleone [117]) and on the premises of the former Waldhaus Klinik Chur [118]. The latter case is particularly interesting because it offers jobs to adults with mental disabilities and provides them with professional training to become florists and gardeners for ornamental plants within the framework of the sheltered workshop of Graubünden Psychiatric Services (Psychiatrischen Dienste Graubünden-PDGR). The company, apart from other activities, maintains the park and the grounds of the historic Waldhaus Clinic in Chur (Figure 6), which includes cultivating 1159 square meters of flower beds on the premises, mowing, trimming, and scarifying the entire lawn, trimming the hedges, maintaining the flower beds' seasonal flower arrangements, watering, or removing weeds. The company is also responsible for many cleaning tasks in the garden and park, such as emptying the rubbish bins on the site, collecting what has been left behind, cleaning the large water basins, and collecting and disposing of leaves and branches from the trees on the site. The company also offers minor gardening tasks on a contractual basis for private customers from the area [119].

This approach marks an important future research direction, but these cases need further structured analysis that takes into account the long-term impact on patients' lives.

(**a**) (**b**)

Figure 6. (**a**) Chur, Waldhaus Psychiatric Clinic, 1.9.1947, aerial photograph by Werner Friedli; source: ETH-Bibliothek Zürich, [120] CC BY-SA 4.0; (**b**) Contemporary view of Klinik Waldhaus (2011), photo by Cloot, CC BY-SA 4.0; source: [121].

Ecotherapy Programmes and Their Relation to a Historic Mental Hospital Therapeutic Landscape

As described above, landscapes and gardens and their use had and still have an important role to play in the construction of the former asylum sites and the well-being of patients. Many connections to nature present in historic mental hospitals still exist, while some of them changed their character because of various reasons, including urban development of the surroundings and significant changes in psychiatric care organisation, to name only two of these reasons.

Table 2 provides a comparison of types of nature connections in historic mental hospitals in the past and nowadays.

Table 2. Comparison of the types of connections to nature and gardens in historic mental hospitals in the past and nowadays; source: author's elaboration.

Connection to Nature in Historic Mental Health Institutions	
In the Past	**Nowadays**
The garden setting of the asylum, located beyond city boundaries, is considered a space of mental escape from illness, and an environment supporting recovery.	Location on the outskirts of cities still considered as pleasant, but often surrounded by residential areas and urban sprawl.
Distant views from the institution showing picturesque landscapes.	Views become limited because of contemporary urban development—nevertheless, the most important can be preserved.
Gardens as places of physical exercise (gymnastics and walks); daily routine in utility and kitchen gardens; places of occupational therapy; terraces in the sun as therapy for tuberculosis.	Meditation and rest in the gardens; gardening therapy in the ward gardens; social activities—talks and therapeutic group meetings in the open air; therapeutic walks in the parks and woods surrounding the institution.
Orchards and arable land for food production as therapeutic work and an economic necessity.	Limited therapeutic farming/agricultural activities and animal tending.
Private gardens of the employees of the institution who lived onsite.	Hospital gardens as sites of regeneration during short breaks for employees.

While contact with nature and activities in the gardens are generally appreciated, there are sometimes concerns about the need for maintenance of the hospital grounds in the aspect of specialist care. The issue of ancient heritage trees and their resistance to climate change-related extreme weather conditions seems particularly demanding [102,111].

Taking this into account, and based on desk research and selected case studies, contemporary ecotherapy programmes can be assigned to particular areas constituting the composition of historic mental hospital gardens today(Table 3). This attribution indicates the potential contemporary use of landscapes for various ecotherapy programmes. While Table 3 presents all possible elements of the typical heritage asylum landscape, implementation of assigned ecotherapy programmes depends on the state of the preservation and integrity of particular sites. In hospital complexes where farming activities were abandoned a long time ago and the arable land was sold, restoring activities aimed at this landscape might be impossible.

Table 3. Distinctive elements of the composition of historic mental hospitals, their sites, and their potential therapeutic use for ecotherapy programmes; source: author's elaboration.

Ecotherapy Programme Opportunities in Historic Mental Hospitals	
Elements of Landscape Composition of Historic Mental Hospital Sites	**Potential Therapeutic Use for Ecotherapy Programmes**
Gardens at the wards.	Social and therapeutic horticulture; nature arts and crafts; green exercise therapy (yoga and open-air gyms).
Ornamental grounds and representative entrance areas.	Social and therapeutic horticulture; nature arts and crafts.
Landscape park.	Environmental conservation; nature arts and crafts; green exercise therapy (walking).
Hospital farm.	Care farming; animal-assisted interventions; animal-assisted therapy.
Former kitchen garden (including historic and new glasshouses), orchards, and arable land.	Social and therapeutic horticulture; nature arts and crafts; care farming.
Private gardens of the employees of the institution who lived onsite.	Social and therapeutic horticulture; nature arts and crafts.

On the contrary, gardens adjoining particular ward buildings are easier to restore and offer many opportunities because of their smaller size and direct link to the building where patients spend time. Their composition was less sophisticated than the outline of historic parks surrounding the institution and their maintenance requires mostly basic gardening skills. There is not much historic planting material, apart from a small number of old trees, that require specialist arborist care. The use of annuals, perennials, and herbaceous plants offers an opportunity for engaging all the senses, and creates sensory gardens, which are an acknowledged type of therapeutic garden [122] and can help provide pleasant and meaningful passive restoration.

While the results of the study enable a rough framework for possible ecotherapy programmes' attribution to the distinctive elements of the historic mental hospital landscape, the study has some limitations resulting from the chosen methodology.

It relies on the researcher's interpretation of data provided by interviewees. Here, conceptualizations are based on the lens of an architect with a special research focus and background in historic conservation and the history of gardens. Therefore, further qualitative and quantitative research involving in-depth psychological and medical perspectives of both patients and therapists is necessary to evaluate the therapeutic effect and clinical significance of ecotherapy conducted in the context of historic mental hospital gardens.

This marks a significant direction for future research, which should be a systematic evaluation of green therapies carried out in institutions located in heritage psychiatric institutions with gardens and parks.

5. Conclusions

Therapeutic gardening might be widely propagated as support for other therapies carried out in historic mental hospitals that nowadays house modern mental healthcare facilities. There is an abundance of evidence that therapeutic gardening may improve physical, psychological, and social health, and contact with nature prevents numerous mental health issues facing today's society. However, much still must be done to include green therapies in psychiatric care and encourage the training of professionals to increase people's opportunities and motivation to engage in gardening activities [123].

The main advantage of historic gardens and parks surrounding psychiatric care institutions located on heritage asylum sites is that they can make use of the existing landscape and gardens and parks, instead of waiting for newly planted material to grow around even the most innovative new hospital units built from scratch. This would be benefit historic parks, which are usually listed in monuments' registers as an integral part of the protected heritage buildings of the hospital. In addition, many sustainable gardening practices, which require manual work and specific individual solutions, can be introduced as green therapies. However, this approach, although promising, has some limitations.

Historically, occupational therapy, which included programmes from the spectrum of contemporary green therapies, was usually imposed on patients. In former asylums, patients had to subordinate to doctors' prescriptions concerning their activities and could not decline the imposed work. What is more, patients did not get paid. For this reason, the institution could maintain its often-extensive grounds by relying mostly on their work. This, naturally, does not prevent possible therapeutic outcomes, but today is considered an ambiguous practice and is associated with the exploitation of the patients treated as an available workforce. Nowadays, it is forbidden to employ patients without remuneration. All activities within the therapeutic treatment model are voluntary and take into account patients' independence and preference. Therefore, it would be difficult to rely only on the patients' sometimes irregular or fluctuating participation in garden maintenance to take appropriate care of the historic garden, which requires certain skills and regularity.

However, the example of a social enterprise (or protected workshops) is a particularly promising opportunity for patients recovering from the acute stage of their illness and on their way to independent living. This example offers the possibility of a meaningful occupation and goes beyond the simplest gardening tasks into the direction of specialization, particularly in the maintenance of historic gardens, which require specialist care. Patients working in social enterprises may have financial motivation combined with a sense of gaining independence and self-development.

Taking into account the beneficial outcomes associated with green therapies relying on restoring human connection to nature, and with maintaining the landscape values of historic mental hospital gardens, implementation of ecotherapy may be considered a step toward improving the well-being of the patients and the care for heritage gardens. Ecotherapy also offers an opportunity to involve local communities from the neighbouring housing areas in nature-based interventions on former asylum sites that could benefit from it, as they are often the nearest green space [124]. This could contribute to a sense of place and place attachment to those sometimes difficult and little-known heritage sites. This way, their mental health and well-being can also be improved, by improving their connection to the heritage [125] park and garden of the historic mental hospital.

Funding: This research received no external funding.

Data Availability Statement: Data available from the author.

Acknowledgments: I direct a special thanks to Elżbieta Filek and Maciej Bóbr, (both from Babiński Clinical Hospital in Kraków Kobierzyn), to Agnieszka Kukułka from Lubiąż Psychiatric Hospital, and to Sophia Lager from Zentrum für Psychiatrie Weinsberg, who were my interviewees for the purpose of this study.

Conflicts of Interest: The author declares no conflict of interest.

References

1. Gerard, D.L. Chiarugi and Pinel considered: Soul's brain/person's mind. *J. Hist. Behav. Sci.* **1997**, *33*, 381–403. [CrossRef]
2. Pinel, P. *Traité Médico-Philosophique sur L'aliénation Mentale ou la Manie*; Caille & Ravier: Paris, France, 1801.
3. Tuke, S. *Description of the Retreat, an Institution near York, for Insane Persons of the Society of Friends: Containing an Account of Its Origin and Progress, the Modes of Treatment, and a Statement of Cases*; Isaac Peirce: Philadelphia, PA, USA, 1813.
4. Rutherford, S. The Landscapes of Public Lunatic Asylums in England, 1808–1914. Ph.D. Thesis, de Montfort University, Leicester, UK, 2003. Available online: https://www.dora.dmu.ac.uk/handle/2086/4783 (accessed on 12 August 2022).
5. Rutherford, S. *The Victorian Asylum*; Shire Publications: Oxford, UK, 2008.
6. Rutherford, S. Landscapes for the Mind and Body, Context 72, December 2001, The Institute of Historic Building Conservation. Available online: http://ihbc.org.uk/context_archive/72/landscapes/hospital.html (accessed on 12 August 2022).
7. Mercier, C. *Lunatic Asylums, their Organisation and Management*; Griffin: London, UK, 1894; pp. 100–114.
8. Allmond, G. The First Garden City? Environment and utopianism in an Edwardian institution for the insane poor. *J. Hist. Geogr.* **2017**, *56*, 101–112. [CrossRef]
9. Mues, A. *Eine Gartenstadt für Psychisch Kranke—Die Baugeschichte der Heil- und Pflegeanstalt Wiesloch, Wiesloch—Beiträge zur Geschichte*; Band 2: Ubstadt-Weiher, Germany, 2001; pp. 290–304.
10. Howard, Ebenezer, To-morrow: A Peaceful Path to Real Reform, London: Swan Sonnenschein & Co., Ltd., 1898. Available online: https://commons.wikimedia.org/wiki/File:Diagram_No.3_(Howard,_Ebenezer,_To-morrow.).jpg (accessed on 12 August 2022).
11. Shorter, E. *A history of Psychiatry: From the Era of the Asylum to the Age of Prozac*; John Wiley & Sons: New York, NY, USA, 1997.
12. Leff, J. *Care in the Community: Illusion or Reality?* John Wiley & Sons: Chichester, UK, 1997.
13. Whitley, R. The Antipsychiatry Movement: Dead, Diminishing, or Developing? *Psychiatr. Serv.* **2012**, *63*, 1039–1041. [CrossRef] [PubMed]
14. Proceedings of a Conference Held at Church House, Westminster, London, on 9m and 10th March, 1961, p. 5. In The Speeches of John Enoch Powell, p. 36. Available online: http://enochpowell.info/wp-content/uploads/Speeches/1957-1961.pdf (accessed on 31 March 2022).
15. Killaspy, H. From the asylum to community care: Learning from experience. *Br. Med. Bull.* **2006**, *79–80*, 245–258. [CrossRef] [PubMed]
16. Taylor, B. The demise of the asylum in late twentieth-century britain: A personal history. *Trans. R. Hist. Soc.* **2011**, *21*, 193–215. [CrossRef]
17. Lowin, A.; Knapp, M.; Beecham, J. Uses of Old Long-Stay Hospital Buildings, *Psychiatr. Bull.* **1998**, *22*, 129–130.
18. *SAVE, Mind over Matter: A Study of the Country's Threatened Mental Asylums*; SAVE Britain's Heritage: London, UK, 1995.
19. Chaplin, R.; Peters, S. Executives have taken over the asylum: The fate of 71 psychiatric hospitals. *Psychiatr. Bull.* **2003**, *27*, 227–229. [CrossRef]
20. Jacobson, D.R. Mapping the location, design and decline of London's psychiatric asylums (1831–2012). *J. Maps* **2016**, *12*, 684–694. [CrossRef]
21. Dumont, M.P.; Dumont, D.M. Deinstitutionalization in the United States and Italy: A Historical Survey. *Int. J. Ment. Health* **2008**, *37*, 61–70. [CrossRef]
22. Lamb, H.R.; Bachrach, L.L. Some perspectives on deinstitutionalization. *Psychiatr. Serv.* **2001**, *52*, 1039–1045. [CrossRef]
23. Joseph, A.; Kearns, R.; Moon, G. Re-Imagining Psychiatric Asylum Spaces through Residential Redevelopment: Strategic Forgetting and Selective Remembrance. *Hous. Stud.* **2013**, *28*, 135–153. [CrossRef]
24. Vermorel, H.; Vermorel, M. De l'asile à la politique de secteur: L'évolution des institutions et des soins psychiatriques à Bassens. *Inf. Psychiatr.* **2012**, *88*, 759–770. [CrossRef]
25. Gusset, S.; Seglias, L.; Lengwiler, M. Versorgen, behandeln, pflegen: Geschichte der Psychiatrie in Graubünden. In *Quellen und Forschungen zur Bündner Geschichte*; Band 38, Herausgegeben vom Staatsarchiv Graubünden; Florian Hitz, Schwabe: Basel, Switzerland, 2021.
26. Wolf, G. (Ed.) *Geschäftsbericht Zentren für Psychiatrie in Baden-Württemberg 2021*; Zentren für Psychiatrie Baden-Württemberg, 2021. Available online: https://docplayer.org/215123331-Geschaeftsbericht-zentren-fuer-psychiatrie-in-baden-wuerttemberg.html (accessed on 10 July 2022).
27. Staniewska, A. Zagrożenia parków i ogrodów towarzyszących dawnym szpitalom psychiatrycznym w Polsce. In *Dziedzictwo Zagrożone, Ogrody Historyczne w Polsce, Hodor, K., Łakomy, K., Eds.*; Politechnika Krakowska: Kraków, Poland, 2015; pp. 102–120.

28. Rosenberg, C.E. Epilogue: Airs, Waters, Places. A Status Report. *Bull. Hist. Med.* **2012**, *86*, 661–670, Project MUSE. [CrossRef] [PubMed]
29. Gesler, W.M. Therapeutic landscapes: Medical issues in the light of the new cultural geography. *Soc. Sci. Med.* **1992**, *34*, 735–746. [CrossRef]
30. Williams, A. Therapeutic landscapes in holistic medicine. *Soc. Sci. Med.* **1998**, *46*, 1193–1203. [CrossRef]
31. Appleton, J. *The Experience of Landscape*; John Wiley & Sons: London, UK, 1975.
32. Kaplan, S. The restorative benefits of nature: Toward an integrative framework. *J. Environ. Psychol.* **1995**, *15*, 169–182. [CrossRef]
33. Kaplan, R.; Kaplan, S. *The Experience of Nature: A Psychological Perspective*; Cambridge University Press: Cambridge, UK, 1989.
34. Ulrich, R.S. View Through a Window May Influence Recovery. *Science* **1984**, *224*, 224–225. [CrossRef]
35. Ulrich, R.S. Visual landscapes and psychological well-being. *Landsc Res.* **1979**, *4*, 17–23. [CrossRef]
36. Ulrich, R.S. Human responses to vegetation and landscapes. *Landsc. Urban Plan.* **1986**, *13*, 29–44. [CrossRef]
37. Dushkova, D.; Ignatieva, M. New Trends in Urban Environmental Health Research: From Geography of Diseases To Therapeutic Landscapes and Healing Gardens. *Geogr. Environ. Sustain.* **2020**, *13*, 159–171. [CrossRef]
38. Saratsi, E.; White, J.; Holyoak, V. *Taking Account of Heritage Values of Urban Parks and Gardens*; Living with Environmental Change Policy and Practice Notes; Note No.36; Living With Environmental Change Polaris House, North Star Avenue: Wiltshire, UK, September 2016. [CrossRef]
39. Gilmore, A. The park and the commons: Vernacular spaces for everyday participation and cultural value. *Cult. Trends* **2017**, *26*, 34–46. [CrossRef]
40. Mell, I.C. Green infrastructure: Reflections on past, present and future praxis. *Landsc. Res.* **2017**, *42*, 135–145. [CrossRef]
41. Swensen, G.; Krokann Berg, S. The 'garden city' in the green infrastructure of the future: Learning from the past. *Landsc. Res.* **2020**, *45*, 802–818. [CrossRef]
42. Cameron, T.; Blanuša, J.; Taylor, E.; Salisbury, A.; Halstead, A.J.; Henricot, B.; Thompson, K. The domestic garden—Its contribution to urban green infrastructure. *Urban For. Urban Green.* **2012**, *11*, 129–137. [CrossRef]
43. Ying, J.; Zhang, X.; Zhang, Y.; Bilan, S. Green infrastructure: Systematic literature review. *Econ. Res. -Ekon. Istraživanja* **2022**, *35*, 343–366. [CrossRef]
44. Hartig, T.; Mitchell, R.; de Vries, S.; Frumkin, H. Nature and health. *Annu. Rev. Public Health* **2014**, *35*, 207–228. [CrossRef]
45. Hartig, T. Restorative Environments. In *Encyclopedia of Applied Psychology*; Spielberger, C., Ed.; Academic Press: San Diego, CA, USA, 2004; pp. 273–279. [CrossRef]
46. Hartig, T. Restoration in Nature: Beyond the Conventional Narrative. In *Nature and Psychology, Schutte, A.R., Torquati, J.C., Stevens, J.R., Eds.*; Nebraska Symposium on Motivation; Springer: Cham, Switzerland, 2021; Volume 67, pp. 89–151. [CrossRef]
47. Annerstedt van den Bosch, M.; Depledge, M.H. Healthy people with nature in mind. *BMC Public Health* **2015**, *15*, 1232. [CrossRef]
48. van den Bosch, M.; Ode Sang, Å. Urban natural environments as nature-based solutions for improved public health—A systematic review of reviews. *Environ. Res.* **2017**, *158*, 373–384. [CrossRef]
49. Twohig-Bennett, C.; Jones, A. The health benefits of the great outdoors: A systematic review and meta-analysis of greenspace exposure and health outcomes. *Environ Res.* **2018**, *166*, 628–637. [CrossRef]
50. Kondo, M.C.; Fluehr, J.M.; McKeon, T.; Branas, C.C. Urban Green Space and Its Impact on Human Health. *Int. J. Environ. Res. Public Health* **2018**, *15*, 445. [CrossRef]
51. Chen, K.; Zhang, T.; Liu, F.; Zhang, Y.; Song, Y. How Does Urban Green Space Impact Residents' Mental Health: A Literature Review of Mediators. *Int. J. Environ. Res. Public Health* **2021**, *18*, 11746. [CrossRef] [PubMed]
52. Maas, J.; Verheij, R.A.; Groenewegen, P.P.; de Vries, S.; Spreeuwenberg, P. Green space, urbanity, and health: How strong is the relation? *J. Epidemiol. Community Health* **2006**, *60*, 587–592. [CrossRef] [PubMed]
53. Berman, M.G.; Kross, E.; Krpan, K.M.; Askren, M.K.; Burson, A.; Deldin, P.J.; Kaplan, S.; Sherdell, L.; Gotlib, I.H.; Jonides, J. Interacting with nature improves cognition and affect for individuals with depression. *J. Affect. Disord.* **2012**, *140*, 300–305. [CrossRef] [PubMed]
54. van den Berg, M.; van Poppel, M.; van Kamp, I.; Andrusaityte, S.; Balseviciene, B.; Cirach, M.; Danileviciute, A.; Ellis, N.; Hurst, G.; Masterson, D.; et al. Visiting green space is associated with mental health and vitality: A cross-sectional study in four European cities. *Health Place* **2016**, *38*, 8–15. [CrossRef]
55. van den Berg, M.M.; van Poppel, M.; van Kamp, I.; Ruijsbroek, A.; Triguero-Mas, M.; Gid-low, C.; Nieuwenhuijsen, M.J.; Gražulevičiene, R.; van Mechelen, W.; Kruize, H.; et al. Do Physical Activity, Social Cohesion, and Loneliness Mediate the Association Between Time Spent Visiting Green Space and Mental Health? *Environ. Behav.* **2019**, *51*, 144–166. [CrossRef]
56. Roscoe, C.; Sheridan, C.; Geneshka, M.; Hodgson, S.; Vineis, P.; Gulliver, J.; Fecht, D. Green Walkability and Physical Activity in UK Biobank: A Cross-Sectional Analysis of Adults in Greater London. *Int. J. Environ. Res. Public Health* **2022**, *19*, 4247. [CrossRef]
57. WHO. Available online: https://www.who.int/news-room/fact-sheets/detail/mental-disorders (accessed on 22 June 2022).
58. Institute of Health Metrics and Evaluation. Global Health Data Exchange (GHDx). Available online: https://vizhub.healthdata.org/gbd-results (accessed on 22 June 2022).
59. Reinert, M.; Fritze, D.; Nguyen, T. The State of Mental Health in America 2022, Mental Health America, Alexandria VA 2021; p. 20. Available online: https://mhanational.org/issues/state-mental-health-america (accessed on 22 June 2022).
60. Halbreich, U. Stress-related physical and mental disorders: A new paradigm. *BJPsych Adv.* **2021**, *27*, 145–152. [CrossRef]

61. World Health Organization. Mental health atlas 2020. World Health Organization. 2021. Available online: https://apps.who.int/iris/handle/10665/345946.License: CC BY-NC-SA 3.0 IGO (accessed on 22 June 2022).
62. Han, A.R.; Park, S.A.; Ahn, B.E. Reduced stress and improved physical functional ability in elderly with mental health problems following a horticultural therapy program. *Complementary Ther. Med.* **2018**, *38*, 19–23. [CrossRef]
63. Overbey, T.A.; Diekmann, F.; Lekies, K.S. Nature-based interventions for vulnerable youth: A scoping review. *Int. J. Environ. Health Res.* **2021**, 1–39. [CrossRef]
64. Siu, A.; Kam, M.; Mok, I. Horticultural Therapy Program for People with Mental Illness: A Mixed-Method Evaluation. *Int. J. Environ. Res. Public Health* **2020**, *17*, 711. [CrossRef]
65. Gonzalez, M.T.; Hartig, T.; Patil, G.G.; Martinsen, E.W.; Kirkevold, M. Therapeutic horticulture in clinical depression: A prospective study of active components. *J. Adv. Nurs.* **2010**, *66*, 2002–2013. [CrossRef] [PubMed]
66. Detweiler, M.B.; Self, J.A.; Lane, S.; Spencer, L.; Lutgens, B.; Kim, D.Y.; Halling, M.H.; Rudder, T.C.; Lehmann, L.P. Horticultural therapy: A pilot study on modulating cortisol levels and indices of substance craving, posttraumatic stress disorder, depression, and quality of life in veterans. *Altern. Ther. Health Med.* **2015**, *21*, 36–41, PMID: 26030115.
67. Wise, J. *Digging for Victory. Horticultural therapy with Veterans for Post-Traumatic Growth*; Routledge: London, UK, 2015.
68. Poulsen, D.V.; Stigsdotter, U.K.; Refshage, A.D. Whatever happened to the soldiers? Nature-assisted therapies for veterans diagnosed with post-traumatic stress disorder: A literature review. *Urban For. Urban Green.* **2015**, *14*, 438–445. [CrossRef]
69. Eichholz, C. Therapeutic Horticulture as a Healing Tool for Veterans. Dissertation, University of Pennsylvania, Philadelphia, PA, USA, 2020. Pennsylvania US. Available online: https://repository.upenn.edu/edissertations_sp2/142 (accessed on 17 February 2022).
70. Murroni, V.; Cavalli, R.; Basso, A.; Borella, E.; Meneghetti, C.; Melendugno, A.; Pazzaglia, F. Effectiveness of Therapeutic Gardens for People with D dementia: A Systematic Review. *Int. J. Environ. Res. Public Health* **2021**, *18*, 9595. [CrossRef] [PubMed]
71. Bourdon, E.; Belmin, J. Enriched gardens improve cognition and independence of nursing home residents with dementia: A pilot controlled trial. *Alzheimer's Res. Ther.* **2021**, *13*, 116. [CrossRef]
72. Cooper-Marcus, C.; Barnes, M. *Gardens in Healthcare Facilities: Uses, Therapeutic Benefits, and Design Recommendations*; The Center for Health Design: Martinez, CA, USA, 1995.
73. Cooper-Marcus, C.; Barnes, M. *Healing Gardens: Therapeutic Benefits and Design Recommendations*; John Wiley: New York, NY, USA, 1999.
74. Gierlach-Spriggs, N.; Kaufman, R.E.; Warner, S.B., Jr. *Restorative Garden: The Healing Landscape*; Yale University Press: New Haven, CT, USA, 1998.
75. Stigsdotter, U.; Grahn, P. What makes a garden a healing garden? *J. Ther. Hortic.* **2002**, *13*, 60–69.
76. Trojanowska, M. *Parki i Ogrody Terapeutyczne*; Wydawnictwo Naukowe PWN: Warszawa, Polandm, 2017.
77. Staniewska, A. *Obłędne Ogrody: Idea Krajobrazu Terapeutycznego w Kompozycji Zespołów Szpitali Psychiatrycznych XIX i Początku XX Wieku*; Wydawnictwo Politechniki Krakowskiej: Kraków, Poland, 2020.
78. Babbie, E. *The Practice of Social Research*; Wadsworth: Belmont, CA, USA, 1992.
79. Kvale, S. *Doing Interviews*; SAGE Publications Ltd.: London, UK, 2007; pp. 2–10. [CrossRef]
80. Jamshed, S. Qualitative research method-interviewing and observation. *J. Basic Clin. Pharm.* **2014**, *5*, 87–88. [CrossRef]
81. Curran, K.; Zimmermann, N. The Dynamics of Collaboration in Heritage Science. *Stud. Conserv.* **2022**, *67*, 136–149. [CrossRef]
82. McGrath, C.; Palmgren, P.J.; Liljedahl, M. Twelve tips for conducting qualitative research interviews. *Med. Teach.* **2019**, *41*, 1002–1006. [CrossRef]
83. Fontana, A.; Frey, J.H. The interview: From neutral stance to political involvement. In *The SAGE Handbook of Qualitative Research*; Denzin, N.K., Lincoln, Y.S., Eds.; Sage: Thousand Oaks, CA, USA, 2005; pp. 695–727.
84. Conolly, J.M. *The Construction and Government of Lunatic Asylums and Hospitals for the Insane*; J. Churchill: London, UK, 1847.
85. Suggestions and Instructions in reference to Sites, Construction, and Arrangement of Buildings, and Plans of Lunatic Asylums, Issued by the Commissioners in Lunacy. *J. Ment. Sci.* **1871**, *16*, 627–631. [CrossRef]
86. Browne, W.A.F. *What Asylums Were, Are, and Ought to Be Being the Substance of Five Lectures Delivered before the Managers of the Montrose Royal Lunatic Asylum*; A.C. Black: Edinburgh, UK, 1837.
87. Parchappe, M. *Des Principes à Suivre Dans la Fondation et la Construction des Asiles d'Aliénés*; Librairie de Victor Masson: Paris, France, 1853.
88. Roller, C.F.W. *Die Irrenanstalt Nach allen ihren Beziehungen dargestellt. Mit Lithographischen und Colorirten Plane*; Müller: Karlsruhe, Germany, 1831; pp. 114–115.
89. Hickman, C. *Therapeutic Landscapes. A History of English Hospital Gardens Since 1800*; Manchester University Press: Manchester, UK, 2013.
90. Zagórski, R. O Zakładzie dla umysłowo i nerwowo chorych w Kobierzynie. Zarys rozwoju Zakładu pod koniec 1924 roku. *Pol. Gaz. Lek.* **1925**, *18*, 420–423.
91. Beveridge, A. What became of Arthur Conan Doyle's father? The last years of Charles Altamont Doyle. *J. R. Coll. Phys. Edinb.* **2006**, *36*, 264–270. Available online: https://www.rcpe.ac.uk/sites/default/files/u_beveridge2.pdf (accessed on 12 August 2022).
92. Montès, J.-F. Le monde clos, ou la raison médico-économique. *Sci. Soc. Et St.* **1993**, *11*, 41–69. [CrossRef]

93. Vanja, C. *Von der herzoglihen Irrenanstalt zum modernen Gesundheitskonzern. Die Geschichte der nassauischen Pscyhiatrie, In Nassauische Annalen, Jahrbuch des Vereins für Nassauische Altertumskunde und Geschichtsforschung*; Bd123_Sonderband: Wiesbaden, Germany, 2021; pp. 603–634.
94. Roszak, T.; Gomes, M.E.; Kanner, A.D. (Eds.) *Ecopsychology: Restoring the Earth, Healing the Mind*; Sierra Club Books: Berkeley, CA, USA.; Los Angeles, CA, USA.; London, UK, 1995.
95. Summers, J.K.; Vivian, D.N. Ecotherapy—A Forgotten Ecosystem Service: A Review. *Front. Psychol.* **2018**, *9*, 1389. [CrossRef]
96. Buzzell, L.; Craig, C. *Ecotherapy: Healing with Nature in Mind*; Sierra Club Books: San Francisco, CA, USA, 2009.
97. Bragg, R.; Wood, C.; Barton, J. *Ecominds Effects on Mental Wellbeing. An Evaluation for Mind*; MIND: London, UK, 2013. Available online: https://www.bl.uk/collection-items/ecominds-effects-on-mental-wellbeing-an-evaluation-for-mind (accessed on 12 August 2022).
98. Ecotherapy, MIND 2021. Available online: https://www.mind.org.uk/media/9661/ecotherapy-2021-pdf-for-download-pdf-version.pdf (accessed on 12 August 2022).
99. Chaudhury, P.; Banerjee, D. Recovering with Nature: A Review of Ecotherapy and Implications for the COVID-19 Pandemic. *Front. Public Health* **2020**, *8*, 604440. [CrossRef]
100. Danilczyk, L.; et al. Kraków—Kobierzyn, ul. Babińskiego 29, Zespół Szpitala Psychiatrycznego. Studium Urbanistyczno-Konserwatorskie. *unpublished material in Babiński Clinical Hospital archive*, Kraków, 1988.
101. Staniewska, A.; Petryshyn, H.; Lukashchuk, H.; Kliusa, T.; Koprynia, K. Values of the historical parks and gardens of psychiatric hospitals in Kulparków and Kobierzyn as the basis for their revalorization, *Wiadomości Konserw. J. Herit. Conserv.* **2021**, *67*, 22–34. Available online: https://repozytorium.biblos.pk.edu.pl/redo/resources/44675/file/resourceFiles/StaniewskaA_WaloryHistorycznych.pdf (accessed on 12 August 2022). [CrossRef]
102. Filek, E.; Babiński Clinical Hospital, Kraków, Poland. Personal communication, 2021.
103. Bóbr, E.; Babiński Clinical Hospital, Kraków, Poland. Personal communication, 2021.
104. Kiejna, A.; Wójtowicz, M. *Prowincjalny Psychiatryczny Zakład Leczniczo-Opiekuńczy w Lubiążu 1830–1912/Provinzial Heil- und Pflegeirrenanstalt zu Leubus/Schlesien: 1830–1912*; Fundacja Ochrony Zdrowia Psychicznego: Wrocław, Poland, 2002.
105. Bresler, J. (Ed.) *Deutsche Heil- und Pflegeanstalten für Psychischkranke in Wort und Bild*; Band 1, Marhold: Halle an der Salle, Germany, 1910.
106. Hohendorf, G.; Rotzoll, M.; Richter, P.; Eckart, W.; Mundt, C. Die Opfer der nationalsozialistischen "Euthanasie-Aktion T4". *Der Nervenarzt* **2002**, *73*, 1065–1074. [CrossRef]
107. Kacalak, J. *Wojewódzki Szpital dla Nerwowo i Psychicznie Chorych w Lubiążu na przełomie wieków*; Oscar Marand P. Spiżak: Warszawa, Poland, 2008.
108. Kukułka, A.; Lubiąż Psychiatric Hospital. Personal communication, 2021.
109. Festschrift „100-Jahre". *Von der Königlichen Heilanstalt Weinsberg zum Klinikum am Weissenhof 1903–2003*; Knaebel, D., Zenth, U., Eds.; Klinikum am Weissenhof: Weinsberg, Germany, 2003.
110. Brochure: Der Grüne Kompass, Klinikum am Weissenhof, Weinsberg 2021. p. 16. Available online: https://www.klinikum-weissenhof.de/fileadmin/user_upload/Weinsberg/News_Veranstaltungen/Gruener_Kompass_KaW.pdf (accessed on 18 November 2021).
111. Lager, S.; Zentrum für Psychiatrie Weinsberg. Personal communication, 2021.
112. AHTA. Definitions and Positions Paper, 2 Mar 2022. Available online: https://www.ahta.org/assets/docs/definitions%20and%20positions%20final%206.17.pdf (accessed on 11 August 2022).
113. Internationalen Gesellschaft Gartentherapie, e.V. Available online: https://www.iggt.eu (accessed on 12 August 2022).
114. Schweizerische Gesellschaft Gartentherapie und Gartenagogik, SGGTA. Available online: https://www.gartentherapie.ch (accessed on 12 August 2022).
115. Grahn, P.; Tenngart Ivarsson, C.; Stigsdotter, U.; Bengtsson, I.-L. Using affordances as a health-promoting tool in a therapeutic garden. In *Innovative Approaches to Researching Landscape and Health: Open Space: People Space 2, Ward Thompson, C., Aspinall, P., Bell, S., Eds.*; Routledge: New York, NY, USA, 2010; pp. 116–154. [CrossRef]
116. Trojanowska, M. A universal standard for health-promoting places. Example of assessment—On the basis of a case study of Rahway River Park. *Bud. I Archit.* **2021**, *20*, 57–82. [CrossRef]
117. La Cooperativa Agricola Monte San Pantaleone. Available online: https://www.agricolamontesanpantaleone.it (accessed on 12 August 2022).
118. Pieren, M.; Zoellner, S.; Tromm, P. Jubiläumspublikation der Psychiatrischen Dienste Graubünden 2017: 125 Jahre Klinik Waldhaus Chur, Hochschule für Technik und Wirtschaft, Psychiatrische Dienste Graubünden, Chur 2017. Available online: https://www.fhgr.ch/fileadmin/news/news-institute_und_zentren/news-2017_10_10-Jubilaeumspublikation_PDGR.pdf (accessed on 12 August 2022).
119. ARBES. Available online: https://arbes.ch/de/arbeiten-bei-der-arbes/anlage-park-chur (accessed on 12 August 2022).
120. Bildarchiv/Stiftung Luftbild Schweiz/Fotograf: Friedli, Werner LBS_H1-010108. Available online: http://doi.org/10.3932/ethz-a-000352242 (accessed on 12 August 2022).
121. Psychiatrische Dienste Graubünden Klinik Waldhaus. Available online: https://upload.wikimedia.org/wikipedia/commons/3/33/Klinik_waldhaus_img2.JPG (accessed on 12 August 2022).

122. Krzeptowska-Moszkowicz, I.; Moszkowicz, Ł.; Porada, K. Evolution of the Concept of Sensory Gardens in the Generally Accessible Space of a Large City: Analysis of Multiple Cases from Kraków (Poland) Using the Therapeutic Space Attribute Rating Method. *Sustainability* **2021**, *13*, 5904. [CrossRef]
123. Soga, M.; Gaston, K.J.; Yamaura, Y. Gardening is beneficial for health: A meta-analysis. *Prev. Med. Rep.* **2016**, *5*, 92–99. [CrossRef]
124. Beyer, K.M.; Kaltenbach, A.; Szabo, A.; Bogar, S.; Nieto, F.J.; Malecji, K.M. Exposure to neighborhood green space and mental health: Evidence from the survey of the health of Wisconsin. *Int. J. Environ. Res. Public Health* **2014**, *11*, 3453–3472. [CrossRef] [PubMed]
125. Spennemann, D.H.R. The Nexus between Cultural Heritage Management and the Mental Health of Urban Communities. *Land* **2022**, *11*, 304. [CrossRef]

land

Article

Historical Water Management Strategies—Case Study of Traditional Villages in Southern China, Hunan Province

Ning DongGe [1], Jinbiao Yan [2], Peilin Liu [3], Martin van den Toorn [4] and Albert Fekete [1,*]

[1] Institute of Landscape Architecture, Urban Planning and Garden Art, Hungarian University of Agriculture and Life Science, 1118 Budapest, Hungary
[2] National-Local Joint Engineering Laboratory on Digital Preservation and Innovative Technologies for the Culture of Traditional Villages and Towns, Hengyang Normal University, Hengyang 421002, China
[3] Institute of Rural Revitalization Research, Changsha University, Changsha 410022, China
[4] Department of Landscape Architecture, Delft University of Technology, 2628 BL Delft, The Netherlands
* Correspondence: fekete.albert@uni-mate.hu; Tel.: +36-205482979

Abstract: Based on the landscape architecture of traditional settlements in southern China, this study takes water as a vital element through field investigation and model analysis to explore the water management strategies of two traditional villages in Xiangjiang River Basin, Hunan Province. We have found that traditional settlements are located between rivers and mountains. The community of the settlement has a strong interaction with the water environment. The water management system consists of two parts: the rainwater collection and storage system of a single building and the settlement' s water collection and drainage system. Through calculation, we found that the amounts of water collected (per year) between the two villages are different: ZhangGuYing (Z village) = 5.73 million L, ShangGanTang (S village) = 1.784 million L, in spite of the fact that water management strategies of the two settlements are similar. Further analysis shows that the difference is related to the adaption of the precipitation and topography of the surrounding areas. The above-mentioned systematic management strategy of water resources has been used until currently, with adaptability, low cost, and sustainability. It has outstanding significance for the current demand for sustainable development from both resource management and cultural aspects.

Keywords: historical commons; human cultural heritage; local development; landscape values; rainwater management

Citation: DongGe, N.; Yan, J.; Liu, P.; van den Toorn, M.; Fekete, A. Historical Water Management Strategies—Case Study of Traditional Villages in Southern China, Hunan Province. *Land* **2022**, *11*, 2107. https://doi.org/10.3390/land11122107

Academic Editors: Jan K. Kazak, Magdalena Wilkosz-Mamcarczyk and Nir Krakauer

Received: 27 September 2022
Accepted: 16 November 2022
Published: 22 November 2022

1. Introduction

Traditional settlements, as part of the landscape, where people have lived in compact communities throughout history [1], are similar to modern settlements. Our lifestyles are based on the development and distribution of resources [2,3]. However, because of the lack of advanced production tools and productivity, traditional settlements are more dependent on a stable and suitable ecological environment [4], including climate, topography, soil, water resources, and other basic natural conditions [5]. Therefore, the traditional ecological knowledge of how to achieve and maintain sustainable development of a settlement has gradually accumulated in different places all over the world [6]. Due to the importance of water for human beings, ancient civilizations around the world have accumulated their own water management technologies and strategies [7] —for instance, the canals in ancient Rome, which realized the spatial distribution of water resources. Ancient Greece also realized the reuse of water resources in the Mediterranean region through the design of urban water pipes and public facilities and solved the problem of uneven time distribution of water resources [8]. Ancient Egypt took advantage of the seasonal changes of the Nile to achieve agricultural prosperity [9]. Although the traditional ecological knowledge is effective, some of it no longer matches the current situation [8]. Therefore, it is particularly important to study the traditional strategies and internal mechanisms of the ecologically sustainable settlements which have survived to this day. Unlike other ancient civilizations, some of the ancient Chinese settlement/facilities are still functioning well until today [10]. The traditional Chinese agricultural landscape is highly dependent on the stable natural environment [11], so the location of settlements [12] and the management and distribution of local resources -especially the water—are prerequisite conditions for the local sustainable development [13].

Based on experience (or observation), traditional knowledge is tentative and probabilistic and will be constantly revised and falsified [14]. Thus, as the ecology of small-scale (settlements) can be continuously adapted iteratively to finally realize a sustainable system of self-sufficiency [15], and at the same time, a highly stable local economy with families as its basic units can be realized [16], which is characterized by high-level social coordination, organized division of labor, and effective autonomy [17]. On the basis of a farming economy that relied on natural resources, ancient Chinese ancestors summed up and developed a set of systematic management theories based on nature for the selection and planning of ecological settlements and management of natural resources [18].

The goal of this study is to reinterpret the accumulated experiences and theories in history from a sustainable perspective for the urban landscape, with a special emphasis on today' s increasingly obvious water resources problems [19]. From the experiences and strategies, we can learn how human–environmental interactions changed through time and space and further discuss how to extend its value [20].

The current research on traditional water resources is mainly qualitative research, which explains the experiences, methods, and technologies of traditional water management in some developing regions [21–23] and tries to apply the simple logic and low-cost strategies to other regions [24]. In addition, there are also studies on the application of rain harvest to agricultural production [25]. The roof rain harvest function in a settlement has potential value [26] but is limited by the data validity [27]; the quantitative analysis based on specific studies cannot be carried out to provide more convincing conclusions for the implementation plans [28]. Some quantitative analysis of rainwater collection has implications for this study [29], especially the analysis and comparison of the changes in precipitation amount and distribution on water resources [30].

In addition to the research on low-cost rainwater harvesting in traditional settlements, the research on traditional settlements' response to floods also has implications for urban stormwater management under today' s climate change situation. [31,32]

Some other studies take the perspective of the functionality of the architecture itself, the water resources management as the main research topic [33], and combine the quantitative analysis and qualitative analysis to research the sustainability and cross-regional application potential of traditional water management [34,35], which can explain the mechanism of traditional ecological knowledge [36], but it does not directly prove the effectiveness of its mechanism and the details of its dynamic changes [37–39].

This study will explore water-related data from the perspective of water management to prove how traditional settlements can achieve adaptability across time and space scales. The research focuses on analyzing the background, logic, and action mechanism of the traditional water management strategies, and finally provides solutions for urban landscape planning and water resources management in the new period.

2. Materials and Methods

Through field investigation, on-site interview, and photogrammetry, this study collected, analyzed, and compared the basic data related to water management in two traditional settlements so as to obtain the operation mechanism of a traditional settlement water management system and further discuss the possibility of its wider application.

2.1. Study Area

The two villages involved in the research are located in Southeast China, Xiangjiang River Basin (Figure 1a).

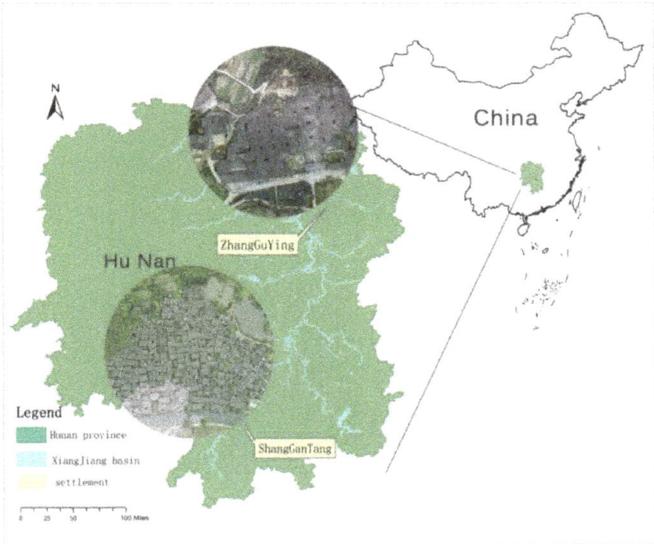

(**a**)

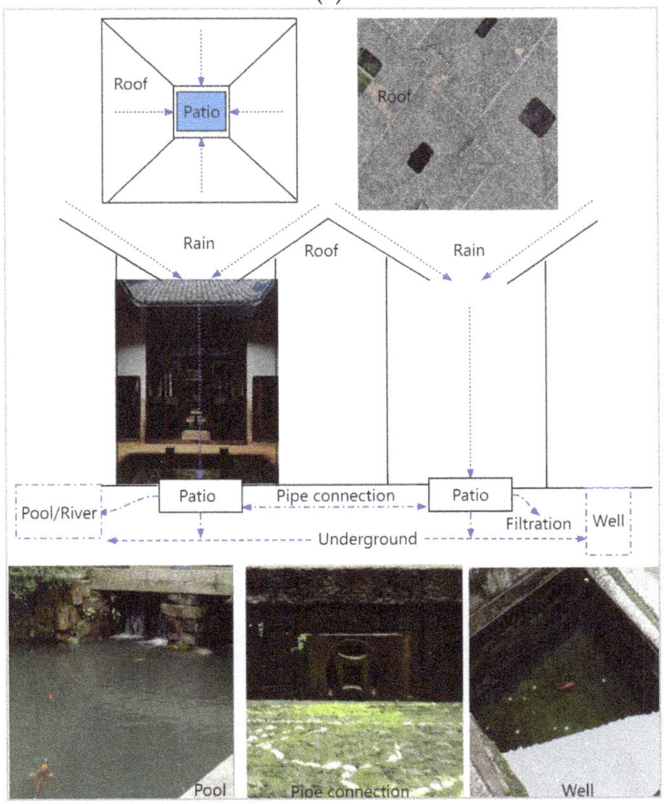

(**b**)

Figure 1. (**a**) Location of the two villages in Xiangjiang River Basin, Southeast China. (**b**) The water collection system used in the villages.

The ZhangGuYing village of YueYang City (29°00′36″ N; 113°128′30″ E) has a population of 2169 inhabitants (data from year 2000, newer official data are not available). The ShangGanTang village of YongZhou City (25°09′09″ N; 111°10′57″ E) has a population of 1865 inhabitants (data from year 2000, newer official data are not available) are located in Hunan Province in terms of administrative division. In terms of geographical location, the two settlements are located in the upper and lower half of the Xiangjiang River Basin. They belong to the subtropical monsoon climate area, but according to the data of the local weather station, the precipitation, humidity, and annual average temperature of the two settlements are different, as shown in Table 1.

Table 1. Meteorological data of the two villages.

Annual Average (1981–2010)	Precipitation	Temperature	Humidity
ZhangGuYing	1353 mm	−5.9–39.2 °C	77%
ShangGanTang	1426 mm	−5.2–40.3 °C	78% [1]

[1] Meteorological Data website [40].

The rainwater collection and storage system mainly have three parts:

1. The rain harvesting roof
2. The patio and underground pipe connection (the patio has an average water capacity of 2.4 m^3)
3. The water storage pool (The pool capacity for Z village is unavailable yet; for S village, the water storage capacity is 73.4^3 [10])

As we can see from Figure 1b, the rainwater was collected by the roof from 4 directions, falling down to the patio, each patio being connected by the underground pipe system, and a bigger amount of water drained into the river; less water was stored in the pool. Some part of the water after infiltration goes down (underground), and the water which was left on the patio can be used for daily purposes (for washing, for microclimate amelioration of the courtyards, etc.). Less is used for drinking, as far as the drinking water obtained from the well. Due to the traditionally used wooden construction materials of the villages, the water from the pool is also used in case of accidental fire.

The villages are located in rural landscapes, representing the organic part of the landscape, which has as its main use paddy rice production. The rice field has a totally independent water supply system called the rice canal, which is connected to the river. For the whole research area, this study will explain the function of the different parts of the water management system used in the village. The data analysis will only focus on the rain harvest part to calculate the amount of water collection and compare the two villages for the similarities and differences in water management strategies.

2.2. Tools and Models

The settlement image data was collected by DJI phantom3, and the 3D model was rendered by the Contextcapter software. Surface measurements and linear regression analysis were performed by Acute3D viewer. Topographic data were collected through Google Earth, and topographic maps and cross-sections of settlements were processed by Surfer software (Figure 2).

2.3. Data Collection and Calculation

Drone photogrammetry was performed on the two urban landscapes in the Xiangjiang River Basin, and the drone routine collection data was entered into the Contextcapture software for processing in order to obtain 3D models of the two settlements. The 3D model was then measured to obtain the catchment area of the patio and roof (Table 2). The regression analysis of the patio and roof area of the buildings showed that there was a linear correlation between the patio and roof area of the two settlements.

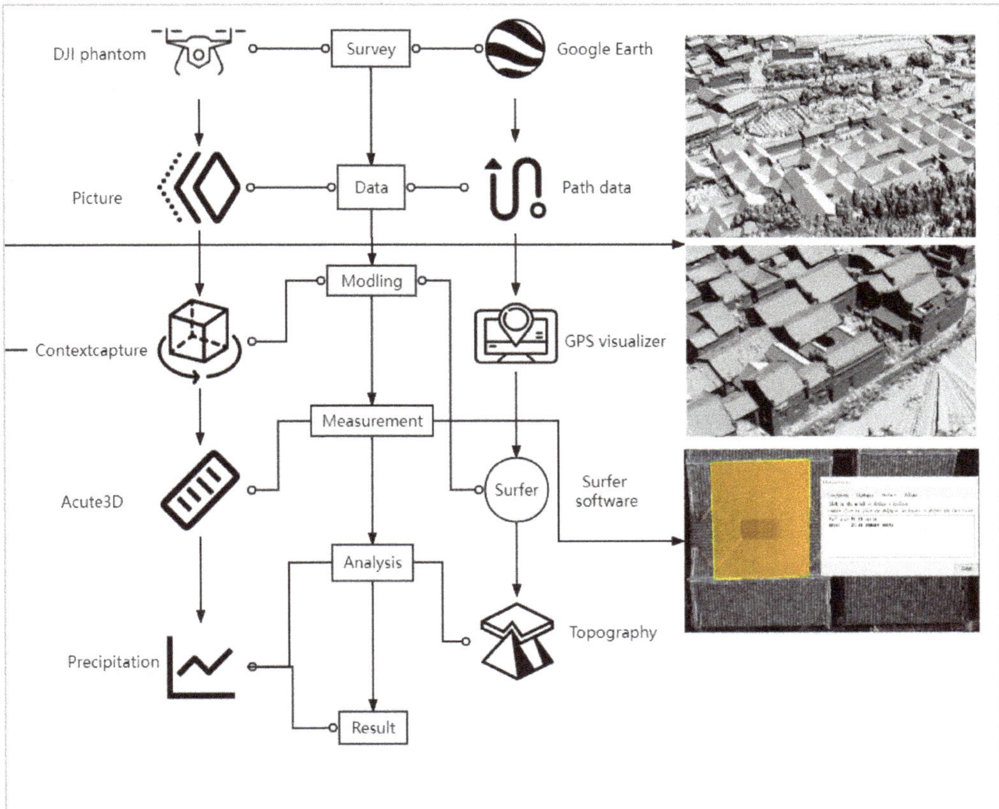

Figure 2. Conceptual framework of the research process.

Table 2. The table of the 3D model water catchment area measurement.

	Patio1 (m^2)	Roof1 (m^2)	P/R1 (ShangGanTang)	Patio2 (m^2)	Roof2 (m^2)	P/R2 (ZhangGuYing)
1	2.78	36.58	0.0759978	3.83	102.15	0.0374939
	5.96	30.73	0.1939473	2.45	112.11	0.0218535
	1.4	13.38	0.1046338	7.5	127.19	0.0589669
4	2.83	33.44	0.0846292	1.22	74.57	0.0163605
	4.67	32.96	0.1416869	12.72	105.62	0.1204317
6	1.4	18.9	0.0740741	2.54	90.09	0.028194
	4.11	25.97	0.1582595	5.79	107.59	0.0538154
	2.77	41.97	0.0659995	5.51	124.05	0.0444176
	3.61	23.92	0.1509197	4.69	149.82	0.0313042
10	4.54	42.82	0.1060252	3.24	116.58	0.0277921
	4.56	38.98	0.1169831	6.64	140.97	0.0471022
12	6.6	64.61	0.1021514	5.75	158.77	0.0362159
13	1.65	26.11	0.0631942	6.45	182.72	0.0352999
	3.32	32.22	0.1030416	6.22	140.33	0.0443241
	7.65	67.93	0.1126159	49.62	277.82	0.1786049
16	2.16	34.76	0.0621404	3.83	104.36	0.0366999
	2.02	28.81	0.0701145	2.79	126.12	0.0221218

Table 2. *Cont.*

	Patio1 (m^2)	Roof1 (m^2)	P/R1 (ShangGanTang)	Patio2 (m^2)	Roof2 (m^2)	P/R2 (ZhangGuYing)
18	1.96	34.41	0.0569602	7.75	137.15	0.0565075
	2.5	27.92	0.0895415	9.62	172.15	0.0558815
	3.18	36.98	0.0859924	8.49	148.07	0.0573377
	1.11	27.07	0.0410048	3.99	96.76	0.041236
22	3.18	24.53	0.1296372	3.93	105.75	0.0371631
	6.71	36.03	0.1862337	5.58	122.95	0.0453843
24	2.36	44.83	0.0526433	3.31	97.23	0.034043
25	5.97	50.87	0.117358	2.24	67.91	0.0329848
	2.49	25.91	0.0961019	2.68	80.23	0.033404
	9.12	42.62	0.213984	7	138.04	0.0507099
28	2.74	27.76	0.0987032	8.67	163.41	0.0530567
	5.09	35.4	0.1437853	8.53	149.18	0.0571792
30	2.85	30.74	0.0927131	2.72	123.34	0.0220529
	1.8	33.37	0.0539407	0.59	89.85	0.0065665
	5.1	29.06	0.175499	5.88	106.79	0.0550613
	5.09	26.37	0.1930224	5.85	105.39	0.0555081
34	28.22	123.09	0.2292631	6.44	97.19	0.066262
SUM	**151.5**	**1251.05**	**0.1210983**	**224.06**	**4242.25**	**0.0528163**
Average	4.4558824	36.795588	0.1130235	6.59	124.77206	0.0470982

Combining precipitation data, the catchment area of the roof, patio, and the measured area were calculated. With data and interviews, the similarities and differences in water management mechanisms of the two settlements were further compared.

Through Surfer software and Google Earth data, the topography of the two settlements was modeled and profiled to obtain the direction of surface runoff, the topographic relief, and a spatial overview of the site.

Finally, the analysis of precipitation and water retention was combined with the logic of village planning and its related adaptive mechanisms (Figure 2).

3. Results

3.1. Regression Result

As we can see from Figure 3, the regression formula is:

$$\begin{aligned} Y1 &= 6.3367X1 + 86.638\ (\text{ZhangGuYingP/R2}) \\ Y2 &= 3.4467X2 + 21.23\ (\text{ShangGanTangP/R1}) \end{aligned} \tag{1}$$

Y1: ZhangGuYing village rainwater harvesting roof area
Y2: ShangGanTang villages rainwater harvesting roof area
X1: ZhangGuYing village patio area
X2: ShangGanTang village patio area
The median of the patio area from 68 dwellings is 3.9 m^2
Bring the median of the patio area into Equation (1):

$$Y1 = 111.35\ m^2 \text{ and } Y2 = 34.67\ m^2$$

As the equations show, the two ratios of the roof catchment area to the patio area are close: in the case of ZhangGuYing (Z) 6.3367 and in the case of ShangGanTang (S) 3.4667. Both the R^2 values are greater than 0, and the area of the roof and patio are positively correlated; that is, the area of the roof (rainwater harvesting) and the area of the two settlements are related to the patio (water storage) area. There is a linear correlation between the roof and the patio (Figure 3).

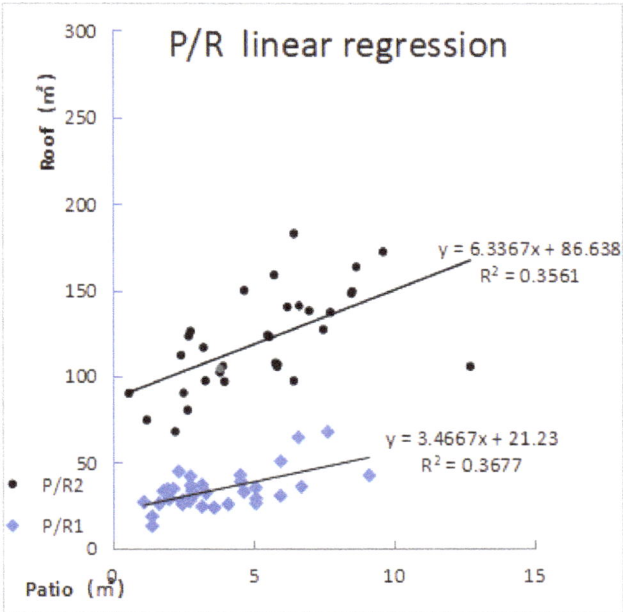

Figure 3. The linear regression of roof and patio surface areas of the two villages.

This is in line with the design language patterns of traditional Chinese architecture; that is, each component of the building is designed according to strict, designated proportions [41]. This design logic has two reasons: one is the standardization of traditional architectural theory, and the other is that it can be combined or adapted to local conditions. It involved adaptive adjustment of the scale and function of the landscape to different situations [42]. From the regression analysis results, we can indirectly conclude that the water management logics of the two traditional settlements are very similar. Meanwhile, the difference is obvious from Table 2: the total roof area of 34 buildings in Z village is 4242.25 m^2, and in S village is 1251.05 m^2. As we can see from Figure 2, the architectural design and structure of the two villages is different: the Z village design is called "DaWu" which means the members of the whole village live as a clan family in one big house. The S village is in the same clan clustering situation, but DaWu is a special form to emphasize the unity between the family members; it is reflected in the design by connecting the roof of each dwelling together. Both villages belong to the logic of clan settlement, but the architectural expression is different. This is the explanation from a cultural perspective.

More importantly, for water management purposes, here we assume that the differences are caused by the adaptation to the environment and topographical conditions of the two settlements. The hypothesis was interpreted in combination with the local interview, topography analysis, and meteorological data.

3.2. Meteorological Analysis

Rainwater amount collection

S village yearly average precipitation (roof) 1426 mm/m^2 * 1251 m^2 =1.784 million L

August (max): 255 mm/m^2 = 0.319million L

Z village yearly average precipitation (roof) 1353 mm/m^2 * 4242 m^2 = 5.739 million L

August (max): 172.9 mm/m^2 = 0.733 million L

Based on the average annual precipitation and the surface area of the settlements, the annual water quantity collected by the roofs of S Village is 1.784 million L, while in the Z village it is 5.73 million L. In Table 3, using the Acute3D measurement tool, we selected the study area including 34 dwellings and a part of the river that flows through the settlement as the total area, and compared the river area with the selection area of the study area.

Table 3. The table of 3D model water catchment area measurement.

	Total Area (Selection)	River Area	River Area Ratio	Max Altitude
ZhangGuYing	16,334.39 m^2	1125.23 m^2	6%	150 m
ShangGanTang	34,174 m^2	3745.28 m^2	10%	280 m

ShangGanTang has 367 traditional dwellings in total; we selected 34 buildings, where the rain harvest function is still working, and this traditional way reusing the rainwater is still used and maintained by the villagers.

ZhanGuYing has 1062 traditional dwellings in total, due to the long-term expansion. The village mainly has three parts: DangDaMen (422,1593DC); WangJiaDuan (468,1802DC), and ShangXinWu (172,1803DC). We only choose 34 buildings, which were built in same period in the WangJIaDuan part of the village, all these buildings having similar rain harvest functions. The total area was selected through surface measurement tools. We used the 367 buildings and the river surface area, which go through the village as the selected area for ShangGanTang village (34,174 $m^{2)}$, and 468 building areas of WangJiaDuan part and river surface area, which go through the village as the selected area for ZhanGuYing village (16,334.39 m^2). We found that the river area ratio in Z village is 6%, much smaller than that in S Village at 10%, and the precipitation in Village Z 1353mm is lower than that in S Village 1426mm, so the theoretical water collection in Z village can increase significantly more than in S Village. Combined with local interviews, the number of floods in the village was also higher in S than in Z village. For S villages, the flood control function is more important, so the patio and roof are smaller than in Z villages; thus, the meteorological analysis supported the hypothesis.

3.3. *Topography Analysis: The Art of Choosing a Location*

According to Figure 4, we can see that both Z and S villages are located on flat ground between hills and rivers.

In the process of the landscape development of the traditional Chinese settlement, the most important step is the positioning of the settlement [43], which can also be understood as the choice of an adequate living environment [44]. Just like in the case of painting [45], the choice of location is equivalent to the painter's first stroke on the canvas, followed by the construction and management of the surrounding environment of the selected location, that is, landscape transformation [46].

In terms of location selection, Chinese landscape architects need to consider many factors, and they also gradually develop a systematic Feng Shui theory in the process of practice [47]. According to the positioning in Feng Shui, the most beneficial settlement location is the "Livable" position [46]. In the Chinese classical philosophy system, "qi" is divided into two kinds: Yin and Yang, which are reflected in the landscape as a mountain (Yang) and water (Yin) [48,49]. Therefore, in ancient times, Feng Shui chose the positions between mountains and rivers to build settlements [50]. From the artistic perspective, this site selection layout is conducive to borrowing scenery, that is, using mountains and rivers to create visual order and a sense of harmony [51,52]. This type of selection and location is also enlightening from the perspective of contemporary science:

1. The thermal circulation generated between the mountains and rivers meets the needs of heat dissipation and ventilation.
2. The wind direction of the monsoon climate is conducive to cooling in summer and avoiding cold air intrusion in winter.
3. Meets the needs for water use and avoids floods.
4. Uses the change of topography to create the layering of the landscape.
5. Layout of the settlement along the river to alleviate spatial and social conflicts.

6. Uses the change of topography combined with the water collection facilities for the sponge effect [10].

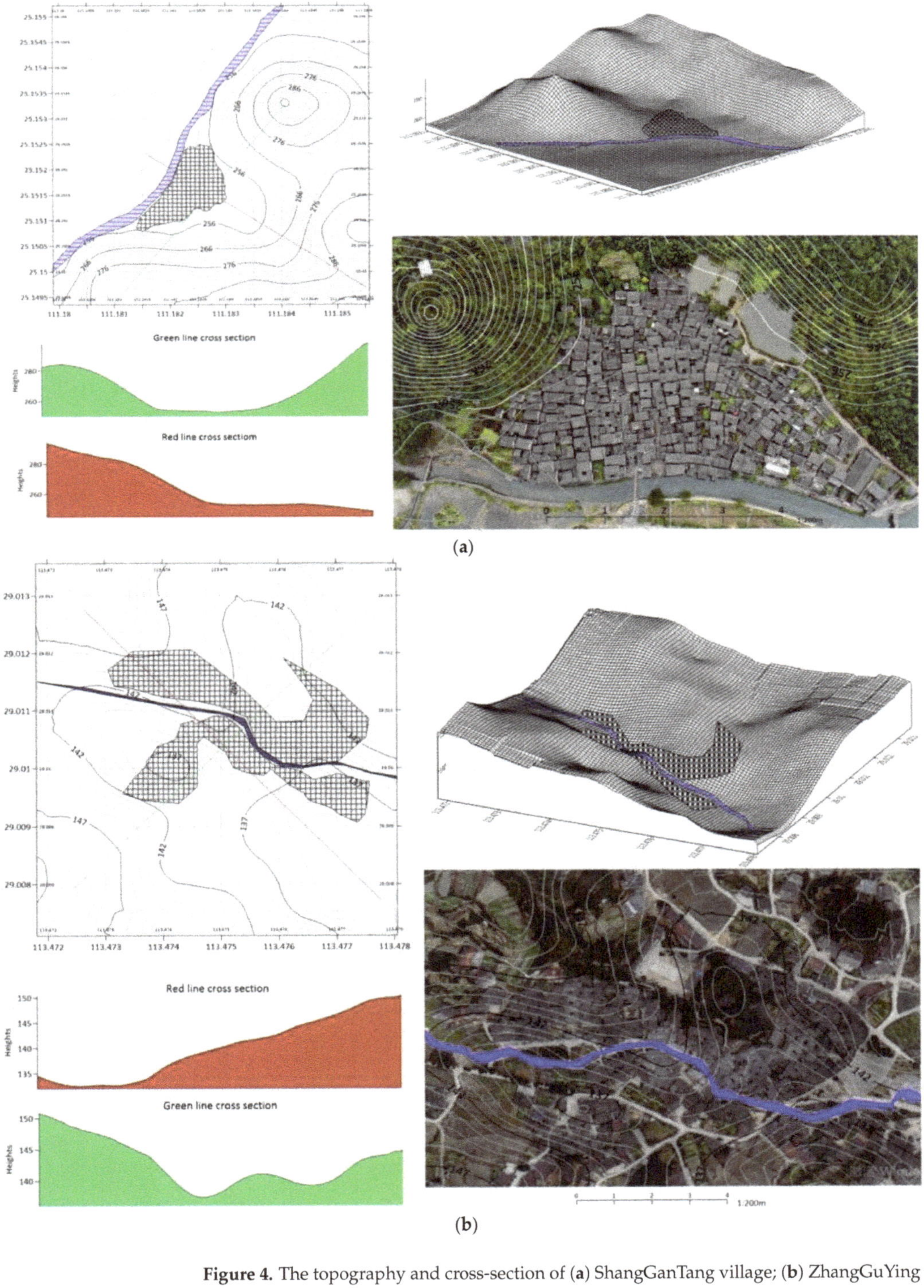

Figure 4. The topography and cross-section of (**a**) ShangGanTang village; (**b**) ZhangGuYing village.

The location of the settlement is part of the adaptive to topography strategies. The surface runoff was considered during the settlement planning. ZhangGuYing Village is relatively flat, while ShangGanTang Village is a typical traditional Feng Shui layout, surrounded by hills on three sides and a river on one side. In this landscape setting, the planners of the traditional settlement fully considered the water management strategies.

For ShangGanTang Village, due to the proximity to the river, the river flow is high, and the surface runoff path of the mountain needs to pass through the settlement. Analyzing Figure 4, it is clearly understandable that the adjustment of the collection capacity of rainwater is low, which is conducive to preventing waterlogging. For Zhanguying Village, the altitude is lower (Table 3). The average annual precipitation is low, the surface runoff is slow, and the settlements are distributed along the two sides of the stream so that the water collection function can be maximized.

3.4. Summary

It can be seen that, according to different environments, the water collection and storage of each building in different settlements has the same logic but with different capacities. The water collection volume is adjusted according to the precipitation, the size of the river, and the topographic characteristics of the landscape. In other words, the strategies are the same but adapted to the specific environment. (Table 4.)

Table 4. The comparative analysis of two village.

	Precipitation	River Size	Topography (Height)	Water Collection Capacity
ZhangGuYing	1353 mm	1125.23 m^2	150 m	5.739 million L
ShangGanTang	1426 mm	3745.28 m^2	280 m	1.784 million L

For Z village, the precipitation is 1353mm, lower than S village, and the river size is 1125.23 m^2. It is also smaller than S village, as the highest point is around 150 m, lower than S village; thus, the Z village adjusted the roof size, maximizing the rainwater collection function.

In addition to the roof–patio rainwater harvesting system based on residential buildings, the location of the settlement also plays an important role in water management. Taking ShangGanTang as an example, the site selection of the settlement conforms to the principle of the traditional Feng Shui theory: the settlement faces the water/river and has the mountain at its back [53]. Such a site selection can combine the water collection and storage systems of single buildings to participate in the water cycle during the formation of precipitation and runoff and has a similar function of regulating water bodies as sponge cities [54].

The integrity of ZhangGuYing Village is more obvious. The roof and patio are linked to each other, which can maximize the collection and storage of rainwater. In addition, the patio and the stream are also connected, and there is a water outlet to control the flow and thus actively adjust the storage capacity. (Figure 4)

4. Discussion: Systematic Water Management Strategies and Sustainable Trends

From the above analysis, we can find that the water management strategies of the two villages are logically similar.

From the point of view of landscape pattern, the two traditional settlements are represented by three parts: dwellings–settlement–landscape (surrounding environment) in terms of water management. The three parts are independent but interconnected. It can be said that the traditional settlement uses landscaping and architecture to achieve less interference with the natural surface runoff and underground runoff.

These functions can be interpreted from an ecological perspective today, but in traditional Chinese society, they are better interpreted in terms of culture, practice, and meaning [55].

According to traditional Chinese culture, man, as a part of nature [56], also has natural attributes [57], while Feng Shui in the traditional landscape literally means Feng: wind (air flows) and Shui: water flows [58]. Air and water are also common fluids in nature, with the attributes of circulation and movement. On the human side, it is more important to enable people and culture to flow continuously, like wind and water. This makes the Chinese traditional water management strategy more disseminated and sustainable at the cultural level [59,60].

In other words, in traditional society, the sustainability of material resources is also culturally combined with the trend of material sustainability. This sustainable strategy has an important impact on today's material resources and spiritual culture. What human beings need is not only the sustainable demand for resources and materials but also the support of sustainable ideas and culture rooted in human consciousness beyond resources and materials.

5. Conclusions

This research proved the importance of the Chinese traditional cultural heritage in rainwater management through the design of the settlement and dwelling architecture (structure, form, and size). From a heritage point of view, sustainable water management strategies also have their cultural roots in the local community. Some future studies are still needed, in order to focus on some practical issues, such as the contemporary adaptabilities of traditional rain harvesting methodologies from a global perspective. The new challenges of climate change and urban sustainability raise the importance of such traditional methods in the collection and use of rainwater, especially in the urban environment. In a European context, there are similar urban and landscape structural systems, which can be improved using the oriental traditions. This is expected to mitigate, among other positive effects, the urban flooding from climate change as well.

Author Contributions: Conceptualization, N.D. and A.F.; methodology, N.D.; software, J.Y.; validation, A.F., N.D. and M.v.d.T.; formal analysis, N.D.; investigation, P.L.; resources, N.D.; data curation, M.v.d.T.; writing—original draft preparation, N.D.; writing—review and editing, N.D.; visualization, N.D.; supervision, M.v.d.T. and A.F.; project administration, N.D.; funding acquisition, N.D. and A.F. All authors have read and agreed to the published version of the manuscript.

Funding: This research received no external funding.

Institutional Review Board Statement: Not applicable.

Informed Consent Statement: Informed consent was obtained from all subjects involved in the study.

Conflicts of Interest: The authors declare no conflict of interest.

References

1. Anagnostopoulos, G.L. Traditional settlements in the landscape. *Plan. Outlook* **1977**, *20*, 3–5. [CrossRef]
2. Wirth, L. Urbanism as a Way of Life. *Am. J. Sociol.* **1938**, *44*, 1–24. [CrossRef]
3. Beatley, T.; Timothy, B.; Manning, K. *The Ecology of Place: Planning for environment, Economy, and Community*; Island Press: Washington, DC, USA, 1997.
4. Meggers, B.J. Environmental Limitation on the Development of Culture. *Am. Anthr.* **1954**, *56*, 801–824. [CrossRef]
5. Priscoli, J.D. Water and civilization: Using history to reframe water policy debates and to build a new ecological realism. *Water Policy* **2000**, *1*, 623–636. [CrossRef]
6. Mays, L. (Ed.) *Ancient Water Technologies*; Springer Science & Business Media: Berlin, Germany, 2010.
7. Grof, S. (Ed.) *Ancient Wisdom and Modern Science*; SUNY Press: Albany, NY, USA, 1984.
8. Angelakis, A.N.; Koutsoyiannis, D.; Tchobanoglous, G. Urban wastewater and stormwater technologies in ancient Greece. *Water Res.* **2005**, *39*, 210–220. [CrossRef] [PubMed]
9. Mays, L.W. Water technology in ancient Egypt. In *Ancient Water Technologies*; Springer: Dordrecht, The Netherlands, 2010; pp. 53–65. [CrossRef]
10. Fekete, A.; Dong Ge, N. Sustainable Water Management Model as Landscape Heritage in Shang Gan Tang Village, China. *IOSR J. Eng. (IO-SRJEN)* **2019**, *10*, 2250–3021. Available online: www.iosrjen.org (accessed on 23 May 2020).
11. Rapoport, A. Environmental quality, metropolitan areas and traditional settlements. *Habitat Int.* **1983**, *7*, 37–63. [CrossRef]
12. Gao, J.; Wang, R.; Huang, J. Ecological engineering for traditional Chinese agriculture—A case study of Beitang. *Ecol. Eng.* **2015**, *76*, 7–13. [CrossRef]

13. Tang, Y.-T.; Chan, F.K.S.; O'Donnell, E.; Griffiths, J.; Lau, L.; Higgitt, D.; Thorne, C. Aligning ancient and modern approaches to sustainable urban water management in China: Ningbo as a "Blue-Green City" in the "Sponge City" campaign. *J. Flood Risk Manag.* **2018**, *11*, e12451. [CrossRef]
14. Huang, P.C. *The Peasant Family and Rural Development in the Yangzi Delta, 1350–1988*; Stanford University Press: Redwood City, CA, USA, 1990.
15. Chang, K.-S. China's rural reform: The state and peasantry in constructing a macro-rationality. *Econ. Soc.* **1992**, *21*, 430–452. [CrossRef] [PubMed]
16. Perkins, D.H. *Agricultural Development in China, 1368–1968*; AldineTransaction: Piscataway, NJ, USA, 2013.
17. Wang, C.; Huang, B.; Deng, C.; Wan, Q.; Zhang, L.; Fei, Z.; Li, H. Rural settlement restructuring based on analysis of the peasant household symbiotic system at village level: A Case Study of Fengsi Village in Chongqing, China. *J. Rural Stud.* **2016**, *47*, 485–495. [CrossRef]
18. Tundisi, J.G. Water resources in the future: Problems and solutions. *Estud. Avançados* **2008**, *22*, 7–16. [CrossRef]
19. Lepofsky, D. The Past, Present, and Future of Traditional Resource and Environmental Management. *J. Ethnobiol.* **2009**, *29*, 161–166. [CrossRef]
20. Yang, L.; Liu, P. The inheritance and its evaluation system of traditional village culture: A case study of traditional village in Hunan province. *Econ. Geogr.* **2017**, *37*, 203–210.
21. Thomas, T. Domestic water supply using rainwater harvesting. *Build. Res. Inf.* **1998**, *26*, 94–101. [CrossRef]
22. Li, L.; Zhang, G. Different purpose of rainwater catchment in China and their environmental effects. In Proceedings of the 8th International Conference on Rainwater Catchment Systems, Tehran, Iran, 21–25 April 1997.
23. Mou, H. Rainwater utilization for sustainable development in north China. *Waterlines* **1995**, *14*, 19–21. [CrossRef]
24. Arnold, P.; Cullis, A. Rainwater harvesting. In *The Collection of Rainfall and Runoff in Rural Areas*; Intermediate Technology Publ.: London, UK, 1986.
25. Wisser, D.; Frolking, S.; Douglas, E.M.; Fekete, B.M.; Schumann, A.H.; Vörösmarty, C.J. The significance of local water resources captured in small reservoirs for crop production—A global-scale analysis. *J. Hydrol.* **2010**, *384*, 264–275. [CrossRef]
26. Singh, O.; Turkiya, S. Assessing Potential for Rooftop Rainwater Harvesting:An Option for Sustainable Rural Domestic Water Supply in Arid Region of Haryana. *J. Rural Dev.* **2017**, *36*, 49. [CrossRef]
27. Patel, U.R.; Patel, V.A.; Balya, M.I.; Rajgor, H.M. Rooftop rainwater harvesting (RRWH) at SPSV Campus, Visnagar: Gujarat—A case study. *Int. J. Res. Eng. Technol.* **2014**, *3*, 821–825.
28. Barthwal, S.; Chandola-Barthwal, S.; Goyal, H.; Nirmani, B.; Awasthi, B. Socio-economic acceptance of rooftop rainwater harvesting—A case study. *Urban Water J.* **2013**, *11*, 231–239. [CrossRef]
29. Palla, A.; Gnecco, I.; Lanza, L. Non-dimensional design parameters and performance assessment of rainwater harvesting systems. *J. Hydrol.* **2011**, *401*, 65–76. [CrossRef]
30. Asadieh, B.; Krakauer, N.Y. Impacts of Changes in Precipitation Amount and Distribution on Water Resources Studied Using a Model Rainwater Harvesting System. *JAWRA J. Am. Water Resour. Assoc.* **2016**, *52*, 1450–1471. [CrossRef]
31. Li, S.; Ding, K.; Ding, A.; He, L.; Huang, X.; Ge, Q.; Fu, C. Change of extreme snow events shaped the roof of traditional Chinese architecture in the past millennium. *Sci. Adv.* **2021**, *7*, eabh2601. [CrossRef] [PubMed]
32. Kazak, J.K.; Dąbrowska, J.; Bednarek, A. Stormwater Management in Urban and Rural Areas. *Water* **2022**, *14*, 3488. [CrossRef]
33. Fekete, A.; Herczeg, Á.; Ge, N.D.; Sárospataki, M. Participatory Landscape Design and Water Management—A Sustainable Strategy for Renovation of Vernacular Baths and Landscape Protection in Szeklerland, Romania. *Land* **2022**, *11*, 95. [CrossRef]
34. Zhou, W.; Matsumoto, K.; Sawaki, M. Understanding the traditional wisdom of harvesting rainwater in household yards: Construction and rainwater usage patterns of settlement water cellars in semi-arid China. *J. Asian Arch. Build. Eng.* **2022**, 1–13. [CrossRef]
35. Song, X.; Mulder, K.; Frostell, B.; Ravesteijn, W.; Wennersten, R. Transition in public participation in Chinese water management. In *Proceedings of the Institution of Civil Engineers-Engineering Sustainability*; Thomas Telford Ltd.: London, UK, 2011; Volume 164.
36. Zheng, X.Y. The ancient urban water system construction of China: The lessons from history for a sustainable future. *Int. J. Glob. Environ. Issues* **2015**, *14*, 187. [CrossRef]
37. Wu, C.; Qiao, M.; Wang, S. Enlightenment from ancient Chinese urban and rural stormwater management practices. *Water Sci. Technol.* **2013**, *67*, 1474–1480. [CrossRef] [PubMed]
38. Xu, Y.-S.; Shen, S.-L.; Lai, Y.; Zhou, A.-N. Design of sponge city: Lessons learnt from an ancient drainage system in Ganzhou, China. *J. Hydrol.* **2018**, *563*, 900–908. [CrossRef]
39. Dong, R.; Zhang, X.; Li, H. Constructing the Ecological Security Pattern for Sponge City: A Case Study in Zhengzhou, China. *Water* **2019**, *11*, 284. [CrossRef]
40. Meteorological Data of the Two Villages. Available online: http://data.cma.cn/data/weatherBk.html (accessed on 3 February 2022).
41. Cody, J.W.; Steinhardt, N.S.; Atkin, T. Nancy Shatzman Steinhardt, and Tony Atkin. In *Chinese Architecture and the Beaux-Arts*; University of Hawaii Press: Honolulu, HI, USA, 2017. [CrossRef]
42. Li, A.I. A Shape Grammar for Teaching the Architectural Style of the Yingzao Fashi. Ph.D. Thesis, Massachusetts Institute of Technology, Cambridge, MA, USA, 2001.
43. Bruun, O. *Fengshui in China: Geomantic Divination between State Orthodoxy and Popular Religion*; University of Hawaii Press: Honolulu, HI, USA, 2003.

44. Marafa, L. Integrating natural and cultural heritage: The advantage of feng shui landscape resources. *Int. J. Herit. Stud.* **2003**, *9*, 307–323. [CrossRef]
45. Zhang, L. Nature and landscape in the Chinese tradition. In *New Perspectives on the Research of Chinese Culture*; Springer: Singapore, 2013; pp. 1–15.
46. Han, K.-T. Traditional Chinese Site Selection-*Feng Shui*: An Evolutionary/Ecological Perspective. *J. Cult. Geogr.* **2001**, *19*, 75–96. [CrossRef]
47. Xu, P. *Feng-Shui: A Model for Landscape Analysis*; Harvard University: Cambridge, MA, USA, 1990.
48. Zhang, Y.H.; Rose, K. *A Brief History of qi*; Paradigm Publications: Boulder, CO, USA, 2001.
49. Wang, R.R. Dong Zhongshu's transformation of Yin-Yang theory and contesting of gender identity. *Philos. East West* **2005**, *55*, 209–231. [CrossRef]
50. Li, X.-D. *Meaning of the Site: A Holistic Approach towards Site Analysis on Behalf of the Development of a Design Tool Based on a Comparative Case-Study between FengShui and Kevin Lynch's System*; Technische Universiteit Eindhoven: Eindhoven, The Netherlands, 1993; 236 p.
51. Xu, J. *A Framework for Site Analysis with Emphasis on Feng Shui and Contemporary Environmental Design Principles*; Virginia Polytechnic Institute and State University: Blacksburg, VA, USA, 2003.
52. Bruun, O. *An Introduction to Feng Shui*; Cambridge University Press: Cambridge, UK, 2008. [CrossRef]
53. Lee, Y.-T.; Yang, H.; Wang, M. Daoist Harmony as a Chinese Philosophy and Psychology. *Peace Confl. Stud.* **2009**, *16*, 68–81. [CrossRef]
54. Qi, Y.; Chan, F.K.S.; Thorne, C.; O'Donnell, E.; Quagliolo, C.; Comino, E.; Pezzoli, A.; Li, L.; Griffiths, J.; Sang, Y.; et al. Addressing Challenges of Urban Water Management in Chinese Sponge Cities via Nature-Based Solutions. *Water* **2020**, *12*, 2788. [CrossRef]
55. Lomas, J.D.; Xue, H. Harmony in Design: A Synthesis of Literature from Classical Philosophy, the Sciences, Economics, and Design. *She Ji J. Des. Econ. Innov.* **2022**, *8*, 5–64. [CrossRef]
56. Li, C. The Confucian ideal of harmony. *Philos. East West* **2006**, *56*, 583–603. [CrossRef]
57. Fang, C.; Wang, Z.; Liu, H. Beautiful China Initiative: Human-nature harmony theory, evaluation index system and application. *J. Geogr. Sci.* **2020**, *30*, 691–704. [CrossRef]
58. Cooper, D.E. Daoism, nature and humanity. *R. Inst. Philos. Suppl.* **2014**, *74*, 95–108. [CrossRef]
59. Li, F.; Cook, S.; Geballe, G.T.; Burch, W.R., Jr. Rainwater harvesting agriculture: An integrated system for water management on rainfed land in China's semiarid areas. *AMBIO A J. Hum. Environ.* **2000**, *29*, 477–483. [CrossRef]
60. Catalani, A. Water & heritage: Material, conceptual and spiritual connections. *Int. J. Heritage Stud.* **2016**, *23*, 174–176. [CrossRef]

land MDPI

Article

Climate Change Mitigation and Preservation of the Cultural Heritage—A Story of the Municipal Park in Rumia, Poland

Monika Trojanowska

Faculty of Civil and Environmental Engineering and Architecture, PBS Bydgoszcz University of Science and Technology, 85-796 Bydgoszcz, Poland; Monika.Trojanowska@pbs.edu.pl

Abstract: Climate change may affect cultural heritage in at least two ways: direct physical effects on the site, building, or structure and effects on social structures. Creating urban parks with therapeutic landscapes can mitigate some of these detrimental effects. This paper presents the revitalization of the former water forge, located in the center of Rumia, near the Tri-City agglomeration. The study focused on the history of the site and the historic manor house called "Dwór pod Lipami" and the preservation efforts. The social engagement, which led to the development of the landscape park and the construction of a talent playground, was an essential factor in the renewal process. The second part of the work presents an assessment of the therapeutic and recreational values of the new urban park using the Universal Standard for Health-Promoting Places, Community Park Audit Tool (CPAT), and mapping the users' preferences. This operation of urban renewal resulted in creating a popular park that helps promote the health and well-being of the local community.

Keywords: climate change mitigation; cultural heritage; urban renewal; urban parks; therapeutic landscapes

Citation: Trojanowska, M. Climate Change Mitigation and Preservation of the Cultural Heritage—A Story of the Municipal Park in Rumia, Poland. *Land* **2022**, *11*, 65. https://doi.org/10.3390/land11010065

Academic Editors: Jan K. Kazak, Katarzyna Hodor and Magdalena Wilkosz-Mamcarczyk

Received: 18 December 2021
Accepted: 30 December 2021
Published: 2 January 2022

1. Introduction

Climate change is a global issue that is important to everyone. The large industrial areas and smaller towns are facing new challenges.

The research question was: What can we do to mitigate climate change in case of the cultural heritage of post-industrial origin located in the town's center to promote the health and well-being of the local community?

The revitalization of a historic post-industrial old village on the banks of Zagórska Struga in the town of Rumia in the vicinities of Tri-City in Poland can serve as a case study. The neglected post-industrial site underwent an urban renewal process to become a favorite recreational site for the city's inhabitants. A public park created around historical monuments and reconstructed water pond became popular among all age groups. The historical building of the "Manor under the linden trees" was reconstructed and today is a local cultural center. The new open-air theatre serves to organize social events and festivals.

A new public park brought cultural and social benefits regarding place identity and place attachment. The local community actively participated in park development. Nivea company funded the new playground because of local citizens' involvement.

The space of the new park with all amenities is an example of a well-developed public space. This study focused on assessing the new park's therapeutic qualities and mapping the users' activity.

The construction of a new urban public park and reconstruction of historic water ponds and waterworks was significant because of local culture and cultural identity, and helped moderate the local climate.

2. Climate Change Risks

Climate change can affect cultural heritage in at least two ways: direct physical effects on the site, building, or structure and the effects on social structures. It may impact communities changing the way they live, work, worship, and socialize in buildings, sites, and landscapes. What can be anticipated is abandonment with the eventual loss of cultural memory [1].

According to UNESCO Report temperature increases ('global warming'), changes in precipitation patterns, and increased frequencies of extreme weather events, i.e., intensity and seasonality of droughts, is exerting considerable impacts on our environment and biodiversity [1–4].

The biodiversity loss is a risk. The growing season of plants is lengthening, and the threat of invasive alien species increasingly impact indigenous species. The small and isolated areas are especially at risk when the grid of green infrastructure connections is broken. Wetlands are especially vulnerable to climate change and have limited adaptive capacity.

The physical substance of cultural heritage buildings is also exposed to threats, like changes in temperature, precipitation and atmospheric moisture, and wind intensity. Historic buildings were designed for a specific local climate. They have a different relation with the ground than modern ones, as they are more porous. Changes in water-table levels and changes in soil chemistry are responsible for destruction of original historic tissue [1]. There is a plethora of research on climate change and its impact on cultural heritage [1–5]. Among those which might affect the case study location in Central and Northern Europe are the following risks:

- Increase in corrosion of built heritage materials
- Increase in salt crystallization
- Material erosion caused by wind and wind-driven rain [2]

Cultural heritage assets such as historical buildings are a legacy that lend a sense of place and identity to local populations. In Rumia, the historic building of the "Manor under the linden trees" is one of only a few historic monuments. It is important to preserve it not only for cultural reasons, but also to promote the sense of place attachment and place identity. The decision to revitalize the historic waterworks and create the urban park helped to preserve the biodiversity and local microclimate.

2.1. Possible Interventions or Adaptive Responses to Counteract Climate Change Threats

UNESCO proposes a specific approach to natural World Heritage sites at risk [1]. A two-pronged approach is required: first, the site's vulnerability should be assessed, and specific site-level mitigation and adaptation strategies should be designed and implemented in partnership with relevant stakeholders. Second, site managers need to look beyond the individual site level and develop and implement regional and transboundary mitigation and adaptation strategies that reduce the site vulnerability in a larger landscape context. That approach may be applied to any cultural heritage site.

The critical question is why the cultural heritage site is sensitive and vulnerable to the pressures of climate change and what could be done to remedy the current situation. Responses in the case of World Heritage sites may include monitoring, maintaining, and managing [1]. In the case of cultural heritage, it might also include urban renewal and adaptation to new functions, if possible, without causing any damage to the original tissue.

When the social and cultural aspects are considered, informing the public about these issues is the most critical strategy. It is vital to emphasize satisfying the public interest and ensuring the participation of the individuals, authorities, and site managers. It is crucial to promote educational activities to raise public awareness [3]. In the case of urban renewal strategies concerning the cultural heritage sites is the entitlement of any given generation to these assets and their right to adapt them to meet contemporary needs with the consent of the larger public [4].

Today, the landscapes are fragmented and subject to human pressures. However, the Kyoto Protocol identifies significant opportunities for mitigating climate change and adapting to climate change while enhancing the conservation of biodiversity. The ecosystem approach is essential [1].

The urban renewal of the historic forge in Rumia was a community-led effort. Numerous uproot initiatives, satisfying the public interest, and ensuring the participation of the individuals, authorities, and site managers was the driving force behind creating this landscape park.

2.2. Climate Change and Human Well-Being. Urban Parks and Therapeutic Landscapes

Some researchers stipulate establishing recreational facilities on poor lands unsuitable for agricultural purposes [3]. The construction of a new urban park can result in the provision and use of new green infrastructure, promoting active forms of travel, thus reducing exposure to air, water, and soil pollutants—climate change impacts in the long run [6]. Public parks offer numerous possibilities of physical and mental restoration, physical activities, and social contacts [7].

Parks, forests, and waterways can be spaces of refuge from brutal and chaotic urbanization. Moreover, public open green spaces offer symbolic ownership of a place to local community members and democratization of access to land, which is crucial to prevent social exclusion, spatial deprivation, and environmental injustice [6]. Engaging residents in the management of urban parks boosts their participation and willingness to use them [5].

Today, we need more complex socio-environmental systems to improve the quality of life in a socially friendly and climate-neutral build environment maintaining achievements of previous generations and cultural heritage [8]. The story of the urban renewal of the public park in Rumia is an excellent example of such an approach.

2.3. Case Study. Park Starowiejski in Rumia, Poland

Rumia (54°34′14″ N 18°23′16″ E) is a city in the Eastern Pomerania region of northwestern Poland in the Kashubian region (Figure 1). The altitude is 7–169.4 m in height above sea level. The area of the city is 30.10 km^2, and the urban density is 1600/km^2. The city has almost fifty thousand inhabitants, but eventually is a prolongation of a linear conurbation of the Tricity—Gdańsk, Gdynia, and Sopot, which has over one million inhabitants. The proximity to the urban center (metro area) is 130,000. The time zone is UTC+1 (CET).

Rumia's climate is classified as Cfb by the Köppen—Geiger system. The average annual temperature in Rumia is 8.7 °C. The average annual rainfall is 550–792 mm, with heavy rain present in all seasons [9].

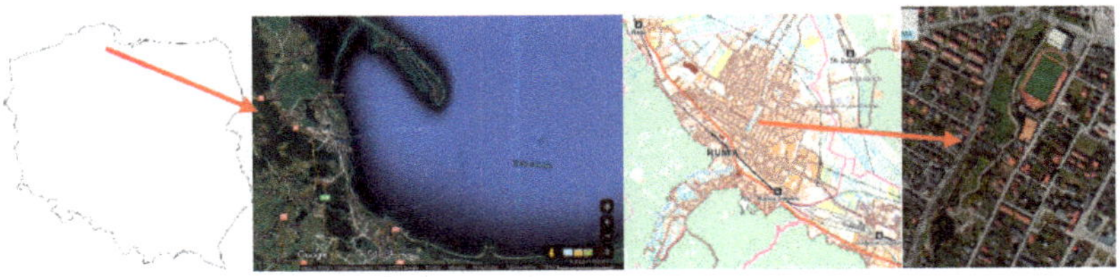

(a) (b) (c) (d)

Figure 1. Case study location—Park Starowiejski in Rumia, Poland (**a**) Map of Poland, source: [10], (**b**) Map of Gdańsk Bay, source [11], (**c**) Map of Rumia, source [12], (**d**) map of Park Starowiejski in Rumia, source [12].

Park Starowiejski is located in the center of Rumia along the banks of Zagórska Struga—a river which flows out of Marchowo Lake, located at an altitude of 154 m above sea level in the Szemud municipality. It then crosses the Kashubian Lake District upland and continues to the valley surrounded by forests of the Tri-City Landscape Park (TPK), to the alluvial fan forming the city of Rumia and the Łyski Canal, and finally enters the Baltic Sea in the Puck Bay. The length of the river is almost 29 km, and the average longitudinal slope is 5.6 ‰ [13].

The streets Starowiejska, Adama Mickiewicza, Dębogórska, and Józefa Wybickiego are marking the borders of the park (Figures 1 and 2).

(a) (b)

Figure 2. Park Starowieski on orthophotomaps, 2021 (**a**) aerial photo, (**b**) 3D map source: [12].

3. The Story of Water Forge—The Most Important Industrial Plant in Rumia

The village of Rumia dates back to ancient times. In 1950 and 1960 in the case study area—Starowiejska and Mickiewicza streets, on both sides of the Zagórska Struga River (Figure 1), over 300 graves were discovered from the ancient time of early Roman influence on the Pomeranian culture [14]. Old stoves for iron ore smelting were found there [15]. The oldest documents mentioning the industrial water facilities in the old village of Rumia date back to the second half of the 16th century. The natural water fault creating a small waterfall was used to move the water wheel of the forge. It is believed that the first water smithy belonged to Cistercian Monks from Oliwa.

The water smithy was the largest industrial plant in Rumia, of great economic importance until the 19th century [15]. (Figure 3)

At the beginning of the 19th century, the Napoleonic wars and the continental blockade disrupted the activity of the forge. It was closed in 1887 and replaced by a woodworking plant that produced thin veneers of wood to make matchboxes. After installing a water turbine to produce electricity, it was powered by electricity [16,17].

During World War II, material to make parachutes was produced in the hall erected on the granary site. This facility was demolished in 1995 due to poor technical conditions. After WWII, the site was owned by PSS Społem and served as a place to store metals and fuel. In the mid-1950s, a butcher's shop in the rebuilt facilities of the plant was organized [17].

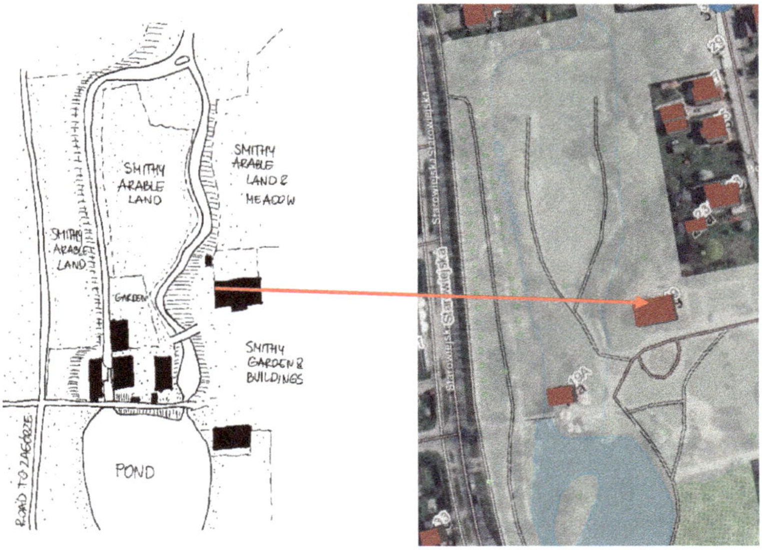

(**a**) (**b**)

Figure 3. The case study area: (**a**) in the XIX century—a manor house from the second half of the 18th century, a residential house from the first half of the 19th century (demolished in the 1970s), a granary, a barn, a pigsty, and a stable next to Mickiewicza Street, and further away—a factory building, a factory office, an oven, a farmhouse, and a pond designated as "public water" Source: author drawing, according to A. Sadłowski [15] and historic map from 1829/66/664—Königl. Domänenrentamta in Brück zu Zoppot, Wojewodzkie Archiwum Państwowe w Gdańsku; (**b**) today—there is only one building remaining from that period—the manor house and a new building constructed next to the pond, designed to resemble historic watermills, orthophoto map, source: [18].

4. Creation of New Identity—Municipal Centre of Culture and Public Park

In the second part of the XVIII century, a residential house was built, later referred to by the local community as a manor house "under the linden trees." After the cease of industrial use in the 1950s, the poorly maintained manor house slowly fell into disrepair until the municipality took it over (Figure 4).

(**a**) (**b**)

Figure 4. The manor house "Dwór pod lipami" from the second half of the 18th century; (**a**) before the renovation, source: [16]; (**b**) and after (picture from the summer of 2021), source: author.

The reconstruction of the manor house started in 1995. Some additions were demolished to restore the original functional plan changed to accommodate the offices on the first floor and apartments on the second floor. Only the foundation, basement walls, exterior walls, and some of the interior partitions were preserved. All the floors, stairs, and roof construction needed to be removed and replaced. New foundations under the new construction walls and a chimney were added. The original walls needed to be cleaned, and the destroyed bricks were replaced.

The monument bears visible traces of arches and vaults preserved in its walls. The arched slab over the basement was renovated. Locally, a layer of reinforced concrete was added from above in inevitable places. The stairs and the slab of the first floor were reconstructed using reinforced concrete. The steel beams support the slab. The second floor and the roof were carefully reconstructed using wood. It is a hipped Mansard roof with a collar-and-purlin wooden structure with a ridge purlin in the axis of the building. The new lintels are made of reinforced concrete or steel. The exterior walls were insulated with mineral wool—7 cm.

The windows were reconstructed with triple glazing. The interior decoration was preserved, renovated, or reconstructed if necessary [19].

The local law protects the manor as cultural heritage. It is inscribed into the communal register of monuments (Gminna ewidencja zabytków -Zarzadzenie Nr 1716/95/2018 z dnia 12 marca 2018 roku) together with the park [20].

In 1996, the manor became the new seat of Municipal Cultural Centre—MDK.

The area between the streets Starowiejska, Adama Mickiewicza, Dębogórska, and Józefa Wybickiego gradually become a public park. The schematic design was prepared by Autorska Pracownia Projektowa "Kwadrat". In 1998, the pond of the former forge and its weir were reconstructed. The historical maps evidence the pond's changes in location and size (Figure 5). Its size and shape vary on different maps, however. That may result from the changes in water level due to seasonal droughts or man-induced enlargement and deepening (Figure 6).

The historic airplane photographs from the WWII period show that the pond was dried. It was more of an open green area until the 90s.

In the place of the historic forge, in 2002, a water mill was built—today, the seat of the Kashubian-Pomeranian Association (Stowarzyszenie Kaszubsko-Pomorskie) and a restaurant with a name in the Kashubian language "Svantopołk" (Figure 7). Protection of Kashubian cultural heritage is vital. Next to the mill, the grinding wheel was reconstructed. The park became the place for education about the local history and heritage.

(a) (b) (c)

Figure 5. *Cont.*

(d) (e) (f)

Figure 5. Location of water smithy and ponds: (**a**) map from XIX century, source: digital archives [21]; (**b**) map from 1938, source [22]; (**c**) historic airplane photo from the IIWW, source [23]; (**d**) historic orthophoto map from 1997, source [23]; (**e**) 2005, digital archives of the city of Rumia, source [23]; (**f**) today, orthophoto map, source: [12].

(a) (b) (c)

Figure 6. The reconstructed pond seasonal changes in water level (**a**) during summer (**b**,**c**) winter. Source: author.

(a) (b)

Figure 7. The water mill—a restaurant in the place of historic forge. (**a**) View from the riverside. Source: author; (**b**) View of the restaurant garden. Source: author.

Gradually the park was developed to become one of the favorite places of local inhabitants. It resembles the XIX century picturesque landscape parks created next to village mansions. The renewal ideology was to restore the references to the past and add new elements. Therefore, new visual connections were added. In 2017, along the Zagórska Struga, from the park to Dębogórska Street, a walking and cycling path was led, which became another attraction of Rumia. It was designated to take part in the competition for the best revitalization project in 2017 [24].

5. Community Engagement

As the park was developed gradually, it became a place for education about local culture, history, and ecological issues—the importance of biodiversity and local insect and bird species. A park is also where children can discover nature and learn using the talent playground.

5.1. A Talent Playground

Over forty-two thousand Rumia inhabitants voted to build a talent playground in the park. They voted for Rumia in a championship organized by the Nivea company, which promised to install a playground called "podwórko Nivea", worth 250,000 PLN in town, which obtained the largest number of votes. It is a fascinating example of social participation and proof of inhabitants' place attachment. The playground has three zones: sports, science, and art. The recreational infrastructure aims to bridge the intergenerational gap and develop various skills [25–27] (Figure 8).

(**a**) (**b**)

Figure 8. The talent playground in the summer of 2021 (**a**,**b**), source: author.

5.2. Educational Gardens

Two educational gardens were installed in the park, a popular place among schoolchildren. One, themed on biodiversity protection, is placed next to fencing and a small parking lot adjacent to the manor house—today MDK (a). It has an insect hotel with educational tables teaching children and adults about the importance of pollinators. The other, themed "from children to children", is placed against the sports hall wall, next to recreational infrastructure and pedestrian path meandering along the banks of Zagórska Struga—local river and the fenced area of a local team soccer field. It offers a flower bed and educational tables (Figure 9).

(**a**) (**b**)

Figure 9. The 1st educational gardens in the summer of 2021 (**a**,**b**), source: author.

5.3. Educational Tables—How to Feed the Birds

The reconstruction of historic ponds created a bird-friendly environment. However, the birds' stomachs are not prepared to digest bread and cakes. Therefore, special vending machines with birds' food were placed strategically where bird feeders usually gather. They are instructions explaining which grains can be given to feed the birds. They are selling a handful of mixed grains for a small amount of money (Figure 10).

(a) (b)

Figure 10. The 2nd educational gardens in the summer of 2021 (**a**,**b**), source: author.

5.4. Educational Tables—Local History

In front of the manor house, next to the main entrance, is a permanent table with a short description of the park's history and a few historical images. The lawn in front of the manor serves temporary exhibitions dedicated to local history (Figure 11).

(a) (b)

Figure 11. Educational tables in front of the manor house (**a**) permanent table with a short description of the history of the park; (**b**) temporary exhibition dedicated to local history, source: author.

6. Methods. Assessment of the Park Therapeutic and Recreational Qualities

Two different methods of assessment were used—the Universal Standard for Health-Promoting Places (Table 1) [28,29] and Community Park Audit Tool (CPAT) [30]. The author—a professional researcher with a degree in architecture and urban design—performed the evaluation during numerous visits to the park at least once a week on various hours and days (weekdays and weekends) over two years in 2020 and 2021. Site observation, mapping the presence of users, and unstructured anonymous interviews were used to evaluate the therapeutic qualities of the park.

6.1. Assessment of the Therapeutic Qualities of the Public Park Using the Universal Standard for Health Promoting Places

The first assessment was performed using the Universal Standard for Health-Promoting Places [28,29,31]. It is a set of criteria connected to health promotion. The tool was developed after a long-term study of over 100 public parks and gardens in Europe and the United States triangulated with literature research [29]. The criteria are divided into five categories: 1. Sustainability, 2. Accessibility, 3. Amenities, 4. Design, and 5. Placemaking. For better clarity, the five categories were split into five separate tables (Tables 1–5) with columns for individual assessment remarks. The detailed assessment required a written explanation of why the researcher thought the attribute was present [28,29,31]. The assessment results were inscribed into the five tables representing five sections of the universal standard. (Tables 1–5).

Both a thick binary and a detailed assessment were performed when applicable.

The thick binary assessment has two categories (0, 1):

No, not observed—0

Yes satisfactory—1

Data n/a stands for data not available.

There are some criteria where the points count is not applicable.

An additional column was added to five tables representing five sections of the universal standard to accommodate the binary assessment. The points were added to represent the score and facilitate comparison with maximal values (Tables 1–5).

Table 1. Assessment of Public Park in Rumia—part 1, source: author.

Public Park in Rumia—part 1		POINTS
1. SUSTAINABILITY		12/12
1.1 Place		Not applicable
Area	Approx. 10 ha (counted with adjacent private gardens)	Not applicable
Location	Rumia is a small town located in the north of Poland, near the cost of the Baltic Sea in the Kashubian Region. The park is located centrally in the oldest part of the town, called old village or old Rumia. The average temperature in January on the coast is −1 °C, the average temperature in July is +18 °C. In addition, there are long transitional periods between summer and winter and clearly cooler spring than autumn. Annual precipitation amounts to about 550–792 mm. The average altitude is approx. 20 m over the sea level [9,32].	Not applicable
Surrounding urban pattern	Low density residential urban tissue with small-scale retail and services	Not applicable
1.2 Environmental characteristics		6/6
Soil quality	Sufficient for recreational use [32].	1
Water quality	The river—Zagórska Struga has qualities similar to mountain creeks with many meanders and cascades. The quality of water is sufficient for recreational use, but no bathing or swimming in the ponds is allowed.	1
Air quality	Good, Measured on 5 October 2021: PM 10; 21.2 $\mu g/m^3$ / PM 2.5; 19.7 $\mu g/m^3$ Source: [33]	1
Noise level	Low to moderate noise level in areas close to traffic routes [3].	1
Forms of natural protection	National heritage registers—manor house, Old monumental trees protected by local law.	1
Green and Blue Infrastructure	Important part of the green and blue infrastructure, especially the corridor along the river—Zagórska Struga.	1

Table 1. *Cont.*

1.3 Biodiversity protection		3/3
Parts of open green space not available to visitors	Artificial island on the pond.	1
Native plants	Planting is a combination of native and non-native species, a couple of patches of invasive plant (Reynoutria japonica) were noted.	1
Native animals	Both native and foreign species were observed.	1
Natural maintenance methods	Data n/a	Data n/a
1.4 Sustainable water management		1/1
Rainwater infiltration	Porous, permeable surfaces.	1
Irrigation with non-potable water	Data n/a	Data n/a
1.5 Parks of Second (New) Generation		1/1
	The park has no boundaries, it spreads to join the green infrastructure and it can be regarded as park of new generation.	1
1.6 Urban metabolism		1/1
	Receptacles for waste segregation and collection are located in the park.	1
1.7 Ecological energy sources		Data n/a
	Data n/a	Data n/a

The park was evaluated as sustainable open green and blue area. The result was 12 out of 12 available points, because there was no evidence about natural maintenance methods, irrigation with non-potable water, or ecological energy sources. However, those issues can be resolved by local authorities and park management.

Table 2. Assessment of Public Park in Rumia, Poland—part 2, source: author.

Public Park in Rumia, Poland—part 2		POINTS
2. ACCESSIBILITY—assessment of walkways to park		24/26
2.1 Distance to park		1/1
	It was observed that most of users either walk or cycle to the park. There are few parking spaces. Bus stops are located next to the park.	1
2.2 Sidewalk Infrastructure-		5/5
Width of sidewalk	Sufficient	1
Evenness of surface	Good	1
Lack of obstructions	Majority of terrain is accessible. Private gardens are fenced.	1
Slope	Flat, no significant slope	1
Sufficient drainage	Sufficient drainage	1
2.3 General conditions of walkways		6/8
Maintenance	The park is well-maintained. No visible traces of litter.	1
Overall aesthetics	Good	1
Street art	None	0/1
Sufficient seating	Some benches along the walking path.	1

Table 2. *Cont.*

Perceived safety	Walkways are perceived as a safe and well-maintained.	1
Buffering from traffic	The planting is providing buffering from the traffic, creating enclosed green interiors.	1
Street activities	Yes, e.g., seasonal decorations	1
Vacant lots	No	1
2.4 Traffic		5/5
Speed	Slow to moderate	1
Volume	Low to moderate	1
Number and safety of crossings	Numerous possibilities for safe crossing of the street.	1
Stop signs	Yes	1
On-street parking	Yes	1
2.5 User Experience		5/5
Air quality	Good	1
Noise level	Low to moderate in places close to traffic lines	1
Sufficient lighting	Sufficient lighting along the walking routes	1
Sunshine and shade	Yes, adequate number of trees providing shade (Figures 3 and 4)	1
Visibility of nearby buildings	Part of area has good visibility of nearby buildings, but there are parts of the green space buffered from the street.	1
2.6 Public transports stops		1/1
	Bus stops next to the park	1
2.7 Sufficient Parking		1/1
	Yes, there are few parking spots and on-street parking.	1

The park was assessed as universally accessible to people in every age and special needs 25/26. The only point which was missing was street art.

Table 3. Assessment of Public Park in Rumia, Poland—part 3, source: author.

Public Park in Rumia, Poland—part 3		POINTS
3. AMENITIES		14/15
3.1. Psychological and physical rejuvenation		5/5
Natural Landscapes	Combination of planting around open green spaces give an impression of a natural landscape.	1
Green open space	Numerous extensive grass-covered grounds (Figures 3 and 4).	1
Presence of water	Zagórska Struga river, artificial ponds (Figure 4).	1
Places to rest in the sun and shade	Multiple places including picnic and play areas.	1
Places to rest in quiet and solitude	Multiple places to rest in quiet and solitude.	1
3.2. Physical Activity Promotion		4/4
Sports infrastructure	Bicycle paths, cross-fit stations, sport field (soccer field of local team)	1
Recreational infrastructure	Recreational infrastructure for all age groups	1
Community gardens	Yes, educational gardens	1
Addressing the needs of people with disabilities	Pathways are wide and even, majority of the park area is accessible (there is small cobbled area).	1

Table 3. *Cont.*

3.3. Catering for basic needs		5/6
Safety and security (presence of guards, cleanliness, maintenance, etc.)	Assessed as a safe place during daytime.	1
Places to sit and rest	Benches in the most popular areas, tables and seats in the restaurant garden, garden pavilions.	1
Shelter	Visitors may find shelter under tree canopies, garden pavilions, or inside buildings, e.g., manor house, reconstructed watermill—restaurant, etc. Provisional temporary structures provide shelter during organized events.	1
Restrooms	Provisional restrooms are placed during organized events.	1
Drinking water	No	0
Food (possibility to buy food in the park or close vicinity)	There is restaurant in the reconstructed watermill.	1

The recreational infrastructure was assessed as satisfactory except for the drinking water fountain. Almost all the possible points were present on 14/15.

Table 4. Assessment of Public Park in Rumia, Poland—part 4, source: author.

Public Park in Rumia, Poland—part 4		POINTS
4. DESIGN		19/20
4.1. Architectural design		9/9
Human scale	Park is designed to fit human scale.	1
Focal points and landmarks	Recognizable landmarks—historic manor, monumental trees, field exhibitions.	1
Structure of interior connections	A clear structure of interior connections	1
Framed views	Natural frames are created by mature trees and meandering pathways.	1
Long vistas (Extent)	Park offers numerous extensive vistas.	1
Pathways with views	Many paths offer interesting views.	1
Invisible parts of the scenery (Vistas which engage the imagination)	Numerous designed vistas which engage the imagination.	1
Possibility to observe other people	Yes, events organised on the stage, festivals, etc.	1
Possibility to observe animals	Plenty of places to see wildlife from a distance. Possibility of bird feeding.	1
4.2. Salutogenic design		5/5
Optimal levels of complexity	Yes, park offers optimal levels of complexity.	1
Engaging features	Multiple elements attract attention: planting, animals, insects, and bridges with colorful handrails. The presence of monuments, temporary exhibitions, colorful playground infrastructure draw the attention of users (Figure 12).	1
Controlled Risk	Several elements offer a subjective feeling of overcoming controlled risk, e.g., bridges.	1
Mystery/Fascination	There are elements of the scenery which are hidden and revealed only when approached, e.g., meandering river views. It provides soft fascination.	1
Movement	Zagórska Struga river, choice of planting which gives impression of movement in the wind.	1

Table 4. *Cont.*

4.3. Sensory stimuli design		5/6
Sensory stimuli: Sight	Colorful leaves in the autumn, flowering trees in the spring	1
Sensory stimuli: Hearing	Sound of meandering river, cascades	1
Sensory stimuli: Smell	Flowering trees in the spring	1
Sensory stimuli: Touch	Trees, water	1
Sensory stimuli: Taste	Restaurant in the reconstructed mill	1
Sensory path	No	0/1

The results of the DESIGN section are excellent 19/20 points.

Only one point was missing—a sensory path, which could enhance the sensory experience.

Table 5. Assessment of Public Park in Rumia, Poland—part 5, source: author.

Public Park in Rumia, Poland– part 5		POINTS
5. PLACEMAKING		14/14
5.1 Social Contact Enhancement		2/2
Organization of events	Multiple events, cultural festivals, sport events, soccer championships, etc.	1
Meeting places for groups	There are multiple places for group meetings—benches, restaurant garden, Cultural Center, garden pavilions. The open-air scene is used by teenagers as a place for gathering (Figure 13).	1
5.2 Human perception—spiritual & symbolic		6/6
Sacred places	Nearby churches	1
Works of Art		1
Monuments	Historic manor (Figure 6)	1
Culture and connections to the past	Reconstruction of water mill in the place of historic forge, temporary open-air exhibitions.	1
Thematic gardens	Yes, educational gardens dedicated to biodiversity.	1
Personalization	Yes, the users can personalize and directly take part in creating the park's landscape. Good examples are educational gardens.	1
5.3 Community Engagement		6/6
Personalising the architectural process	Yes, the names of people responsible for the park design and maintenance are publicly known, e.g., persons who organized the competition for talent playground.	1
Participation of all stakeholders, including inhabitants and users	Yes, e.g., the competition for talent playground.	1
Determining the rules of conduct and self-management	Yes, the rules of conduct are determined, written, and put on public display. There are tables in the park with rules of conduct.	1
Space for social contact	Yes, numerous places with equipment can facilitate social contacts (friendly engagement between strangers), e.g., gazebos, playgrounds, benches.	1
-third places	Some places can become the third most important place after home and work, e.g., MDK Municipal Cultural Centre, Restaurant, and park.	1
-fourth places	Park is being used as a fourth place—to spend time in between other engagements, e.g., open green space, seating places.	1

The aspect of placemaking is very important. The maximum of 14 out of 14 points were awarded.

(**a**) (**b**)

Figure 12. (**a**) There are some points of interest to facilitate the creation of mental maps and wayfinding. The bridge railing cutouts resemble the Kashubian embroidery style. Source: author; (**b**) The paths in the park are comfortable, have even surfaces, and provide interesting views with long vistas. Source: author.

(**a**) (**b**)

Figure 13. Open-air scene. (**a**) During organized event. Source: author; (**b**) As seen every day in the summer. Source: author.

6.2. Assessment with Community Park Audit Tool (CPAT)

The Community Park Audit Tool (CPAT) is a user-friendly tool for evaluating the ability of parks to promote youth physical activity that would enable diverse community stakeholders to audit local parks. The CPAT is a comprehensive yet user-friendly instrument that is six pages long and contains four sections: park information, access and the surrounding neighborhood, park activity areas, and park quality and safety. It has demonstrated a high degree of inter-rater reliability with percent agreement for the vast majority of the items in the tool between 80–90% [26].

The park was evaluated using the CPAT tool and scored in all categories, except for the off-leash dog park and skate park. The swimming pool and tennis courts are within walking distance from the park [26]. Therefore, it may be presumed that the park Starowiejski in Rumia is a place that promotes youth physical activity.

6.3. Mappig the Users Activity

The users' activity was observed over summer 2021 by one researcher during numerous visits conducted during various times of day on weekdays and weekends. Main gathering points identified included:

1. pedestrian path along the stream,
2. sports field,
3. benches along the pedestrian path,
4. open-air stage,
5. talent playground,
6. tot lot.

Among the other most popular gathering spots, the open-air stage was a favorite gathering spot for teenagers on weekdays when it was not used for events (Figure 13).

7. Results and Discussion

These study results are highly satisfactory. When assessed with Universal Standard for Health-Promoting Places, the park scored 84 of 87 points. There were only a few missing points, such as a sensory path and street art, which the local community could add if needed.

The assessment with CPAT demonstrated a well-functioning and well-equipped public park which may promote youth physical activity. The park scored in all categories, except for the off-leash dog and skate parks, which are possible to construct if the local community wishes to.

The mapping of user activity demonstrated that activity pockets were aligned with the recreational infrastructure. Main gathering points were: the well-frequented pedestrian path along the stream, sports field of a local soccer team which gathered spectators and players on assigned days, wooden benches along the pedestrian path, open-air stage used for organized events, as well as by local teenagers during weekdays as a favorite gathering point, talent playground used by the children and their parents, tot lot frequented by the youngsters with their babysitters.

7.1. Preventing the Nocive Effects of Climate Change and Urban Heat Islands

The revitalization of the post-industrial site brought back and revived the historic pond, which is a valuable asset to mitigate heat waves. The green cover of open green areas and the green canopy of mature trees are helping to prevent the detrimental effects of climate change. Existing trees are well-maintained, and new ones are added. People choose to stroll along the stream or cycle to do their everyday errands, and some of them opt to do so and not to use their cars. Creating the public park was an important step to improve the local microclimate. People can reach this area to escape the heat urban island effect.

7.2. Promotion of Social Life

The revitalization of the former industrial site brought additional benefits. The engagement of the local community was manifested during the voting for the talent playground. Place identity is enhanced by events organized in the park and open-air exhibitions. A park is also a place for ecological education. Numerous events which promote social contacts are held in the park. It is a favorite public open green space.

8. Conclusions

The case study of the park in Rumia constructed on the post-industrial site is an example of good practices. On a previously neglected industrial lot, a public park provides joy to the people. A necessary revitalization process was done, with excellent results, providing green space with adequate infrastructure and maintenance for people's enjoyment. The difference between the previous and the current situation lies in how people can now safely and meaningfully enjoy the park [34].

The revitalization of the old manor house, which was turned into the local cultural center, was essential for success. The old monument houses a new function and therefore is well-maintained as it has a new owner.

The assessment of therapeutic qualities and using the CPAT tool demonstrated that the post-industrial cultural heritage site could be turned into a health-promoting urban place. Mapping the users' activity confirmed that a park is a popular place. The presence of people was noted all day long, and the pockets of activity were formed in places with recreational infrastructure.

Additional benefits of creating a well-functioning public park were attachment and identity enhancement demonstrated during voting for the talent playground. The public park in an old village in Rumia demonstrates that it is possible to create a health-promoting place in place of a post-industrial site, revitalize the cultural heritage, and mitigate climate change at the same time.

Funding: This research received no external funding.

Informed Consent Statement: Not applicable.

Acknowledgments: Special thanks to Irena Gaj from Miejski Dom Kultury MDK Municipal Cultural Centre in Rumia and Paweł Dąbrowski from Rumia Municipal Office.

Conflicts of Interest: The authors declare no conflict of interest.

References

1. UNESCO. World Heritage Centre. *Climate Change and World Heritage Report on Predicting and Managing the Impacts of.Climate Change on World Heritage and Strategy to Assist States Parties to Implement Appropriate Management Responses*; UNESCO: Paris, France, 2007.
2. Sesana, E.; Gagnon, A.S.; Ciantelli, C.; Cassar, J.A.; Hughes, J.J. Climate change impacts on cultural heritage: A literature review. *WIREs Clim. Chang.* **2021**, *12*, e710. [CrossRef]
3. Çolakoğlu, E.O.; Mil, Z. Effects of climate changes on tourism. *Int. J. Contemp. Econ. Admin. Sci.* **2011**, *1*, 234–271. Available online: http://www.ijceas.com/index.php/ijceas/article/view/40 (accessed on 18 December 2021).
4. Sikora, D. Threats to historic parks arising from the conflict of cultural and natural values. *Tech. Trans./Czasopismo Techniczne* **2015**, *5-A*, 275–290.
5. Jay Mar, D.Q.; Uchiyama, Y.; Kohsaka, R. Local perceptions of blue carbon ecosystem infrastructures in Panay Island, Philippines. *Coast. Eng. J.* **2021**, *63*, 227–247. [CrossRef]
6. Miciukiewicz, K. Can a park save the city?: Hopes and pitfalls of the London national park city. *Cult. Stud. Rev. /Przeglad Kulturoznawczy* **2020**, *4*, 424–434.
7. Cooper-Marcus, C.; Francis, C. *People Places. Design Guidelines for Urban Open Space*; John Wiley & Sons: New York, NY, USA, 1998.
8. Kazak, J.K.; Hodor, K.; Wilkosz-Mamcarczyk, M. Natural environment and cultural heritage in the city, a sustainability perspective. *Sustainability* **2021**, *13*, 7850. [CrossRef]
9. Available online: https://pl.climate-data.org/europa/polska/pomeranian-voivodeship/rumia-10057/ (accessed on 18 December 2021).
10. Available online: https://pixabay.com/pl/vectors/rumunia-mapa-europa-kraj-nar%C3%B3d-29529 (accessed on 18 December 2021).
11. Google Maps. Available online: https://www.google.pl/maps/@54.5753342,18.3966939,784m/data=!3m1!1e3 (accessed on 18 December 2021).
12. Geoportal.gov.pl. Available online: https://mapy.geoportal.gov.pl/imap/Imgp_2.html?gpmap=gp0 (accessed on 18 December 2021).
13. Szydłowski, M.; Szpakowski, W. Impact of polder flood plain of the Cisowska Struga on the transformation of flood waves between Gdynia and Rumia. *Acta Sci. Pol. Form. Circumiectus* **2019**, *18*, 113–126. [CrossRef]
14. Paner, H. Pradzieje. In *Historia Rumi od pradziejów do 1945*; Wydawnictwo Region: Gdynia, Poland, 2012.
15. Sadłowski, A. Kuźnica Rumska I Jej Dworek. 1998. Available online: https://www.rumia.edu.pl/dr/dr03.asp (accessed on 9 September 2021).
16. *Historia Rumi, od Pradziejów do 1945 Roku*; Wydawnictwo Region: Gdynia, Poland, 2012.
17. *Rumia. Miasto w Pradolinie*; Muzeum piśmiennictwa i muzyki kaszubskiej w Wejherowie, Wydawnictwo Diecezji Pelplińskiej Bernardinum: Pelplin, Potland, 2004.
18. Geoportal. Available online: https://mapy.geoportal.gov.pl/imap/Imgp_2.html?gpmap=imap3d (accessed on 20 November 2021).
19. Sulikowski, G. The Architectural Project for the Renovation of the Manor. stored in the archives of the MDK Municipal Cultural Centre.
20. Gminna Ewidencja Zabytków w Rumi. Available online: https://bip.rumia.pl/Article/get/id,33583.html (accessed on 18 December 2021).
21. Digital Archives. Available online: https://maps.arcanum.com/en/map/europe-19century-secondsurvey/?bbox=2045621.5971180573%2C7277813.864416852%2C2051655.3450383185%2C7280933.450633741&layers=158%2C164 (accessed on 20 November 2021).

22. Wojskowy Instytut Geograficzny, Warszawa, Digital Archives. 1938. Available online: http://maps.mapywig.org/m/WIG_maps/series/025K/P30-S27-G_RUMIA_1938_LoC_G6520_s25_.P6.jpg (accessed on 20 November 2021).
23. Digital Archives of the City of Rumia. Available online: https://rumia.e-mapa.net/ (accessed on 16 December 2021).
24. Konkurs na Modernizację Roku. Available online: https://www.modernizacjaroku.org.pl/pl/node/762 (accessed on 9 September 2021).
25. Podwórko Nivea. Available online: https://podworko.nivea.pl/nagrody#strefa3 (accessed on 9 September 2021).
26. Tymczasem w Rumi. Available online: https://tymczasemwrumi.pl/2019/09/19/przyjdz-na-oficjalne-otwarcie-podworka-talentow-nivea-w-rumi/ (accessed on 10 September 2021).
27. Tymczasem w Rumi. Available online: https://tymczasemwrumi.pl/2019/05/21/nowy-ogrod-edukacyjny-w-rumi/ (accessed on 10 September 2021).
28. Trojanowska, M. Therapeutic qualities and sustainable approach to heritage of the city. The coastal strip in Gdańsk, Poland. *Sustainability* **2020**, *12*, 9243. [CrossRef]
29. Trojanowska, M. A universal standard for health-promoting places. Example of assessment—On the basis of a case study of Rahway River Park. *Bud. I Archit.* **2021**, *20*, 57–82. [CrossRef]
30. Kaczynski, A.T.; Wilhelm Stanis, S.A.; Besenyi, G.M. Development and testing of a community stakeholder park audit tool. *Am. J. Prev. Med.* **2012**, *42*, 242–249. [CrossRef] [PubMed]
31. Krzeptowska-Moszkowicz, I.; Moszkowicz, Ł.; Porada, K. Evolution of the concept of sensory gardens in the generally accessible space of a large city: Analysis of multiple cases from Kraków (Poland) using the therapeutic space attribute rating method. *Sustainability* **2021**, *13*, 5904. [CrossRef]
32. Madejska, A. *Prognoza oddziaływania na środowisko do Miejscowego Planu Zagospodarowania Przestrzennego dla obszaru położonego w Rumi w pasie pomiędzy ul.Starowiejską, a ul.Mickiewicza*; Archives of the Rumia Municipal Office; 2019.
33. Jakość Powietrza w Rumi. Available online: https://powietrze.rumia.eu/map (accessed on 10 September 2021).
34. Jaszczak, A.; Pochodyła, E.; Kristianova, K.; Małkowska, N.; Kazak, J.K. Redefinition of park design criteria as a result of analysis of well-being and soundscape: The case study of the Kortowo Park (Poland). *Int. J. Environ. Res. Public Health* **2021**, *18*, 2972. [CrossRef] [PubMed]

land

Article

Landscape Fragmentation in Qinling–Daba Mountains Nature Reserves and Its Influencing Factors

Yingzhuo Zhang [1,2], Haoran Yin [2], Lianqi Zhu [2] and Changhong Miao [1,2,*]

1 Key Research Institute of Yellow River Civilization and Sustainable Development & Collaborative Innovation Center on Yellow River Civilization Jointly Built by Henan Province and Ministry of Education, Henan University, Kaifeng 475001, China; 104752180027@vip.henu.edu.cn
2 College of Geography and Environmental Science, Henan University, Kaifeng 475004, China; 104753190112@henu.edu.cn (H.Y.); lqzhu@henu.edu.cn (L.Z.)
* Correspondence: chhmiao@henu.edu.cn

Citation: Zhang, Y.; Yin, H.; Zhu, L.; Miao, C. Landscape Fragmentation in Qinling–Daba Mountains Nature Reserves and Its Influencing Factors. *Land* **2021**, *10*, 1124. https://doi.org/10.3390/land10111124

Academic Editors: Jan K. Kazak, Katarzyna Hodor and Magdalena Wilkosz-Mamcarczyk

Received: 22 September 2021
Accepted: 20 October 2021
Published: 22 October 2021

Abstract: Climate change and intensified human activity have altered the landscape pattern of nature reserves and are expected to induce persistent changes in habitat quality. Using GIS technology and landscape ecological theories, we quantitatively analyzed landscape fragmentation characteristics and the driving factors for the interior and peripheries of the Qinling–Daba Mountains nature reserves during 2010–2017. Using spatial principal component analysis, landscape pattern indices, and Geodetector, we evaluated the habitat quality status of different nature reserve types in different regions and the impacts of human disturbance on these areas. The results are as follows: (1) Most national nature reserves in the Qinling–Daba Mountains were moderately or highly fragmented during 2010–2017, and the fragmentation degree of a few reserves exhibited a decreasing trend. (2) The fragmentation degree of landscape patches from the core areas to the experimental areas of the inner nature reserves showed a trend of being low in the middle and high in the surrounding area; the level of landscape fragmentation gradually decreased from the edge of 1 km (M-1) to 5 km (M-5). (3) There was spatial differentiation in the intensity of landscape fragmentation among the nature reserves; human activity intensity, land-use degree, elevation, slope gradient, and topographic relief were the factors influencing the spatial differentiation of landscape fragmentation, and the contribution of anthropogenic factors was significantly greater than that of natural factors. Human activities, such as the construction of network infrastructures, irrational partition management, expansion of agricultural and industrial production activities, were the main reasons for the spatial differentiation of landscape fragmentation in the nature reserves. These results can provide significant scientific support for ecological restoration in the nature reserves and contribute to the coordinated development between socio-economic system and ecological environment in the exceedingly impoverished areas.

Keywords: nature reserves; landscape fragmentation; north–south transitional zone; spatial differentiation; Qinling–Daba Mountains; habitat quality

1. Introduction

Landscape fragmentation is an intrinsic driver of habitat quality deterioration and a major cause of ecosystem degradation and biodiversity reduction [1,2]. The reduced patch size, increased habitat isolation, greater edge distances, and lower connectivity caused by this fragmentation leads to changes in interactions among ecosystem components and alters species' growth environments. The effects of landscape fragmentation on habitats are continuous, detrimental, and unpredictable. The process and effects of landscape fragmentation can be identified by the reactions of many organisms [2,3]. When habitats are highly exposed to anthropogenic activity, their structures and function exhibit long-term changes [4], which is of great significance for research on the spatial structure and characteristics of landscape fragmentation under different land-use scenarios and their impact on habitat quality.

Nature reserves have been regarded as an important tool for protecting habitat integrity and species diversity. The landscape integrity and ecological benefits of reserves play an essential role in maintaining the human–land relationship [5]. Although systematic research on the effectiveness of nature reserves is still in its infancy, the effectiveness and suitability of the range of nature reserves remain to be explored, as only 20–50% of global nature reserves are effectively managed [6]. The landscape ecosystems of nature reserves bordering densely populated areas have undergone significant changes, thus affecting the balance of ecosystem diversity; landscape fragmentation is the main cause of these changes [7]. Effective and precise conservation is a critical method for avoiding biodiversity loss and maintaining ecosystem balance [8]. In the face of potential threats due to the presence of multiple internal and external environmental factors, the timeliness of the scope of nature reserves and the sustainability of the related policy and regulations requires substantial research support, especially in developing areas where coordinated development and ecological protection are the top priorities. In recent years, the number of nature reserves in China has increased rapidly, but conservation benefits have not improved [9]. Therefore, it is necessary to study the fragmentation of landscape in typical national nature reserves in depth.

Island biogeography theory is fundamental for studying the ecological succession pattern of landscapes and has been widely used by domestic and international scholars in the research of habitat fragmentation and landscape renovation [10,11]. Landscape-based habitat quality evaluation is mainly applied to regions, such as small watersheds, mountains, plateaus, and cities [12–14], with relatively broad research areas. The study of terrestrial surface landscapes and habitat fragmentation based on land cover is currently mainstream. Land cover change can directly affect species' habitats and the ecosystem's functions and services [15], leading to significant alterations in the ecological linkage between habitat quality and land-use intensity [14,16]. In addition, the loss, degradation, and fragmentation of natural habitats caused by land-use change are the dominant anthropogenic drivers of biodiversity change [17]. Studies have shown that 70% of remaining forests are located within 1 km of forest edges, 50% of forest fragments are within 500 m of the forest edges, 20% of the remaining forests are located within 100 m from the edges, and most remaining forest patches are less than 10 hm^2 in size [18]. This edge effect of vegetation primarily reflects the degree of landscape fragmentation within the ecosystem, the effects of which are persistent [3,19–21]. Therefore, investigating landscape fragmentation and its effects on habitat status based on land cover was the focus of this study. We selected ten national nature reserves of different types in China's north–south transitional zone and the Qinling–Daba Mountains as the research objects because this area is relatively sensitive to changes in the surrounding ecological environment. Based on land use and land cover data, we explored the variation characteristics of habitat fragmentation in different landscapes and their spatial differentiation using landscape patch indices and the spatial principal component analysis method. Then, we investigated the critical drivers of fragmentation using Geodetector. This study aimed to provide a reference for the precise management of nature reserves in China's north–south transitional zone affected by climate change and anthropogenic activities. Moreover, by revealing the habitat quality status and landscape ecological patterns of the nature reserves in the Qinling–Daba Mountains, we can provide theoretical support for these nature reserves' sustainable development and ecosystem optimization.

2. Study Area and Methods

2.1. Overview of the Study Area

The Qinling–Daba Mountains are located between 30°–36° N and 101°–114° E, spanning Sichuan, Gansu, Shaanxi, Chongqing, Hubei, and Henan provinces. This region has a width of approximately 300 km from north to south and a length of approximately 1000 km from east to west, with a total area of around 300,000 km^2 and an altitude range of 13–5528 m (Figure 1). The Qinling–Daba Mountains sit in the transitional region from the northern subtropical zone to the warm temperate zone, serving as an essential eco-geographical transition zone and a natural dividing line of climate in China. Within this transition zone, the topography varies significantly, and its natural geographical location has led to complex horizontal and vertical distribution patterns of vegetation coverage. The northern slope of the Qinling Mountains is mainly covered by warm–temperate deciduous broadleaved forests, while evergreen–deciduous, broadleaved mixed forests cover the southern slope. The northern slope of the Daba Mountain is connected with the southern slope of the Qinling Mountain, and the vegetation types on these slopes are similar. The southern slope of the Daba Mountains is covered by subtropical evergreen broadleaved forests. The regional vegetation and temperature change with altitude, showing obvious vertical zonality [22,23]. The Qinling–Daba Mountains are one of the 17 vital Biodiversity Ecological function zones identified in the National Major Function Zone Planning of China. The region has a large population with a scattered distribution, and the mountainous area between the Qinling and Daba Mountains is economically under-developed and has prominent problems with poverty, making it an important concentrated contiguous special hardship area in China. Effective management by local governments has led to the integrated development of land resources, but the partition management mode has made the natural habitat more fragmented. By the end of 2015, there were 26 national nature reserves in the Qinling–Daba Mountains, with a total area of approximately 1.47×10^4 km^2 accounting for 5% of the entire land area of the Qinling–Daba Mountains, and 43 local nature reserves with a total area of 1.09×10^4 km^2 accounting for 3.6% of the land area, which is lower than the global average (12%). In this study, we selected ten national nature reserves according to the division of land-use types and their indicative levels in the north–south transitional zone [24,25] as shown in Figure 1.

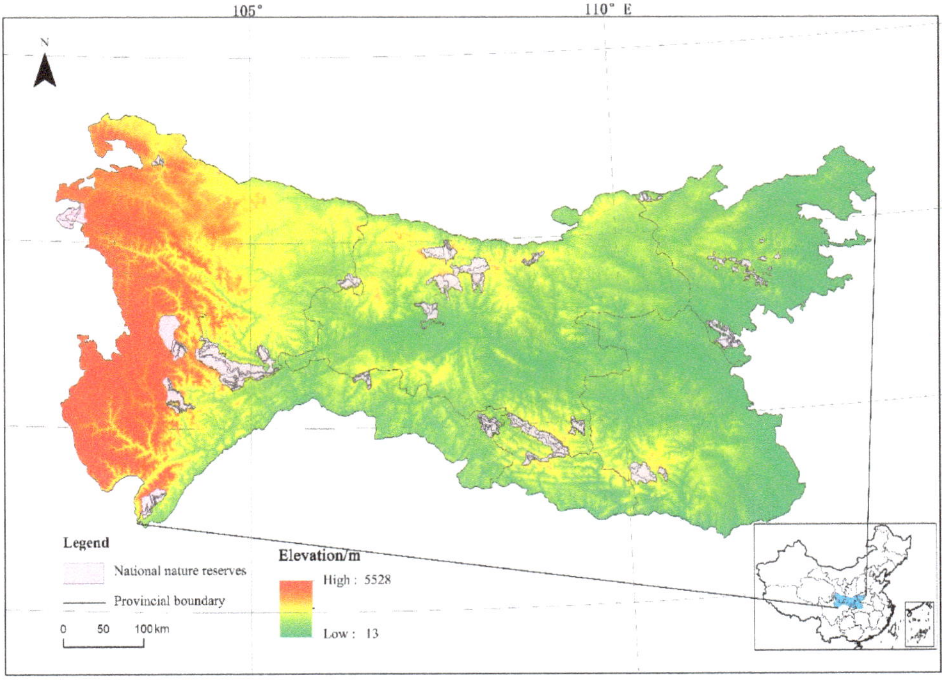

Figure 1. Distribution of national nature reserves in the Qinling–Daba Mountains.

2.2. Data Sources and Processing

The 2010 and 2017 land-use type data used in this study were obtained from the Globeland30 data set developed and improved by the National Geomatics Center of China (http://ngcc.sbsm.gov.cn/ngcc/) (accessed on 20 September 2020) for the National High-Tech Research and Development Program (863 Program), with a spatial resolution of 30 m. Supported by ArcGIS 10.3 software, we preprocessed the data by cropping, mosaicking, and resampling and reclassified the land-use types into nine categories including agricultural land, grassland, forest, shrubland, wetland, water area, land for facilities, bare land, and snowfield, according to the land-use/cover type classification standards of the Resource and Environment Science and Data Center (RESDC), Chinese Academy of Science (http://www.resdc.cn/) (accessed on 16 September 2020). Then, we verified the distribution pattern of these land-use types with reference to the 1 km land-use data set from the RESDC website. Information from the nature reserves was obtained from the functional zoning maps provided on the website of the Ministry of Environmental Protection of China (http://www.zhb.gov.cn/stbh/zrbhq/) (28 July 2020), and the nature reserves' boundaries in the maps were vectorized to obtain layers for the whole study area.

DEM data were derived from the NASA_SRTM data set with a 30 m spatial resolution. The elevation and slope gradient information were first extracted using ArcGIS, and the topographic relief of the entire Qinling–Daba Mountains was obtained by the mean change-point analysis method. Referring to the other region extraction method applied to the study area [26], we classified the topographic relief of the Qinling–Daba Mountains into six levels: flat ground (<30 m), terrace (30–70 m), hills (70–200 m), low-relief mountains (200–500 m), medium-relief mountains (500–1000 m), and high-relief mountains (>1000 m) [26].

2.3. Study Methods

2.3.1. Construction of a Composite Landscape Fragmentation Indicator

To investigate the overall landscape fragmentation pattern of the selected ten nature reserves in the Qinling–Daba Mountains, in this study, we calculated the landscape indices using the Patch metric module in FRAGSTATS 4.2 software to analyze the landscape fragmentation with the comparative method quantitatively. Considering the local situation of the Qinling–Daba Mountains, we selected the patch area (PA), patch shape index (PSI), fractal dimension index (FRAC), contagion index (CONTIG), and Euclidean nearest neighbor distance (ENN) to characterize the degree of landscape fragmentation in different regions [27,28]. The algorithm and ecological significance of each landscape index are explained in related literature [29]. Because the information of landscape fragmentation reflected by the selected landscape patch indices may exhibit overlap, we used the principal component analysis method to reorganize the selected indices into a set of independent synthetic variables. By characterizing the degree of regional landscape fragmentation using this method, redundancy can be avoided. These synthetic variables were then used as the analysis variables of the Geodetector to explore the influencing factors. The results of the principal component analysis are listed in Table 1.

The formula for a composite indicator of landscape fragmentation is as follows:

$$Z_i = PC_n\{f(PA, PSI, FRAC, CONTIG, ENN)\} \quad (1)$$

where Z_i represents the degree of landscape fragmentation; i denotes the nature reserve type in the study area; PC_n represents the principal component scores of n indices; PA, PSI, $FRAC$, $CONTIG$, and ENN represent the patch area index, patch shape index, patch fractal dimension index, contagion index, and Euclidean nearest neighbor distance, respectively.

Table 1. Results of principal component analysis of landscape indices.

Year	KMO	Bartlett's Test (*sig*)	Nature Reserve	Common Factor Variance	Cumulative Variance Contribution Rate /%	Number of Principal Components
				Initial Value		
2010	0.590	0.000	Baishuijiang	1.000	87.196	3
	0.589	0.000	Dabashan	1.000	72.862	2
	0.620	0.000	Shennongjia	1.000	73.957	2
	0.552	0.000	Micangshan	1.000	72.815	2
	0.569	0.000	Taibaishan	1.000	72.905	2
	0.538	0.000	Xuebaoding	1.000	73.773	2
	0.623	0.000	Danjiang Wetland	1.000	71.791	2
	0.597	0.000	Hanzhong Crested Ibis	1.000	90.134	3
	0.578	0.000	Baishuihe	1.000	76.367	2
	0.630	0.000	Lianhuashan	1.000	72.279	2
2017	0.578	0.000	Baishuijiang	1.000	72.176	2
	0.550	0.000	Dabashan	1.000	74.043	2
	0.596	0.000	Shennongjia	1.000	71.768	2
	0.561	0.000	Micangshan	1.000	73.891	2
	0.586	0.000	Taibaishan	1.000	72.162	2
	0.576	0.000	Xuebaoding	1.000	72.302	2
	0.586	0.000	Danjiang Wetland	1.000	71.797	2
	0.621	0.000	Hanzhong Crested Ibis	1.000	73.027	2
	0.547	0.000	Baishuihe	1.000	74.583	2
	0.592	0.000	Lianhuashan	1.000	71.319	2

The path metrics and Class metrics modules in the FRAGSTS 4.2 software were used to calculate the landscape indices of different nature reserve types in the Qinling–Daba Mountains in 2017 with the aim of generating synthetic variables to characterize the degree of fragmentation. The results are shown in Figure 2. To investigate the spatial differentiation pattern of landscape fragmentation from the core to the edges of each nature reserve, we selected the functional partitions of the nature reserve and a 5 km buffer zone on its periphery as the research objects, which are (from inside to outside): the interior (core area, buffer area, and experimental area) and edge areas of 1–5 km (M-1, M-2, M-3, M-4, and M-5). The edge areas were divided according to the Euclidean distance of 1–5 km from the boundary to the core area, respectively, and the landscape fragmentation of the nature reserve was studied on this basis.

2.3.2. Determination of Factors Influencing Landscape Fragmentation

Factor detection and interactive detection methods in the Geodetector model were employed to investigate the relationship between landscape fragmentation in the Qinling–Daba Mountains nature reserves and its driving factors to quantitatively analyze the factors influencing regional landscape fragmentation [30]. First, the explanatory variables were constructed with the ArcGIS platform, and 2000 random points were created to extract the synthetic variables of landscape fragmentation, topographic relief, land-use type, human activity intensity, elevation, and slope gradient. Then, these explanatory variables were discretized into categorical variables, and the natural breakpoint method and land-use classification criteria were applied for partitioning. The basis of Geodetector is that if an independent variable had a significant effect on a dependent variable, then the spatial distribution of these two variables should be similar [31]. The model is as follows:

$$q = 1 - \frac{1}{W{\alpha_i}^2} \sum_{i=1}^{n} Wi{\alpha_i}^2 \quad (2)$$

where q is the spatial heterogeneity of each indicator, and $q \in [0,1]$; W is the sample size of all indicators in the study area; $\alpha^2{}_i$ is the variance of sub-regional indicators; $i = 1, 2, \ldots, n$, i denotes the partition of each indicator, and n represents the number of all partitions. The magnitude of q reflects the degree of spatial heterogeneity of each nature reserve in the Qinling–Daba Mountains. The larger the q value, the higher the spatial heterogeneity, and the greater the influence of partitioning factors on the spatial distribution of regional landscape fragmentation. The smaller the q value, the stronger the randomness of spatial distribution. Specifically, $q = 0$ indicates that the landscape fragmentation in the study area is absent of spatial heterogeneity, and $q = 1$ means that the indicators have perfect spatial heterogeneity.

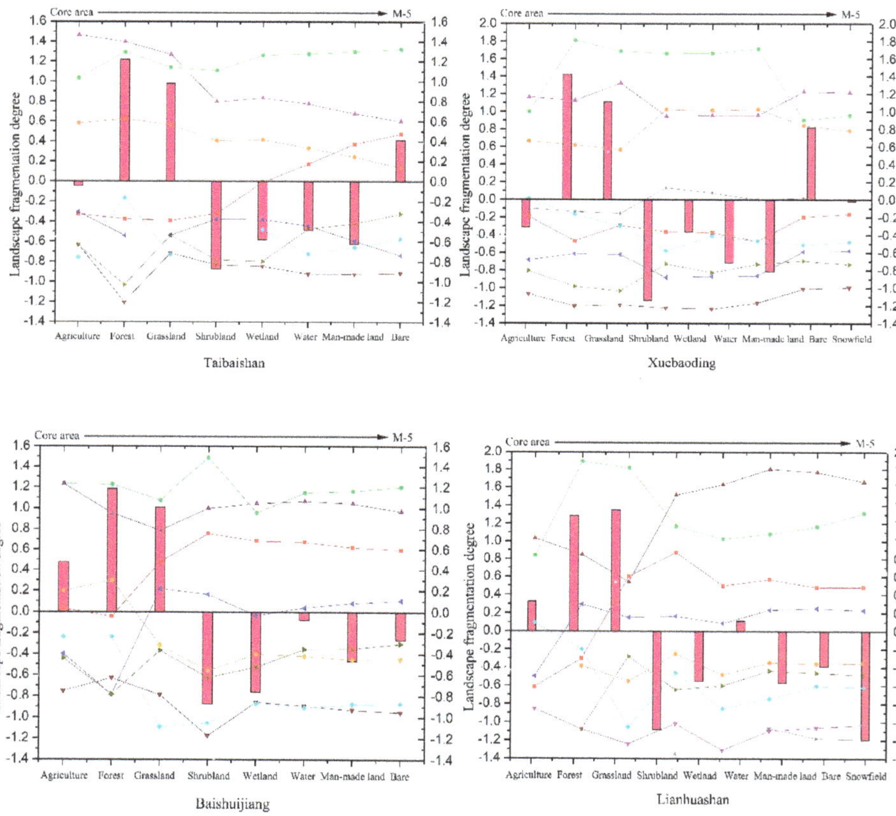

Figure 2. *Cont.*

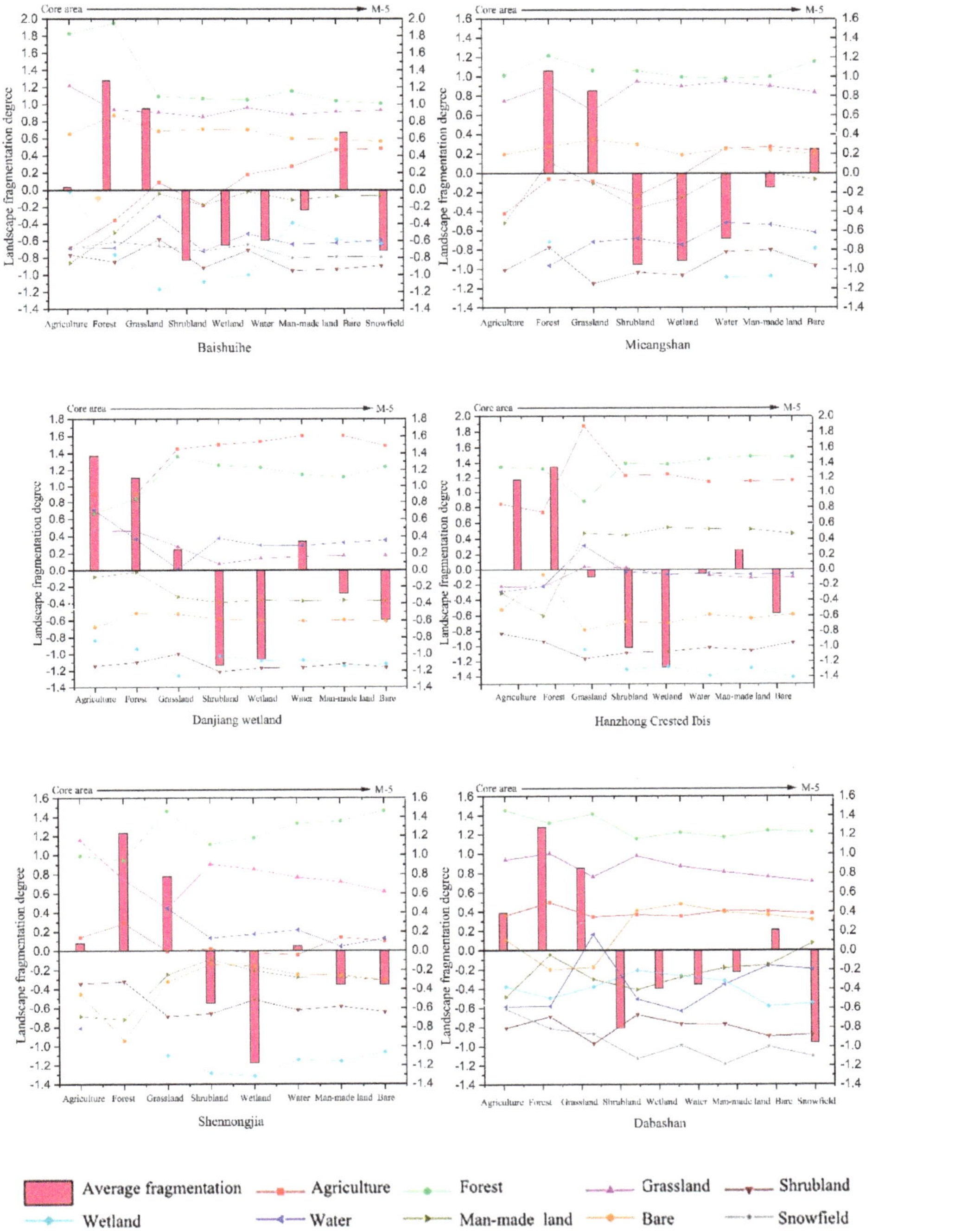

Figure 2. Fragmentation of core–edge landscapes in Qinling–Daba Mountains nature reserves.

In this study, topographic relief, elevation, slope gradient, land-use degree, and human activity intensity were chosen as the explanatory variables reflecting the landscape fragmentation. The factor detection method can reveal the extent to which each driving factor explains the spatial variation of the dependent variable in the Qinling–Daba

Mountains nature reserves [30]. Interactive detection allows us to explore whether the interaction among multiple driving factors has increased or decreased the explanatory power of synthetic variables on the dependent variables in the nature reserve [30].

Based on the results of principal component analysis for landscape patch indices, we extracted synthetic variables to characterize the degree of landscape fragmentation and chose the Hanzhong Crested Ibis nature reserve, which has a large KMO and cumulative variance contribution rate, as the research object to explore the influencing factors of landscape fragmentation in the nature reserves. The synthetic variables were input into the Geodetector for the analysis. In the partitioning process, which was conducted simultaneously, topographic relief classification was performed following the specified criteria, but the other driving factors were subjected to the systematic classification method of ArcGIS and the natural breakpoint method.

2.3.3. Quantitative Indices of Human Activity Intensity

To characterize the impact of regional human activity intensity on the degree of landscape fragmentation, we took the artificial land surface as the basic feature and used the intensity of land surface development and utilization and the level of human disturbance as the quantitative standard for human activity intensity considering the land-use intensity data from ten national nature reserves in the Qinling–Daba Mountains. Based on the regional land-use/cover variation data, we estimated the quantitative disturbance intensity of each nature reserve as the intensity of human activities [32]. The quantitative model of human activity intensity can be calculated by the following equations:

$$HMI = \frac{S_{CV}}{S} \times 100\% \tag{3}$$

$$S_{CV} = \sum_{i=1}^{n} SV_i \times G_i \tag{4}$$

where HMI is the intensity of human activities in a nature reserve; S_{CV} is the converted equivalent area of impervious surface; S is the total area of the region; SV_i is the area of the land-use/cover type i; G_i is the conversion coefficient for the impervious surface equivalent of the land-use/cover type i; n is the number of land-use/cover types in the region. S_{CV} is a quantization unit that compares the degrees of influence of different human activity modes on surface land use and their reflections in each land-use type. G_i is the index weight of each land-use type and surface vegetation characteristic in response to human activities. It is converted into a conversion coefficient for the artificial impervious surface equivalent according to human activity intensity.

For land-use/cover types with different impact degrees, the variation characteristics of the surface's natural attributes can serve as a basis for determining G_i values. With reference to the relevant study of Xu, Y., et al. [32], the conversion coefficients were determined by considering the conservation objectives of different nature reserves. The results are shown in Table 2.

Table 2. Conversion coefficients of impervious surface equivalents for different land-use types.

Land-Use/Cover Type	Agricultural Land	Woodland and Shrubland	Grasslands	Wetland and Water Areas	Land for Man-Made Facilities	Bare Land and Snow-fields
Conversion coefficient	0.2	0.0	0.067	0.6	1.0	0.0

3. Analysis of Results

3.1. Variation Characteristics of Landscape Fragmentation in Different Types of Nature Reserves

The variations in landscape fragmentation in the studied nature reserves in the Qinling–Daba Mountains from 2010 to 2017 show that landscape fragmentation had an overall decreasing trend. The degree of landscape fragmentation declined in seven nature reserves and increased in the remaining three reserves. From the results of landscape indices presented in Table 3, we can see that the PA and CONTIG of the Baishuijiang, Dabashan, Xuebaoding, and Lianhuashan nature reserves had an obviously decreasing trend from 2010 to 2017, indicating that the landscape patch areas in these nature reserves were becoming smaller, and the patches were gradually dispersed. Moreover, the increasing trend in FRAC suggests that the regional landscape patch structure tended to be complicated, and the degree of landscape fragmentation exhibited a rising trend. The PA and CONTIG indices of the Shennongjia and Micangshan nature reserves also had a downward trend, indicating that the area of the patches in these regions shrank and the degree of spatial aggregation decreased; the PSI and FRAC indices slightly increased, showing that the shape of the landscape patches developed from regular to irregular, and the degree of landscape fragmentation in these regions had a slight rise. The variation trends of PA, FRAC, and CONTIG of the Baishuihe nature reserve were consistent with those of Shennongjia and Micangshan, with larger magnitudes. Their ENN index increased, indicating that the patches became more dispersed and poorly connected, and the regional landscapes were increasingly fragmented. The Danjiang Wetland and Hanzhong Crested Ibis nature reserves had significant changes in CONTIG and ENN, and their variation trends were consistent; their PSI index decreased, indicating that the degree of landscape fragmentation in these two regions improved from 2010. Similarly, the Taibaishan nature reserve also had gradual improvement in its landscape fragmentation degree. Its PA and FRAC values increased, and there were more large patches which had shapes that became more regular.

Table 3. Landscape type index of each nature reserve in the Qinling–Daba Mountains.

Year	Patch Index	Baishuijiang	Dabashan	Shennongjia	Micangshan	Taibaishan	Xuebaoding	Danjiang Wetland	Hanzhong Crested Ibis	Baishuihe	Lianhuashan
2010	PA/ha	143.55	111.20	50.70	21.15	18.40	47.22	4.77	11.06	21.98	7.20
	PSI	1.38	1.34	1.13	1.23	1.22	1.60	1.43	1.42	1.59	1.54
	FRAC	1.24	1.25	1.20	1.18	1.23	1.13	1.17	1.25	1.12	1.16
	CONTIG	0.96	0.94	0.96	0.96	0.91	0.99	0.79	0.74	0.99	0.94
	ENN/m	65.79	71.84	67.02	66.18	67.93	676.98	97.76	100.85	69.26	76.43
2017	PA/ha	132.97	107.5	49.63	14.33	35.62	30.20	3.95	12.20	2.62	5.77
	PSI	1.23	1.22	1.31	1.25	1.20	1.21	1.30	1.17	1.25	1.25
	FRAC	1.25	1.30	1.22	1.27	1.20	1.28	1.14	1.27	1.18	1.22
	CONTIG	0.87	0.87	0.94	0.77	0.83	0.80	0.80	0.79	0.49	0.75
	ENN/m	59.75	64.19	64.75	68.95	80.21	68.35	55.65	76.93	74.73	70.34

The synthetic variables of the landscape indices of the Qinling–Daba Mountains nature reserves from 2010 to 2017 show that most nature reserves were significantly fragmented (Figure 3). The degree of fragmentation varied among the different types of nature reserves. Its order from high to low in 2010 was Baishuihe > Xuebaoding > Lianhuashan > Danjiang Wetland > Micangshan > Hanzhong Crested Ibis > Shennongjia > Taibaishan > Dabashan > Baishuijiang, and Baishuihe and Xuebaoding exhibited more severe landscape fragmentation than the other nature reserves. In 2017, the greatest to least fragmentation was observed as follows: Danjiang Wetland > Lianhuashan > Baishuihe > Hanzhong Crested Ibis > Micangshan > Shennongjia > Taibaishan > Xuebaoding > Baishuijiang > Dabashan. The landscape fragmentation was particularly significant in Danjiang Wetland and Baishuihe, and the fragmentation degree was the lowest in Dabashan. After 2010, the degree of fragmentation decreased in Taibaishan, Danjiang Wetland, and Hanzhong Crested Ibis in the order of Danjiang Wetland > Hanzhong Crested Ibis > Taibaishan. The fragmentation in their different types of landscapes was improved to a certain extent. In summary, the overall landscape fragmentation in the Qinling–Daba Mountains nature reserves became more severe during the seven years, but the fragmentation in some nature reserves was mitigated.

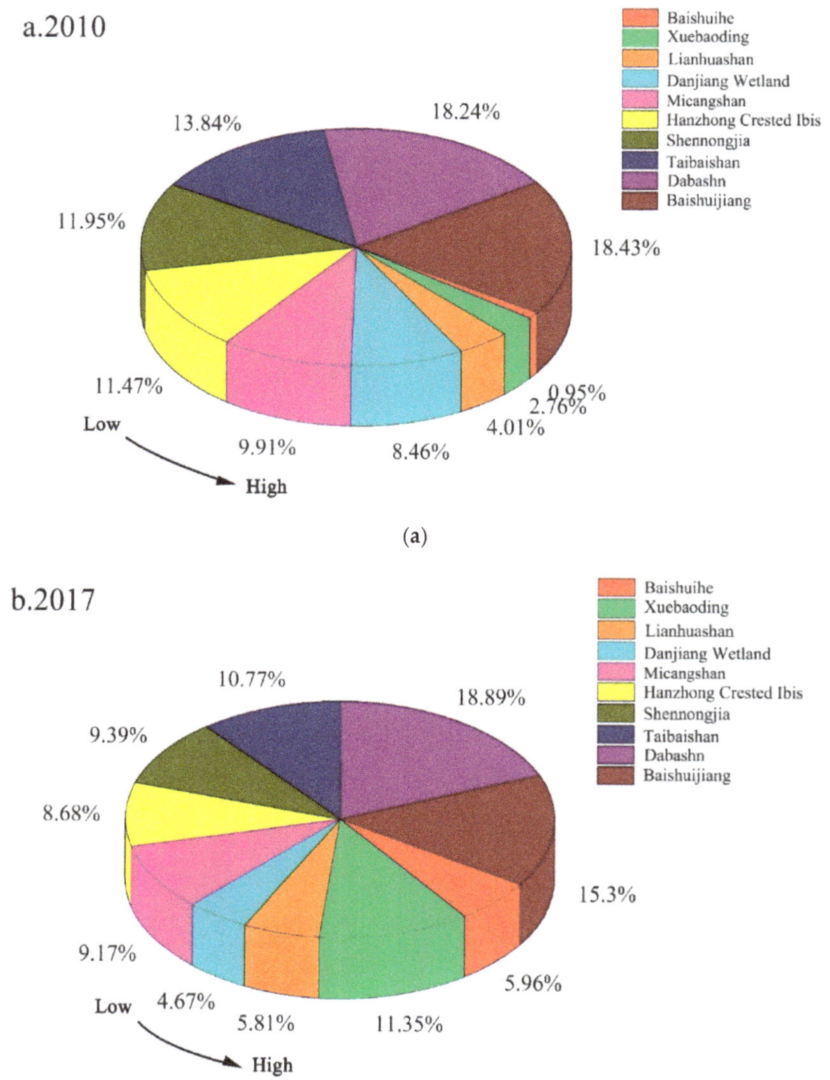

(b)

Figure 3. Proportion of the comprehensive variable scores of the landscape patch index in 2010 (**a**) and 2017 (**b**).

The spatial distribution of landscape fragmentation in major nature reserves in the Qinling–Daba Mountains is demonstrated in Figure 4. In 2010, the highly fragmented areas of the Baishuijiang nature reserve were mainly concentrated in its northwest, southwest, and some marginal areas in the northeast; in 2017, the range of the highly fragmented regions remained the same, but the degree of fragmentation became higher, and the fragmentation gradually spread to the core area. The spatial distribution pattern of fragmentation in the Dabashan nature reserve during 2010–2017 was not apparent. Sporadic fragmentation mainly occurred in the north and southwest and was rarely distributed in the central area; the broken topography made the landscape more isolated and fragmented. The moderately and highly fragmented regions in Shennongjia were dispersed in its interior, and the fragmentation degree of the edges was relatively low. In 2010, the landscape of the Micangshan nature reserve was dominated by low to moderate fragmentation, and the highly fragmented regions were scattered in the central and western areas; by 2017, fragmentation became more severe at the edges of the entire nature reserve and had spread to the interior. The distribution of landscape fragmentation in Micangshan is closely related to topography and rivers. The dramatic topographic relief and convergence of river tributaries have created diverse landscapes. In 2010, the highly fragmented regions in the Xuebaoding nature reserve were concentrated in its western, northern, and southern border areas; by 2017, the fragmented regions had gradually developed to the central area with an increased number of patches, forming a continuous belt. The landscape fragmentation in the Baishuihe nature reserve is particularly significant. Baishuihe exhibited the most serious landscape fragmentation in the watershed. In 2017, the moderately and highly fragmented regions were concentrated in its edges, and the mildly fragmented regions were alternated with moderately to highly fragmented regions in its central area. In the Lianhuashan nature reserve, the moderately and highly fragmented regions were distributed from the southern periphery to its edges. The fragmentation at its northern borders was more serious than in its southern area, and the fragmented regions gradually expanded from the edges to the interior of the nature reserve. In comparison, the Taibaishan, Danjiang Wetland and Hanzhong Crested Ibis nature reserves were the focus of national environmental policies [33], and human disturbance was gradually reduced in these regions. From 2010 to 2017, the degree of fragmentation gradually decreased in the entire region, and most of this area was dominated by medium to high fragmentation, with the highly fragmented regions concentrated at the edges. In summary, from the spatial and temporal distribution characteristics of landscape fragmentation in the ten nature reserves in the study area from 2010 to 2017, we can find that the fragmentation degree had an upward trend in most nature reserves. Moreover, the moderately and highly fragmented regions were concentrated at the edges, and the fragmentation degree was reduced in only a few nature reserves. The highly fragmented nature reserves were all subjected to different levels of disturbance in their interior. The anthropogenic disturbance was relatively significant, and the fragmentation tended to develop from the edges to the core areas.

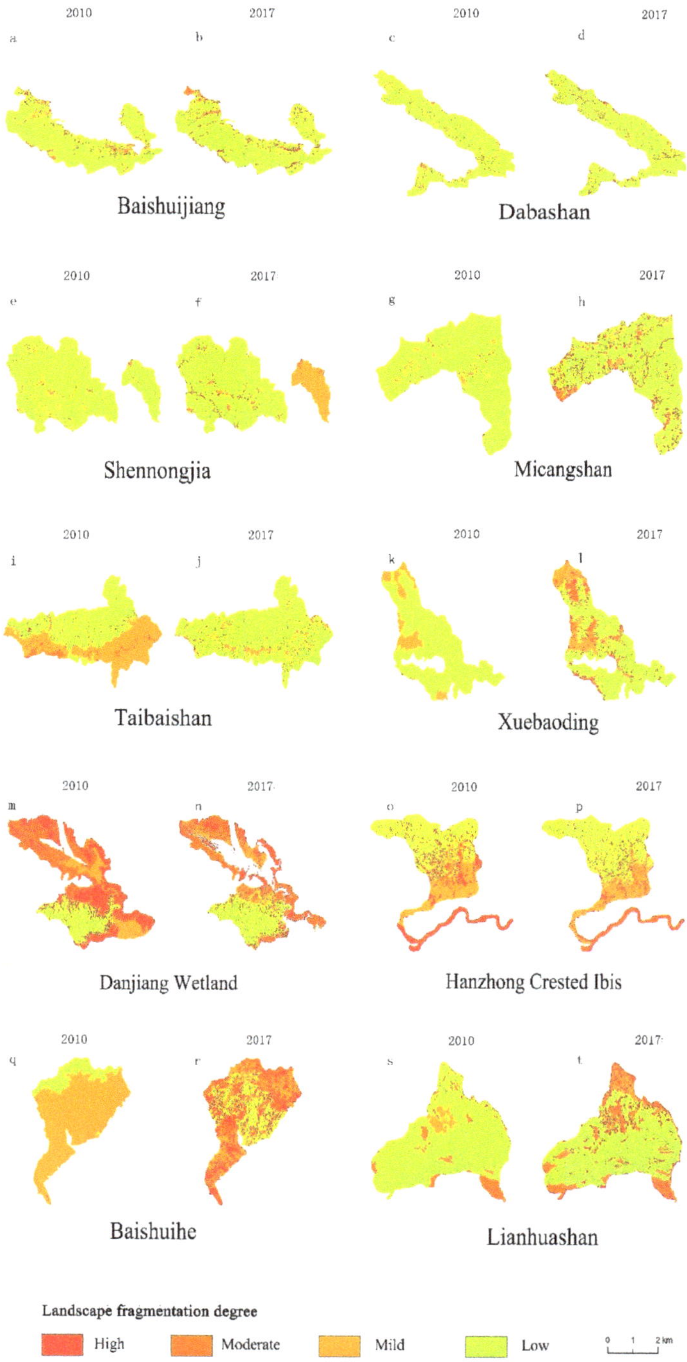

Figure 4. Spatial distribution of landscape fragmentation in different types of nature reserves.

3.2. Core–Edge Spatial Differentiation of Landscape Fragmentation among Nature Reserve Types

3.2.1. Analysis of Landscape Fragmentation in the Interior of the Nature Reserves

The kappa coefficient was 0.83, and the classification results were applied to the entire study area. Most nature reserves in the study area exhibited a high degree of landscape fragmentation in their interior, with significant differences in spatial distribution. The fragmentation degree mainly decreased from the core area to the experimental area. According to Figures 2 and 4, the fragmentation was the most severe in the Danjiang Wetland nature reserve, with an overall fragmentation score of less than 1.5 for the entire nature reserve. The average scores for shrubland and wetland were much lower than 0, the scores for grasslands, wetland, and water areas gradually decreased from the core area to the experimental area, and the general fragmentation level was high. Changes in the fragmentation degree of the internal functional areas were the most drastic in the Lianhuashan nature reserve, where the average fragmentation degree was the highest for shrubland, wetland, and snowfield landscapes, and the average scores for shrubland and snowfield were lower than 0.8. The fragmentation variations in the landscape patches of the internal forest and water areas were relatively large. From the core area to the experimental area, the fragmentation degree increased first and then decreased, and the spatial differentiation of landscape fragmentation was evident in the three main functional areas. The variations of the fragmentation curves of the Baishuihe nature reserves were generally consistent with few fluctuations. The fragmentation levels of the forest and shrubland patches in the core area were higher than the average values of 1.28 and 0.95, indicating a low degree of fragmentation. However, the fragmentation degree of the forest and wetland significantly increased in the experimental area, showing a prominent edge effect. The Hanzhong Crested Ibis nature reserve had dramatic changes from the buffer area to the experimental area. The fragmentation degree showed a downward trend in the forest, bare land, and shrubland. The fragmentation of the shrubland patches was the most severe, and the fragmentation levels of the agricultural land and man-made facility patches rose rapidly in the experimental area. The overall fragmentation of patches in the Micangshan nature reserve first decreased and then increased, with relatively significant variations observed in the patches of shrubland and man-made facilities. The fragmentation degree of shrubland, wetland, and water area landscapes was below −0.6, and the overall fragmentation was significant. The variations of landscape fragmentation in the Shennongjia nature reserve were also drastic, especially for the forest, grassland, and bare land, and the magnitude of the changes was between 0.4 and 0.6. The fragmentation degree of the wetland patches was generally high, with intense human activities on their peripheries. The landscape fragmentation characteristics and variations of the interior of the Taibaishan and Xuebaoding nature reserves were generally consistent. The fragmentation degree of shrubland was significantly higher than that of other types of patches, but the overall fragmentation magnitude of Xuebaoding tended to level off. The Baishuijiang and Dabashan nature reserves had a relatively low fragmentation degree, and the landscape fragmentation in these two regions was the most prominent in shrubland and snowfield, respectively. The overall score of the water area patches had great variations, and the fragmentation level of the internal man-made facility patches decreased first and then increased. The comparison of the nature reserves revealed that although the matrix landscapes were generally forest and grassland, the number of matrix patches increased while the core area was reduced. The matrix at the edges of most nature reserves became more fragmented, and a trend in habitat loss emerged. Some of the agricultural land patches of protected areas appear in the core area and even have a tendency to expand towards the core area, and the expansion of construction land also follows this pattern. The region is mostly a concentration of poor people, and the pattern of cultivation is rougher, and the negative impact of the management model accelerates the trend in landscape fragmentation within the protected areas. These findings indicate that ecological protection is essential in the Qinling–Daba Mountains. Some landscape ecological types that urgently require conservation are not protected. On the contrary, their area is shrinking under the disturbance of intense human activities, and the landscapes have been further fragmented.

3.2.2. Analysis of Landscape Fragmentation at the Edges of Nature Reserves

Variations in the landscape fragmentation on the peripheries of the nature reserves were relatively minimal across the study area, excluding those around M-1, and the fragmentation level of the landscape patches around M-1 was relatively high. The magnitude of landscape fragmentation variations on the peripheries of the Baishuijiang, Xuebaoding, and Lianhuashan nature reserves was also relatively high. Specifically, the forest patches around Xuebaoding had the most prominent variation magnitude of 0.8. The fragmentation degree of forest patches gradually increased from M-1 to M-5, while the variation trends of grassland, water areas, and shrubland were the opposite. All landscape types between M-1 and M-2 of Baishuijiang and Lianhuashan had the most remarkable variations (0.55). Particularly, the cross impacts between agricultural land and vegetation landscapes were significant, and the patches of major habitats were unstable. The greatest landscape fragmentation variations on the peripheries of the Shennongjia and Hanzhong Crested Ibis nature reserves were observed in M-1 (0.67), and as the distance from the nature reserves increased, the variation curves flattened. For the Taibaishan, Micangshan, and Baishuihe nature reserves, the fluctuations in the variation curves were mildest within M-5, but the agricultural land and man-made facilities exhibited significant variations. The degree of landscape fragmentation in this area gradually decreased, the patches were concentrated, and the anthropogenic disturbance was greater than natural disturbances. The fragmentation fluctuations of the forest, grassland, and wetland landscapes on the periphery of the Taibaishan nature reserve were lower than other areas, and the level gradually increased from M-5 to the boundary of the nature reserve, which was considered to be closely related to human activity at the edge of this area. Landscape fragmentation at the edges was more obvious than at the interior, where the landscape matrix gradually loses its dominance, biological habitats undergo potential migration, and external patches gradually replenish the missing habitats, causing the destruction of the original habitats, leading to fragmentation in terms of biological processes and, more importantly, the addition of external human factors dominates this fragmentation process.

In summary, the degree of landscape fragmentation in the interior and exterior of all nature reserves in the study area differed significantly. The most significant differences occurred at the edges of the nature reserves and their peripheries, and the fragmentation degree of the core areas rose under the impact of the edge areas. With the increasing severity and breadth of human impact, anthropogenic directional selection has led to the clustered distribution of agricultural land and man-made facility land patches. These land-use types are interspersed among natural habitats, increasing the degree of landscape fragmentation. Changes to landscape patches, forest, grassland, shrubland, and wetland areas were generally reduced, especially in the core areas of nature reserves and at the edges, while that of agricultural land and man-made facility areas demonstrated increasing trends. For different types of landscapes, the area of some anthropogenic landscapes is expanding, and the patches were distributed in a contiguous region, with a decreased degree of fragmentation. Particularly, the area of agricultural land was expanding, and the old and new farmlands were overlapping and complementing each other, forming a contiguous area, making agricultural land the dominant landscape type in certain regions. Internal landscape habitat processes and external environmental changes provide the conditions for landscape fragmentation in protected areas. Agricultural production activities and zoning management of protected areas continually partition essential ecological landscapes, resulting in the fragmentation of native forests, grassland, shrubland, and wetland. These landscape patches exhibited a mosaic distribution pattern. The dominance of these landscapes is gradually decreasing, and their fragmentation degree is rising. Thus, species have to migrate due to the shrinking habitats, and biodiversity is reduced.

3.3. Analysis of Influencing Factors

The detection results for the driving factors in the Hanzhong Crested Ibis nature reserve are listed in Table 4. It is shown that the q-statistics of land-use type and human activity intensity were 0.478 and 0.384, respectively, both indicating significant spatial differentiation and high influence strength; the explanatory power of slope factor on regional landscape fragmentation was 0.254, with the least influence. The strength of the influencing factors on landscape fragmentation in the nature reserve ranks in the following order: land-use intensity > human activity intensity > elevation > topographic relief > slope gradient. All influencing factors passed the significance test. Although the overall intensity of human activity in nature reserves exhibited a decreasing trend during 2010–2017 (Figure 5), the explanatory power of human disturbance factors on landscape fragmentation was still more dominant than that of natural factors. The results reveal that the gradual reduction of the natural vegetation area of the forest and grasses was accompanied by an increase in the amount of built-up land and population in the area that had affected the spatial differentiation of the landscape pattern. Therefore, lowering the intensity of human activity within and around nature reserves is crucial for protecting ecological landscape patterns and reducing landscape fragmentation.

Table 4. Factor detection results.

Influencing Factor	Human Activity Intensity	Topographic Relief	Land-Use Type	Slope Gradient	Elevation
q-statistic	0.384	0.323	0.478	0.254	0.351
Significance q	0.000	0.000	0.000	0.000	0.000

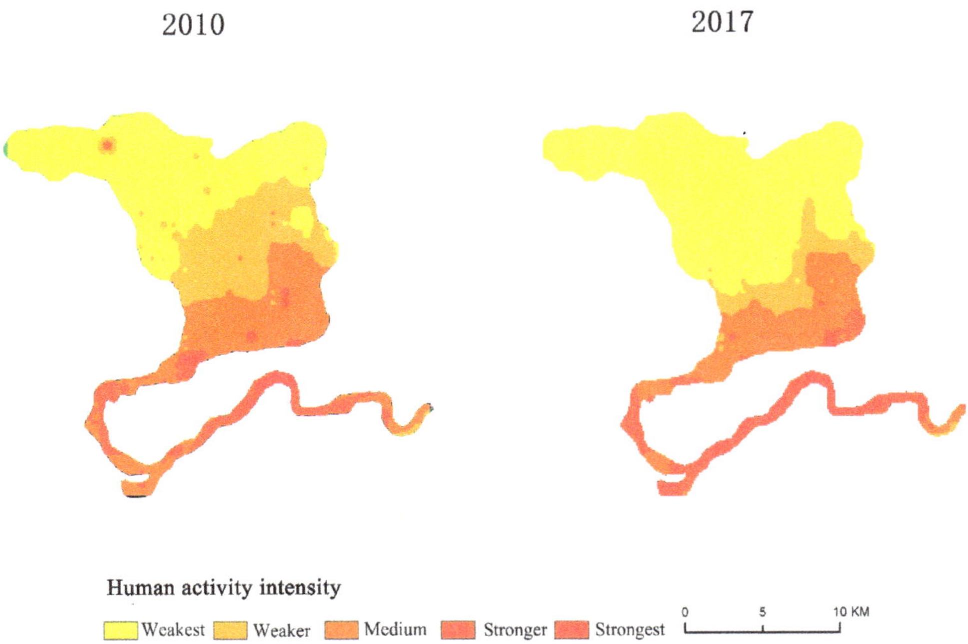

Figure 5. Intensity of human activities in the Hanzhong Crested Ibis nature reserve.

Table 5 shows that the combination of any two factors had a more significant impact on the degree of landscape fragmentation in the nature reserve than any individual factor, and the factors with relative significant influence exhibited a relatively significant effect of double-factor enhancement. In terms of the interaction degree of the influencing factors of fragmentation, the order of their explanatory power is as follows: human activity intensity ∩ land-use type > topographic relief ∩ land-use type > elevation or slope gradient ∩ land-use type > elevation ∩ human activity intensity > topographic relief or slope gradient ∩ human activity intensity > topographic relief or slope gradient ∩ elevation > topographic relief ∩ slope gradient. Moreover, the explanatory power of the interaction of any two factors is greater than that of a single factor. Therefore, it can be seen that the effect of a single natural factor on landscape fragmentation is limited, but after interacting with an anthropogenic factor, the influence can become more prominent. This means that the synthetic effect of anthropogenic factors had a more significant impact, which is crucial for explaining the landscape fragmentation in nature reserves.

Table 5. Interactive detection results.

Influencing Factor	Human Activity Intensity	Topographic Relief	Land-Use Type	Slope Gradient	Elevation
Human Activity Intensity	0.384				
Topographic Relief	0.442	0.323			
Land-Use Type	0.681	0.576	0.478		
Slope Gradient	0.450	0.364	0.527	0.254	
Elevation	0.471	0.392	0.549	0.373	0.351

4. Conclusions and Discussion

4.1. Conclusions

In this study, we took ten typical national nature reserves within the Qinling–Daba Mountains as research objects to explore the degree of landscape fragmentation and its influencing factors in the Qinling–Daba Mountains from 2010 to 2017. The results are as follows.

The landscape pattern of most nature reserves in the Qinling–Daba Mountains exhibited a trend of increasing fragmentation, although the degree of landscape fragmentation in some of the evaluated nature reserves was decreasing. From 2010 to 2017, the landscape fragmentation level was the highest in the Danjiang Wetland and Baishuihe nature reserves, and Dabashan had the lowest degree of fragmentation. The fragmentation degree in the Taibaishan, Danjiang Wetland, and Hanzhong Crested Ibis nature reserves gradually decreased, and the habitat conditions improved.

The degree of landscape fragmentation from the core areas to the experimental areas in the interior of the Qinling–Daba Mountains nature reserves was low in the middle and high in the surrounding areas. The level of landscape fragmentation gradually decreased from the edge of 1 km (M-1) to 5 km (M-5). Human activities were relatively intense around the edges at 1 km (M-1), and the degree of landscape fragmentation in these areas was found to be high. The spatial distribution of fragmented landscape patches in the nature reserves were differentiated. The high-value areas appeared in the regions with intense human activities, and the low-value areas were located in the regions with mild topographic relief.

The land-use degree, human activity intensity, elevation, slope gradient, and topographic relief significantly affected the spatial differentiation of landscape fragmentation in the Qinling–Daba Mountains nature reserves, and anthropogenic activities play a leading role. Human disturbance was the main cause of the fragmentation of ecological landscapes in the Qinling–Daba Mountains nature reserves. In particular, the construction of network infrastructures, irrational partition management, and expansion of agricultural and industrial production activities are the three most important factors that lead to the fragmentation of landscapes in nature reserves.

4.2. Discussion

The global climate is undergoing unprecedented changes. The distribution of species diversity is restricted by the changing climate and is sensitive to climate change's effects. The impact of climate change on the functional structure of landscape ecosystems will be a major cause of biodiversity loss in the future [34]. Many studies on the influence of climate change on habitat status and biodiversity have revealed the significant effect of climate change on the reduced suitability and spatial distribution of nature reserves. Climate change also has considerable potential impacts on habitat quality [35–37]. In this study, we analyzed the landscape fragmentation in nature reserves from the aspects of topographic and anthropogenic influencing factors, not involving the effects of climate variables on the suitability of nature reserves. The Qinling–Daba Mountains are a climatic transition zone. It may be necessary to explore the impact of climate change on habitat fragmentation in nature reserves in further studies to elucidate climate change-induced variations in the ecological landscape patterns in the Qinling–Daba Mountains.

In addition, since most nature reserves are small, with varying shapes and distribution scales, the use of low-resolution data may lead to the mismatch of spatial scales, thus affecting the accuracy of the findings presented herein. Therefore, it is necessary to reveal the landscape fragmentation of each nature reserve from the entire region of the Qinling–Daba Mountains. The present study's results also indicate that the interiors of some nature reserves are still affected by human activities, especially the nature reserves which have peripheries that are densely populated and highly urbanized. The influence of anthropogenic factors on landscape fragmentation in nature reserves is greater than that of natural factors, basically in agreement with Zhang, J.Q., et al. [38].

The expansion of rural development may damage nearby nature reserves. Human activities inside nature reserves are becoming increasingly frequent, mainly including the networking of tourist facilities and transportation infrastructure, illegal deforestation and land clearing, inappropriate production methods, and destructive living habits (Figure 6). Various types of anthropogenic disturbances are combined, jointly leading to significant edge effects in nature reserves, basically in agreement with Jin, C.P., et al. [39].

Human activities are more intense within 1 km of the peripheral area. The expansion of agricultural economic development has made land-use types more differentiated and landscape patterns more complex and fragmented. There are a large number of poor populations in the region, and slow economic development hinders the transformation and upgrading of the management model of nature reserves. Excessive intervention by policy measures and unreasonable zoning management measures have accelerated the trend of landscape fragmentation in nature reserves [40]. From the perspective of sustainable development, strict protection measures suitable for the local area should be formulated to properly control the human activities in the national nature reserves and precisely manage the ecological and environmental problems in these areas.

Landscape fragmentation is often accompanied by widespread habitat fragmentation, whereas the effects of habitat fragmentation are two-sided. Although it can cause habitat loss through species migration lag [41] and extinction debt [42], it also strengthens the functional linkage among species habitats, brings positive edge effects, decreases population competition, and enhances landscape complementarity [41]. Habitat restoration is most effective when carried out by natural mechanisms, but anthropogenic conservation measures can also improve the overall habitat quality. Therefore, although the impacts of habitat fragmentation within the landscapes are negative in some specific cases, these impacts cannot be treated as the same in all contexts. The results of these impacts must be analyzed in detail with species distribution models, such as ecological niches, to provide theoretical support for the effectiveness of nature reserves based on biodiversity and eco-environmental quality.

At present, nature reserves are established to conserve internal biodiversity and protect the natural environment. Conservation means have also been increasing in recent years. For example, through the reconstruction of specific habitats and corridors, the connectivity among the fragments increases, and the negative effects of fragmentation are thus mitigated [43]. However, the habitats within nature reserves are simultaneously affected by natural factors and human activities. It is impossible to assess the rationality of the nature reserve design based on available technical tools, and the time lag of nature reserves cannot be evaluated [44]. Distinguishing between the contribution of natural and anthropogenic disturbances is the key to judging the suitability of nature reserves. This can help develop conservation policies flexibly, thereby contributing to the sustainable development of nature reserves. In this study, we used land-use data to analyze habitat fragmentation among different landscape types. Our work was limited in that we were neither able to study habitat quality nor evaluate the suitability of the studied reserves. As climate change and human activity continue to impact global ecosystems, the spatial changes in nature reserves ought, with time and the functional partitioning from core to edge, to warrant deeper investigation [45]. Different types of nature reserves have different protection objects and roles, and research on habitats should be specific in its focus to yield suitably detailed findings. Multi-dimensional and multi-perspective exploration of internal habitat statuses will be a key approach for future research.

Field photos taken in 2020 showing the trend in landscape fragmentation in the inner and marginal areas of the Qinling–Daba Mountains nature reserves:

Figure 6. Field research photos of the Qinling–Daba Mountains nature reserves.

Author Contributions: Conceptualization, L.Z., H.Y. and Y.Z.; methodology, Y.Z. and H.Y.; software, Y.Z. and H.Y.; validation, L.Z., Y.Z. and C.M.; formal analysis, Y.Z.; investigation, Y.Z. and H.Y.; resources, L.Z. and C.M.; data curation, H.Y.; writing—original draft preparation, Y.Z. and H.Y.; writing—review and editing, Y.Z., H.Y., L.Z. and C.M.; supervision, L.Z. and C.M.; project administration, L.Z. and C.M.; funding acquisition, L.Z. and C.M. All authors have read and agreed to the published version of the manuscript.

Funding: This research was funded by the National Natural Science Foundation of China, grant number 41430637, and the Special Investigation of Basic Science and Technology Resources, grant number 2017FY100902. The National Natural Science Foundation of China is funded by the National Natural Science Foundation of China (NSFC), and the Special Investigation of Basic Science and

Technology Resources is funded by Institute of Geographic Sciences and Natural Resources Research (CAS).

Institutional Review Board Statement: Not applicable.

Informed Consent Statement: Not applicable.

Data Availability Statement: Data sharing not applicable.

Acknowledgments: The authors are particularly grateful to Baiping Zhang, researcher and PhD supervisor of the University of Chinese Academy of Sciences, who presides over the "Comprehensive Scientific Investigation of China's North–South Transition Zone" project. As members of the research group for the project (funding number 2017FY100902), we thank Quantong Chen, Pingping Zhang, and Qingdong Dong for their contributions during field trips and raw data collection. The authors are also very grateful to the editors and reviewers for their comments and suggestions on improving this study.

Conflicts of Interest: The authors declare no conflict of interest.

References

1. Ding, L.; Xu, G.; Lu, J.; Zhang, D.; Fang, B. Landscape fragmentation and its effect on biodiversity. *J. Jiangsu For. Sci. Technol.* **2005**, *32*, 45–49.
2. Evans, M.J.; Banks, S.C.; Driscoll, D.A.; Hicks, A.J.; Melbourne, B.A.; Davies, K.F. Short- and long-term effects of habitat fragmentation differ but are predicted by response to the matrix. *Ecology* **2017**, *98*, 807–819. [CrossRef] [PubMed]
3. Haddad, N.M.; Brudvig, L.A.; Clobert, J.; Davies, K.F.; Gonzalez, A.; Holt, R.D.; Lovejoy, T.E.; Sexton, J.O.; Austin, M.P.; Collins, C.D.; et al. Habitat fragmentation and its lasting impact on Earth's ecosystems. *Sci. Adv.* **2015**, *1*, e1500052. [CrossRef] [PubMed]
4. Lindenmaver, D.; Fischer, J. Habitat fragmentation and landscape change: An ecological and conservation synthesis. *Austral Ecol.* **2010**, *32*, 477–478.
5. Geldmann, J.; Barnes, M.; Coad, L.; Craigie, I.D.; Hockings, M.; Burgess, N.D. Effectiveness of terrestrial protected areas in reducing biodiversity and habitat loss. *Biol. Conserv.* **2013**, *161*, 230–238. [CrossRef]
6. Watson, J.; Dudley, N.; Segan, D.B.; Hockings, M. The performance and potential of protected areas. *Nature* **2014**, *515*, 67–73. [CrossRef]
7. Zhao, R.; Jiang, P.; Zhao, H.; Fan, J. Fragmentation process of wetlands landscape in the middle reaches of the Heihe River and its driving forces analysis. *Acta Ecol. Sin.* **2013**, *33*, 4436–4449. [CrossRef]
8. Pressey, B.; Mccauley, D.J.; Morgan, L.; Possingham, H.P. A to-do list for the world's parks. *Nature* **2014**, *515*, 28–31.
9. Wang, Y.; Gao, J.; Wang, J.; Gao, F. Study on ecosystem service values in national nature conservation areas in Guangdong Province from 2000 to 2010. *Res. Environ. Sci.* **2014**, *27*, 1157–1163.
10. Langlois, L.A.; Drohan, P.J.; Brittingham, M.C. Linear infrastructure drives habitat conversion and forest fragmentation associated with Marcellus shale gas development in a forested landscape. *J. Environ. Manag.* **2017**, *197*, 167–176. [CrossRef]
11. Collins, C.D.; Holt, R.D.; Foster, B.L. Patch size effects on plant species decline in an experimentally fragmented landscape. *Ecology* **2009**, *90*, 2577–2588. [CrossRef]
12. Moreiraa, M.; Fonsecabc, C.; Vergílioc, M.; Caladod, H.; Gile, A. Spatial assessment of habitat conservation status in a Macaronesian island based on the InVEST model: A case study of Pico Island (Azores, Portugal). *Land Use Policy* **2018**, *78*, 637–649. [CrossRef]
13. Zhu, J.; Gong, J.; Li, J. Spatiotemporal change of habitat quality in ecologically sensitive areas of eastern Qinghai-Tibet Plateau: A case study of the Hehuang Valley, Qinghai Province. *Resour. Sci.* **2020**, *42*, 991–1003.
14. Yin, K.; Zhao, Q.; Wen, M.; Hua, L.; Lin, T.; Shi, L. Assessment of landscape pattern effect and ecosystem service of island urban forest. *Remote. Sens. Land Resour.* **2014**, *26*, 128–133.
15. Cerrillo, R.M.N.; Vieira, D.J.E.; Ochoa-Gaona, S.; De Jong, B.H.J.; Delgado-Serrano, M.D.M. Land cover changes and fragmentation in mountain neotropical ecosystems of Oaxaca, Mexico under community forest management. *J. For. Res.* **2019**, *30*, 143–155. [CrossRef]
16. Wu, J.; Cao, Q.; Shi, S.; Huang, X.; Lu, Z. Spatio-temporal variability of habitat quality in Beijing–Tianjin–Heibei Area based on land use change. *Chin. J. Appl. Ecol.* **2015**, *26*, 3457–3466.
17. May, F.; Rosenbaum, B.; Schurr, F.M.; Chase, J.M. The geometry of habitat fragmentation: Effects of species distribution patterns on extinction risk due to habitat conversion. *Ecol. Evol.* **2019**, *9*, 2775–2790. [CrossRef]
18. Pfeifer, M.; Lefebvre, V.; Peres, C.; Banks-Leite, C.; Wearn, O.; Marsh, C.; Butchart, S.H.M.; Arroyo-Rodríguez, V.; Barlow, J.; Cerezo, A.; et al. Creation of forest edges has a global impact on forest vertebrates. *Nature* **2017**, *551*, 187–191. [CrossRef]
19. Pfeifer, M.; Lefebvre, V.; Gardner, T.A.; Arroyo-Rodriguez, V.; Baeten, L.; Banks-Leite, C.; Barlow, J.; Betts, M.G.; Brunet, J.; Cerezo, A.; et al. Supporting information for "BIOFRAG—A new database for analyzing biodiversity responses to forest fragmentation". *Ecol. Evol.* **2014**, *4*, 1524–1537. [CrossRef]

20. Barlow, J.; Lennox, G.D.; Ferreira, J.; Berenguer, E.; Lees, A.C.; Nally, R.M.; Thomson, J.R.; Ferraz, S.F.d.; Louzada, J.; Oliveira, V.H.F.; et al. Anthropogenic disturbance in tropical forests can double biodiversity loss from deforestation. *Nature* **2016**, *535*, 144–147. [CrossRef]
21. Rosa, I.M.D.; Smith, M.J.; Wearn, O.R.; Purvis, D.; Ewers, R.M. The environmental legacy of modern tropical deforestation. *Curr. Biol.* **2016**, *26*, 2161–2166. [CrossRef]
22. Zhang, B.; Zhou, C.; Chen, S. The geo-info-spectrum of montane altitudinal belts in China. *Acta Geogr. Sin.* **2003**, *58*, 163–171.
23. Zhou, Z.; Ren, Z.; Li, J. Study on the ecological value of soil conservation of vegetation in the Qinling-Daba Mountains. *Arid. Zone Res.* **2006**, *1*, 144–148.
24. Li, Y.; Liu, G.; Liu, D.; Qin, Y. Geographical expression and quantitative exploration of the China's north-south transitional zone. *Geogr. Res.* **2021**, *40*, 1857–1869.
25. Zhao, F.; Zhang, B.; Zhu, L.; Yao, Y.; Cui, Y.; Liu, J. Spectra structures of altitudinal belts and their significance for determining the boundary between warm temperate and subtropical zones in the Qinling-Daba Mountains. *Acta Geogr. Sin.* **2019**, *74*, 889–901.
26. Zhang, J.; Li, Y.; Zhu, L. Relief amplitude based on county units in west Henan Mountain area and its correlation with distribution of population and economic activities. *Areal Res. Dev.* **2019**, *38*, 55–60.
27. Tang, J.; Tan, F.; Zhou, L. Analysis on the influence of landscape fragmentation on species diversity. *J. Green Sci. Technol.* **2018**, *24*, 158–161.
28. Dong, Y.; Liu, S.; An, N.; Yin, Y.; Wang, J.; Qiu, Y. Landscape pattern in Da'an City of Jilin Province based on landscape indices and local spatial autocorrelation analysis. *J. Nat. Resour.* **2015**, *30*, 1860–1871.
29. Wu, J. *Landscape Ecology: Pattern, Process, Scale and Hierarchy*; Higher Education Press: Beijing, China, 2007.
30. Wan, J.; Xu, C. Geodetector: Principle and prospective. *Acta Geogr. Sin.* **2017**, *72*, 116–134.
31. Xie, Y.; Gong, J.; Qi, S.; Zhang, Y.; Guo, J. Assessment of ecosystem food production and its spatiotemporal heterogeneity in the Bailong River Watershed, Gansu. *Acta Ecol. Sin.* **2017**, *37*, 1719–1728.
32. Xu, Y.; Xu, X.; Tang, Q. Human activity intensity of land surface: Concept, method and application in China. *J. Geogr. Sci.* **2016**, *26*, 1349–1361. [CrossRef]
33. Sun, B.; Duan, W.; Ding, H.; Feng, Y.; Wen, Y. Preference analysis of household ecological compensation in Crested Ibis protected area in Hanzhong, Shaanxi based on choice experiments. *Resour. Sci.* **2017**, *39*, 1792–1800.
34. Pereira, H.M.; Leadley, P.W.; Proençarob, V.; Scharlemann, A.P.W.; Fernandez-Manjarrés, J.F.; Araújo, M.B.; Balvanera, P.; Biggs, R.; Cheung, W.W.L.; Chini, L.; et al. Scenarios for global biodiversity in the 21st century. *Science* **2010**, *330*, 1496–1501. [CrossRef] [PubMed]
35. Staudinger, M.D.; Carter, S.L.; Cross, M.S.; Dubois, N.; Duffy, J.E.; Enquist, C.; Griffis, R.; Hellmann, J.J.; Lawler, J.J.; O'Leary, J.; et al. Biodiversity in a changing climate: A synthesis of current and projected trends in the US. *Front. Ecol. Environ.* **2013**, *11*, 465–473. [CrossRef]
36. Morgine, M.; Jetz, W. The effect of range changes on the functional turnover, structure and diversity of bird assemblages under future climate scenarios. *Glob. Chang. Biol.* **2015**, *21*, 2917–2928.
37. Araújo, M.B.; Alagador, D.; Cabeza, M.; Nogués-Bravo, D.; Thuiller, W. Climate change threatens European conservation areas. *Ecol. Lett.* **2011**, *14*, 484–492. [CrossRef]
38. Zhang, J.Q.; Gong, J.; Liu, D.Q. Dynamics and driving factors of landscape fragmentation based on Geodetector in the Bailongjiang watershed of Gansu Province. *Sci. Geogr. Sin.* **2018**, *38*, 1370–1378.
39. Jin, C.P.; Wang, C.; Wen, R.H.; Hou, P. Analysis of land cover change in Danjiang Wetland National Nature Reserve and its inner and outer areas from 2000 to 2015. *Environ. Monit. Early Warn.* **2019**, *11*, 8–13.
40. Yan, W.W. Study on the status quo and management measures of ecotourism in Taibai Mountain Nature Reserve. *J. Green Sci. Technol.* **2020**, *21*, 214–217.
41. Jackson, S.T.; Sax, D.F. Balancing biodiversity in a changing environment: Extinction debt, immigration credit and species turnover. *Trends Ecol. Evol.* **2010**, *25*, 153–160. [CrossRef]
42. Tilman, D.; May, R.M.; Lehman, C.L.; Nowak, M.A. Habitat destruction and the extinction debt. *Nature* **1994**, *371*, 65–66. [CrossRef]
43. Damschen, E.I.; Drudvig, L.A. Landscape connectivity strengthens local–regional richness relationships in successional plant communities. *Ecology* **2012**, *93*, 704–710. [CrossRef]
44. Zhu, P.; Huang, L.; Xiao, T.; Wang, J. Dynamic changes of habitats in China's typical nature reserves on spatial and temporal scales. *Acta Geogr. Sin.* **2018**, *73*, 92–103. [CrossRef]
45. Dobrowski, S.Z.; Littlefield, C.E.; Lyons, D.S.; Hollenberg, C.; Carroll, C.; Parks, S.A.; Abatzoglou, J.T.; Hegewisch, K.; Gage, J. Protected-area targets could be undermined by climate change-driven shifts in ecoregions and biomes. *Commun. Earth Environ.* **2021**, *2*, 198. [CrossRef]

land

Article

Urban Cemeteries—Places of Multiple Diversity and Challenges. A Case Study from Łódź (Poland) and Leipzig (Germany)

Andrzej Długoński [1], Diana Dushkova [2,*] and Dagmar Haase [2,3]

1 Institute of Biological Sciences, Faculty of Biology and Environmental Sciences, Cardinal Stefan Wyszynski University in Warsaw, 01-815 Warszawa, Poland; a.dlugonski@uksw.edu.pl
2 Helmholtz Centre for Environmental Research—UFZ, 04318 Leipzig, Germany; dagmar.haase@ufz.de
3 Institute of Geography, Humboldt University of Berlin, Rudower Chaussee 16, 12489 Berlin, Germany
* Correspondence: diana.dushkova@ufz.de

Abstract: This article presents a pilot study investigating the multidimensional diversity of cemeteries as an important element of cultural heritage and green infrastructure within the urban landscape. We studied the state and diversity of nature, perceptions, and activities of visitors. As religion is an important aspect that differentiates cemeteries from each other, we studied a sample of four multi-confessional urban cemeteries in Łódź (Poland) and Leipzig (Germany) by using site observation and a questionnaire survey. We found that cemeteries are far undervalued as public green resources that can perform important functions in sociocultural life and the mental well-being of the general public, as the perceptions of silence- and contemplation-seeking visitors tell us. The perception of cemeteries depends on the level of secularization, varying from a sacrum sphere up to specific recreational and touristic opportunities; findings that should be considered by town planners when optimizing the cultural ecosystem services of green spaces.

Keywords: ecosystem services (ES); urban cemeteries; biocultural diversity (BCD); urban green infrastructure (UGI); questionnaire survey; Leipzig; Łódź

Citation: Długoński, A.; Dushkova, D.; Haase, D. Urban Cemeteries—Places of Multiple Diversity and Challenges. A Case Study from Łódź (Poland) and Leipzig (Germany). *Land* **2022**, *11*, 677. https://doi.org/10.3390/land11050677

Academic Editors: Jan K. Kazak, Katarzyna Hodor, Magdalena Wilkosz-Mamcarczyk and Francesca Cigna

Received: 26 March 2022
Accepted: 30 April 2022
Published: 3 May 2022

Publisher's Note: MDPI stays neutral with regard to jurisdictional claims in published maps and institutional affiliations.

1. Introduction

The increasing number and size of urban areas and the continuous transformation of urban landscapes pose big challenges for urban environments, such as the decline in the quality and quantity of green space, loss of biodiversity, ecosystem degradation, and citizens' disconnection from nature, which affect their health and well-being [1,2]. In this regard, it becomes highly important that green spaces in cities are appropriately designed to become multifunctional to meet the various needs of the urban population. A review of existing research on urban green spaces and ecosystem services (ES) shows that in most cases, scientists focus on urban forests, street trees, parks, and gardens, while cemeteries remain overlooked. Moreover, a vast literature deals extensively with the definition of green space that is rarely based on urban atlas classification, namely, "forest" and "green urban areas". It includes different land uses in one class, and important urban green spaces such as cemeteries are not included. As cemeteries are defined differently in different areas (or entirely overlooked), there is no shared understanding of their role and function in wider built environment networks. In addition, current research on cemeteries has rarely used an interdisciplinary approach, and there are few studies that are based on both concepts of ES and biocultural diversity (BCD) that are of an interdiscip, linary nature. This is because cemeteries, even when included in the concept of urban green infrastructure (UGI), are not ascribed qualities similar to those ascribed to public green spaces such as parks, forests, street trees, green roofs, and gardens; instead, most qualities are related to cultural history [3–5]. To date, research on cemeteries has been preliminarily restricted to biodiversity and conservation aspects, including research on their vegetation and fauna [6–10], restorative components and compatibility [5,11], historical and cultural (e.g., cultural heritage and cultural encounters) [12–15], and aesthetic, spiritual and recreational [5,14,16–22] aspects. However, some current studies have also highlighted the big potential of cemeteries as an element of a city's greenery, being of multifunctional use with different uses and meaning [3–5,14,16,23].

In this context, this paper aims to fill the knowledge gap by examining the contribution of the urban cemetery as neglected green space to UGI and ES. We suggest the application of a multidimensional set of indicators to assess urban cemeteries as an important element of UGI, which can provide a specific range of ES. We empirically studied this new phenomenon by using a sample of four multi-confessional cemeteries in Leipzig (Germany) and Łódź (Poland) through site observation and a questionnaire survey. Additionally, this study also applied the concept of biocultural diversity (BCD) which gives acknowledgement to the diversity of understandings and appreciation of urban nature, and promotes alternative ways of living and being within cities [24,25]. Because there is neither specific literature on the large BCD of cemeteries nor any empirical studies, we had to work creatively with different sets of literature to produce our own synthesis. To the best of our knowledge, no study has addressed this overlap of processes (ES provided by urban cemeteries as an element of UGI of multifunctional use and meaning, resulting from a BCD approach) to date, in a spatially explicit manner. In the four different sets of literature (e.g., biodiversity aspects of cemeteries, cemeteries as cultural heritage including its historical and cultural values, recreational and restorative values of cemeteries, cemeteries as an element of a city's greenery) that we analyzed, there was a lack of research of the following perspectives: (a) diversity in a double sense, (b) multifunctionality, (c) multiculturality, and (d) integration of neglected social groups. From these perspectives, we developed our research questions to explore cemeteries as repositories of natural and cultural diversity, highlighting their huge potential for UGI and ES provision, and to introduce a framework for enhancing ecosystem processes and human uses in urban cemeteries. In doing so, we focused on viewpoints of utilization of the cemetery space and visual impact of the cemetery to people. We aimed to learn the perceptions of people who visit cemeteries, because they can provide everyday observations and perspectives related to those cemeteries. In order to achieve this goal, a pilot study was carried out to observe the sites (determination of the extent of ES and visitor's activities in given cemeteries) and by conducting a questionnaire survey among users.

We proposed five working research questions and objectives that help to better conceptualize possible future trends in the development and use of cemeteries as elements of UGI:

1. Do cemeteries have specific socio-ecological potential and fulfil ecological functions (e.g., climate regulation, habitat sphere, niche, original species niche) that differ from other urban green spaces?
2. What ES are provided by urban cemeteries? Do they differ in selected cases from Łódź and Leipzig?
3. Do cemeteries serve specific groups of urban population (neglected people) that are significantly different from those groups that use other green spaces within the city?
4. Is there a difference in the opinions of cemetery visitors within nations (German and Poles)?
5. What are the specific and informed recommendations for better integration of cemeteries in urban GI planning?

These five questions are of an integrative heuristic nature because, to date, theoretical considerations or solid empirical research findings on this topic barely exist. They are derived from the recent literature mentioned above and from pre-surveys conducted by the authors in April and June 2017 and 2018, in both case study cities [26,27]. These questions not only describe trajectories of future development of cemeteries as an element of UGI, but also provide an important conceptual starting point for our empirical research.

2. Materials and Methods

Methodically, the study used a four-step process, starting with preliminary research and identification of the research gap, research design and methodology for data collection (step 1), collecting data using site observation and a questionnaire survey among cemeteries' visitors (step 2), processing and analyzing data by applying spatial analysis and statistical analysis of the data obtained (step 3), and finally, data interpretation and alignment with previous studies (step 4).

Considering the five working research questions and objectives mentioned above, we developed a research framework which incorporates all steps 1–4 (Figure 1). It consisted of three main elements: cultural diversity, biological diversity, and biotic features and grey infrastructure, which aimed to explore and better conceptualize cemeteries as elements of UGI, reveal a set of ES provided by cemeteries, and investigate different ways of using cemeteries in daily life and visitors' activities, so as to understand people's opinions about these activities. This should enable a better understanding of the current role of the cemeteries within the urban fabric, not only for GI planners and scientists, but also for neglected groups of people who need other types of recreation which cannot be provided by certain crowded public parks or gardens.

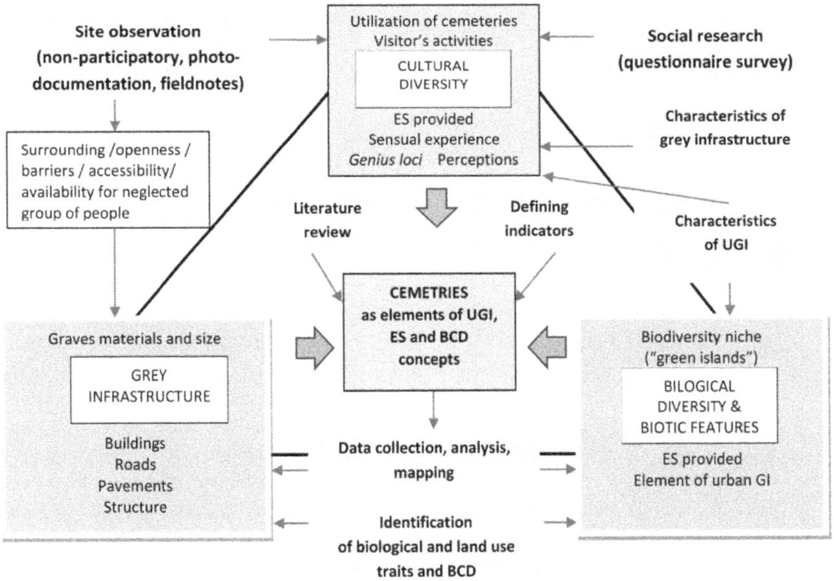

Figure 1. Research framework. Source: authors.

2.1. Study Area

We studied a sample of four multi-confessional cemeteries—two in Leipzig (Saxony, Eastern Germany), and two in Łódź (Łódzkie Voivodeship, Central Poland). The cities are similar due to their historical background as cities with factories which developed in the 19th century (Figure 2).

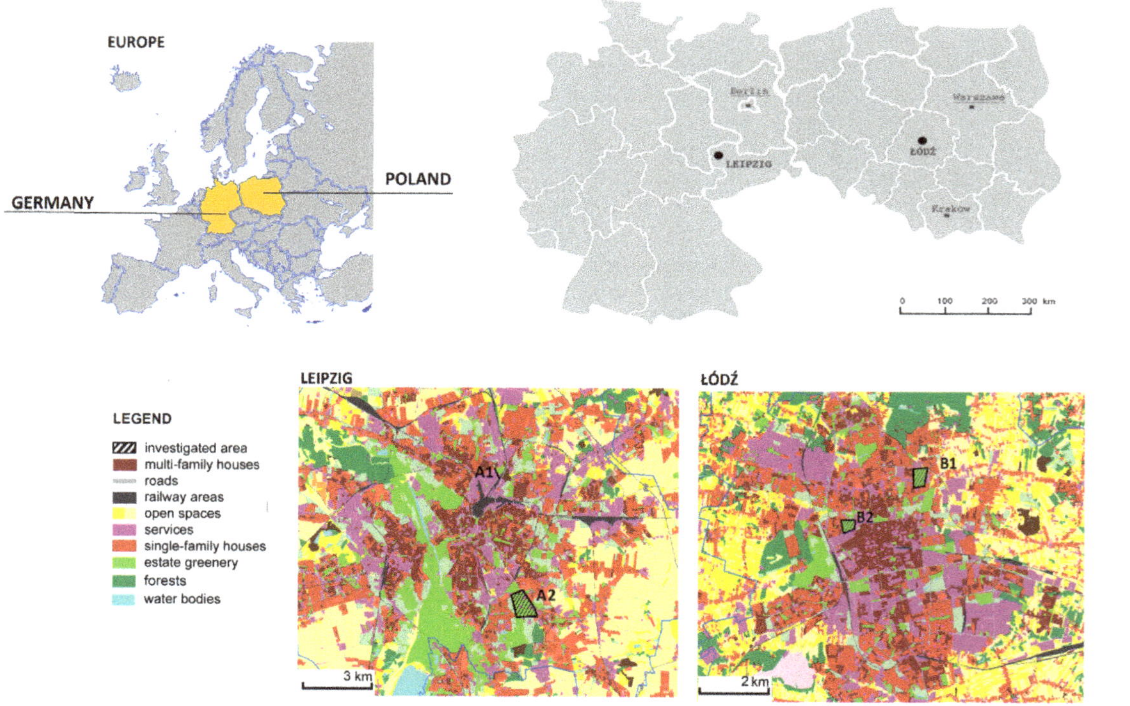

Figure 2. Study areas and site locations in Leipzig, Germany (A1 and A2) and Łódź, Poland (B1 and B2). Author: A. Długoński, based on [28,29].

Two different cemeteries in Leipzig, South Cemetery (Südfriedhof) and Old Jewish Cemetery (Der alte jüdische Friedhof), were chosen for the pilot study, to compare with Ogrodowa Street Cemetery (Stary Cmentarz) and Jewish Cemetery (Cmentarz Żydowski) situated in Łódź (Figure 3). The rationale for case sites' selection relates to several reasons. Firstly, these cemeteries have similar confessions (Christianity and Judaism), the highest recreational and tourist potential, and the most representative function of existing cemeteries in both cities. Secondly, they have strong traditions derived from other cultures and different genius loci (the spirit of the place). Thirdly, the size and shape of the areas, in addition to the same annual seasons and comparable bio-regions, provided important constants, which are key in such a comparison. For each city, two cemeteries were selected which are different in terms of their form and size, and canopy cover (high/low), and thus show a maximum variability. The selected cemeteries are unique, have multicultural characteristics, and are places of memory or remembrance of important people who were directly involved and have contributed to the creation of these two similar cities in Central Eastern Europe in their time. Moreover, the selection of the study sites included practical aspects such as good accessibility for the researchers and the experience of working in this study area, and thus the discussion benefited from this embedded knowledge. Cultural familiarity of the researchers with the religions of the cemeteries enabled the field study to be sensitively conducted, acknowledging all ethical aspects and tacit behavioral rules at the sacred places.

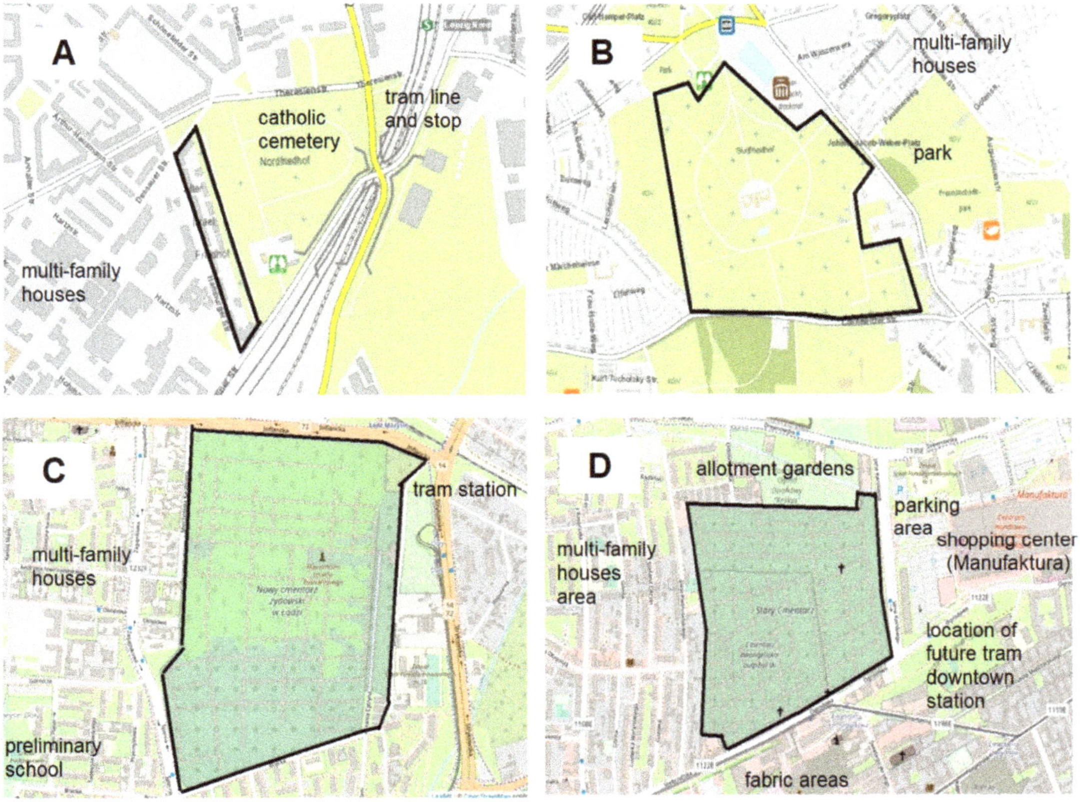

Figure 3. Maps with location of case studies (black line—boundary of given cemetery) with neighborhood: (**A**)—Old Jewish Cemetery in Leipzig (Germany), (**B**)—Südfriedhof Cemetery in Leipzig (Germany), (**C**)—Jewish Cemetery in Łódź (Poland), (**D**)—Ogrodowa Street Cemetery in Łódź (Poland). Author: A. Długoński, based on [28].

The analyzed cemeteries are located in densely built-up cities. Südfriedhof Cemetery in Leipzig is located in the southern part of the city, and the Old Jewish Cemetery in the city center. Ogrodowa Street Cemetery is in the central part of Łódź city, and the Jewish Cemetery is in the northeastern part of this city. Although in the past, cemeteries were located on the outskirts of cities, with time the cities grew, and cemetery facilities began to adjoin multi-family buildings, industrial and communication infrastructure, and less often with open areas. Currently, all analyzed objects are surrounded by multi-family buildings, some of which are located in close proximity to other green areas (allotment gardens—Ogrodowa Street Cemetery; or parks—Südfriedhof; or other religious cemeteries—Old Jewish Cemetery), in addition to important communication routes (main city roads—both Jewish or railroad cemeteries—Old Jewish Cemetery, Ogrodowa Street Cemetery) thus providing easy access or travel to visiting people (Figures 2 and 3).

2.2. Data Collection

On the basis of TEEB classification of ecosystem types [30] and the framing provided by Common International Classification of Ecosystem Services (CICES) V5.1 [31], we elaborated the classification scheme of cemeteries' ecosystem services' research, extending it by our own developed approach which was applied by us in previous studies of ES of different urban green areas [1,25,32,33]. Because this study is geared toward fundamental research in the (new) overlapping field that incorporates the concepts of UGI, BCD, and

ES [1,3,24,25,30,31,34] for green spaces, we used a literature review on ES provided by urban cemeteries and analyses of data on UGI based on different techniques developed and applied by us in previous research [35–42]. Other methodological approaches such as site observation, photo documentation, field notes, and non-participatory observation, were also applied [43–45].

2.2.1. Literature Review on Ecosystem Services Provided by Cemeteries

Using the above-mentioned framework adapted from [32,33], we explored the potential ecosystem services that may be provided by cemeteries in general, and later linked them to the ES supply by the selected cemeteries. It focused on provisioning (e.g., water for drinking, cultivated plants), regulating (e.g., mediation of wastes, flood protection) and cultural ecosystem services (e.g., experiential and physical use, education), although acknowledging that these are underpinned by "supporting" conditions (e.g., primary production). We examined the potential ecosystem services provided by cemeteries through a review of the literature related to cemetery design, management and use, further spatial analysis of the OS Open Greenspace data, and surveys of four studied cemeteries in Leipzig and Lodz. The results of the literature review are further provided in the results section and linked to the supply of ecosystem services by the selected cemeteries (Table 1).

2.2.2. Site Observation Using Non-Participatory Observation, Photo Documentation and Field Notes

Based on the approach of Bryman 2016 [43], site observation was first used to form a research hypothesis, and later, to gain data concerning a burial practice and visitors' behavior. This method was chosen as most probable to provide the initial understanding of cemeteries' functions. Furthermore, observation of the sites provided an understanding of how selected burial sites looked, and observation of people visiting the sites provided information on how these burials were treated within German and Polish societies. Observation of these cemeteries was conducted as non-participatorily and non-obstructively as possible. Field notes were taken immediately on the site, and were expanded upon in the evening of the same day, in addition to later being discussed. The notes included a description of location, structure of visitors, the purposes and length of visits, date and time of the day, characteristics of burial ground and its design, etc.

Photo documentation and field notes were also used to explore the local context before starting the site observation and survey, but later they helped to support (and in some cases better interpret or provide more details to) the results from the questionnaire survey and spatial analysis. In particular, we applied the approach of Emerson et al. (2011) [44] to writing ethnographic field notes in order to produce data that could be analyzed to gain insight and advance our knowledge. Following the approach of Holm (2014) [45], we conducted visual documentation by creating images that documented and answered specific research questions.

2.2.3. Questionnaire Survey and Pre-Survey

Survey data were collected using an on-site questionnaire distributed simultaneously in four selected cemeteries of Leipzig (Südfiedhof Cemetery and Old Jewish Cemetery) and Łódź (Ogrodowa Street Cemetery and Jewish Cemetery) in June–August 2019. As a starting point for the survey, pre-surveys were conducted by the authors in April–September 2017 and 2018 in both case study cities [26,27]. The sample was restricted to visitors of the selected cemeteries. This pre-survey not only helped to explore the key issues of cemeteries as an element of UGI, but also provided an important conceptual starting point for our empirical research and to describe trajectories for future development.

A total of 122 individuals (37 men and 41 women from Łódź; 18 men and 26 women from Leipzig) participated in the questionnaire survey in the summer of 2019.

The survey consisted of two parts. In the first part, the respondents were asked their basic data (gender, age, education, marital status) in order to assign them to a group of respondents. In the second part of the questionnaire, consisting of five main questions written in the respective local language (German or Polish), respondents were asked about the uses of cemetery spaces. This part of the questionnaire focused on their viewpoints of the utilization of the cemetery space and the visual impact of the cemetery on people. However, the survey also aimed to explore the perceptions of people who lived in the surrounding area of those cemeteries, because they could provide everyday observations and perspectives to those cemeteries. All questions were provided with several fixed responses (from three to five options to select) and explanation where needed. The questions included the issue of cemetery use, its further development and management, potential for the cemetery to become a space for recreation, and opportunities cemeteries can provide for recreation (Figure 4).

Questionnaire survey about the uses of cemetery spaces **Date of survey**________________

Site ID__________ **Name of cemetery)**________________ **Surveyor**________________

- We focus on viewpoints of utilization of cemetery space and visual impact of cemetery to people;
- We want to learn the perceptions of people who live in the surrounding area of those cemeteries because they can provide everyday observations and perspectives to those cemeteries.

1. What was the reason for their your visit to the cemetery?

☐ visit the graves of their relatives ☐ practicing tourism, walking and admiring given cemetery area

☐ practicing tourism, walking and admiring given cemetery area ☐ making other things there: "participation in a funeral, cleaning the graves, observing nature and sculptures, re-search etc."

2. What is your opinion on management in cemeteries?

☐ Lack of knowledge about management of urban cemeteries ☐ Improper plant species selection

☐ Lack of organizational efficiency ☐ They are well managed ☐ Lack of integration with other organizations or communities of interest ☐ They are neglected places

3. What kind of recreational uses can cemeteries provide?

☐ ***Potential recreational Uses*** " I observe many people use cemetery spaces but I never go there. "Maybe anniversary events...it been around a long time...I think we should really start to think to put another building or houses in there....Maybe a church."

☐ ***Not sure:*** "it does have historical significant that definitely can be... the cemetery is like a preserved area that they can built buildings on it."

☐ ***No recreational use:*** "I rather disagree with the idea that cemeteries should be used for recreational uses, since it's not about recreational. ...It is a place of absolutely silence. Places between the party zone and where I live...it creates some kind of buffer in the middle of the town.

4. Should cemeteries be part of town's recreational inventory?

☐ **No:** "I don't think they would. In our culture, we think cemetery is a place to respect and designated only for funerary uses. If we go there, it would be a reason to see our relatives. But never thought about it can be a place to go have types of activities or celebrations."

☐ **Dobts:** "I don't know what can be there. This is a place to respect and sacrum sphere ...."

☐ **Not sure:** "I would rather deny as cemetery can be a part of recreational inventory as long as it is used as a cemetery but for walking purpose it is good for people live nearby.

☐ **Not needed anymore:** "I would rather deny as cemetery can be a part of recreational inventory as long as it is used as a cemetery but for walking purpose it is good for people live nearby.

☐ **Yes, due to lacking other recreation sites:** "I think this cemetery should be relocated or removed in this area, he notes that "for the type of people live here, I didn't think it is a place that people will go there have activities..."

5. Is this cemetery accessible to you?

☐ **Yes:** "you can see a lot of cars driving through it, also people walking"

☐ **Not sure:** "Cemeteries had difficulties with finding not sure with entrance to the cemetery, ...It it problem to find place for parking or way how to getting by public transport... I do not expect all rules for being in the area of cemetery..., I am aware about theand open hours problematic."

☐ **No:** "The cemetery in not accessible form me..." I do not want to go there because the given cemetery is unpleasant or horrible and overwhelming space for me"

Figure 4. Questionnaire template (Source: authors).

2.3. Data Analysis

2.3.1. Statistical Analyses

Statistical analysis using the chi-quadrat test was undertaken to analyze the data of the questionnaire survey conducted during the pilot study, in order to determine future spatial and societal trends in urban development, and to draw conclusions. All selected data are expressed in tables showing the number and percentage of responses by visitors in each nation (separately for Poles and Germans). Statistical comparisons of response rates between respondents by nationality in selected questions (Q1–Q5) of the questionnaire survey were performed using the chi-quadrat test. Values of $p \leq 0.05$ were considered to be statistically significant. All statistical analyses were performed with Statistica 13.0 (statSoft) software [46].

2.3.2. Spatial Analyses

Maps illustrating the location of selected cemeteries in the neighborhood and foreground of a given city were elaborated based on geographic information systems (GIS) in QuantumGIS (version 3.12) software (Mountain View, CA, USA), and OSMStandard by OpenStreetMap GIS portal [28] and Urban Atlas [29].

3. Results from Pilot Study

3.1. Ecosystem Services Provided by Urban Cemeteries

The observations showed that the selected cemeteries from Leipzig and Łódź have specific recreational, cultural and natural potential that differ from other urban green spaces. The explanation for this is their urban location: the selected cemeteries are typically located within the cities and are (as in the case of Südfriedhof and Ogrodowa Street cemeteries) larger and older than many municipal parks. From this urban perspective, cemeteries play a key role in contributing to the UGI and delivery of a wide range of ES (Table 1). Not underestimating the value of urban cemeteries to deliver provisioning ES, in our pilot study we mostly focused on their supporting, regulating and cultural ES. Cemeteries have multiple specific biodiversity, and serve as habitats for unique species of urban flora and fauna (supporting ES) (see Section 3.2). This is mostly due to the fact that, in comparison with other green spaces in Leipzig and Łódź, selected cemeteries are not such dynamic landscapes: once constructed, they remain as burial space with appropriate low-impact management and maintenance in naturalistic style, without radical change (they were never transferred to a built-up or sealed area). The selected cemeteries also make substantive contribution to delivering a range of regulating ES (e.g., climate regulation, habitat sphere, and green corridors) that help to mitigate the effects of urban heat island, flooding, air and water quality, and loss of biodiversity, which was revealed within the previous research of the authors [36–39,41,42]. Our pilot study also revealed a wide scope of cultural ES delivered by cemeteries (see Section 3.3). Among them are the therapeutic role of cemeteries as places for psychological regeneration and for rebuilding the body spiritually, which seems to be especially important in hectic or dense urban spaces such as Leipzig and Łódź, where there is a high stress level during/after working hours and lack of place for (self-)reflection. They are areas with historical significance, telling the history of culture and religion and people who developed and built the historical heritage objects in both cities. This creates an opportunity for developing thanatourism (sightseeing of cemeteries), observed especially strongly in both Jewish cemeteries, as a narrow range of tourism. The selected cemeteries were also characterized as sites with multiple diversity, with a cultural display for visitors and offering selected recreational activities. Within the pilot study, we realized that cemeteries are important spaces for "passive" recreation (e.g., to accommodate low-impact activities such as walking or bird watching, or provide an area for visual release) for some specific (neglected) social groups who need places for silence and self-reflection which cannot be provided by urban parks, gardens, etc.

Table 1. Characteristics of ES provided by selected cemeteries in Leipzig and Łódź considering the results of a literature review on the respective ES of cemeteries. Legend: A—Old Jewish Cemetery in Leipzig (Germany), B—Südfriedhof Cemetery in Leipzig (Germany), C—Jewish Cemetery in Łódź (Poland), D—Ogrodowa Street Cemetery in Łódź (Poland).

ES Category	ES Characteristics	Studies Confirmed	A	B	C	D
supporting	provide habitat for unique species of urban flora and fauna	[6–10,12]	x	x	x	x
	soil formation, photosynthesis, primary production, nutrient and water cycling	[16,47]	x	x	x	x
provisioning	landscape of consumption and use which is incrementally changing with each new burial	[5,13,20]	x	x	x	x
regulatory	help to mitigate the effects of the urban heat island, flooding, air and water pest and disease	[16,47,48]	x	x	x	x
cultural	provide special type of recreation	[5,11,14,18,19,22]	x	x	x	x
	potential benefits for mental health and well-being	[14,16,22,47]	x	x	x	x
	role as sacred places	[5,17,20,49]	x	x	x	x
	help in preserving and enhancing the character and cultural identity of the cemetery landscape	[14,15,47,50]	x	x	x	x
	tell diverse stories of the city and represent intangible notions of the character of giving place	[15,19,51–54]	x	x	x	x
	cognitive development, spiritual enrichment, educational-civic function	[17,48]	x	x	x	x
	aesthetic value and experience	[17,47,52,55,56]	x	x	x	x
multiple	type of urban green providing a variety of ES, place of biological and cultural diversity	[3,4,47,50,56]	x	x	x	x
ES disservices	"disservices" such as allergens, invasive/dangerous/poisonous species, and the degradation of groundwater quality negative social perception of cemetery as a space for recreation and the ecosystem in which the cemetery exists	[16,47,50] [51–58]	x	x	x	x

Legend: x—the occurrence of a given feature was observed in the given cemetery.

To sum up, all four cemeteries in Łódź and Leipzig, regardless of their multidimensional diversity and several characteristics (size, history, use intensity, structural diversity, etc.), deliver all categories of ES, which leads to the conclusion that they offer fundamental ecological functions in terms of the benefits they provide to society as human well-being, which are similar, in this term, to other areas of UGI (Table 1).

3.2. *Wild Nature in Cemeteries*

All four selected cemeteries contain nature elements, including: old symbolic and valuable trees and shrubs (e.g., birch, lime, ash, spruce, yews, etc.) and creepers (e.g., ivy); different animals (e.g., squirrels, bugs, wildcats, hedgehogs, etc.), birds (e.g., thrushes, woodpeckers and towers, etc.) and insects (e.g., beetle, ants, etc.); and cultural–historical elements (gravestones, sculpture). Because of the high potential to serve as refuge areas (e.g., surface for retreat) for flora and fauna in terms of biodiversity, they provide cities with habitat for native wildlife (Figure 4). The landscape of the Südfriedhof is structured, and resembles a city park. It is similar to the Ogrodowa Street Cemetery, where trees are planted with avenues modelled on the park cemetery, such as the French Père-Lachaise Cemetery. A different situation was observed in both Jewish cemeteries. Cmentarz Żydowski in Łódź is dilapidated and densely overgrown, with an undergrowth of old trees and ruderal vegetation that overgrow cemetery quarters. Only a small part of the Jewish Cemetery in Łódź is covered with grass (ghetto area), which is closer to the design of the Old Jewish Cemetery in Leipzig (Figures 5 and 6).

Figure 5. Wildlife and tombstones in Jewish cemeteries in Łódź and Leipzig. Legend: (**A**)—sequel, (**B**)—two examples of insect groups, (**C**)—creepers *Hedera helix*. Photographs: A. Długoński.

Figure 6. Views on case studies (site observation): (**A**)—Old Jewish Cemetery in Leipzig (Germany), (**B**)—Südfriedhof Cemetery in Leipzig (Germany), (**C**)—Jewish Cemetery in Łódź (small photo: ghetto area), (**D**)—Ogrodowa Street Cemetery in Łódź. Photographs: A. Długoński and D. Dushkova.

The pilot study revealed that all four cemeteries have a large variety of tree species, especially when compared with the adjacent/neighboring area. Both young and old, single and groups of trees were recorded. Tree-lined paths were also mapped. Many different hedges and shrubs were also revealed. In some parts of the Südfriedhof and Ogrodowa Street cemeteries, a densely wooded area similar to relict nature was recorded. On the partly cultivated and partly natural meadows, a large number of different ground covers and meadow plants, together with diverse spontaneous vegetation, was revealed. Along with such natural hotspots of biodiversity, in several parts of Südfriedhof and Ogrodowa Street cemeteries, a wide variety of plantings on the graves and flowerbeds was revealed, that additionally contributed to a great diversity of flora within the cemeteries' landscape. According to site observation, there was no infrastructure revealed for active recreation, for example, cycle paths, playgrounds, sports fields or tennis tables. Only infrastructure elements for low-impact recreation such as benches, pavilions, monuments, statues, memorial plaques and a mourning café were identified. Public sanitary facilities and a lighting fixture were also present in the southern cemetery. Altogether, it again underlined the contribution of burial sites to biodiversity support in urban landscapes.

3.3. Utilization of Cemeteries: Visitors' Activities and Potential for Recreation

3.3.1. Results from Site Observation

Site observation analyses of the four cemeteries showed that most activities that occurred were for visiting the graves of relatives, sightseeing (for tourism purposes or observing places of historical value such as thanatourism), walking and contemplating, spending time with children, and relaxing (Figure 7).

For the Südfriedhof Cemetery, additional activities included reading books and enjoying nature, because this landscape is similar, as stated earlier, to an urban park. Activities such as cleaning graves, cycling, sitting on benches, resting, and participation in funerals were rarely observed at Südfriedhof and Ogrodowa Street cemeteries.

We observed that Leipzig visitors mostly went to the cemetery for walking, running, inhaling fresh air, meeting friends, or contemplating. Łódź visitors, however, mostly went to the cemetery for a concrete reason (visiting the grave of relatives), or sometimes the reason for visiting was to contemplate or observe wild nature. This is due to urban planning associating the idea of cemeteries to "city parks." This issue was also apparent in Südfriedhof Cemetery in Leipzig and Ogrodowa Street Cemetery in Łódź. Germans often treat cemeteries as part of urban green public places (Südfriedhof), compared with Polish people who regard cemeteries as sacred/holy places that need to be separated from the city and protected, by maintaining silence and using only for the purpose of resting.

We also observed different activities occurring at the studied Jewish cemeteries (Cmentarz Żydowski and Old Jewish cemetery) of both cities. Observations showed that these places were visited mostly by tourists, rather than by relatives. This is also due to the historical background of citizens and past times of these sites. Therefore, it can be concluded that Jewish cemeteries are more often visited for thanatourism due to their historical and cultural meaning, or human curiosity.

In Łódź' cemeteries, some activities (cycling and running) were prohibited, whereas in Germany, more of these physical activities were not allowed but they are sometimes practiced (Südfriedhof). However, physical activities were strictly prohibited in the Old Jewish Cemetery because of culture and religion. This difference is not only due to the different cultural approaches of Christian and Jewish communities but also because of their nationalities.

The above-discussed examples of activities at cemeteries could be divided into three groups in terms of their level of effort: as active recreation, semi-active recreation, and passive recreation; and undiscovered activities that could be added to each group in terms of religion and specific development of the given cemetery Active recreation included cleaning graves, cycling, or running (not allowed but observed only in Südfriedhof cemetery as a "park place" in main transit roads of the object); semi-active recreation included thanatourism and observation of sculptures, while passive recreation included reading books, contemplating, sitting on the bench, and walking (Figure 8).

Figure 7. The most common activities carried out in the cemeteries in Łódź and Leipzig. Legend: **1**—visit to the graves of relatives, **2**—sightseeing (for tourism purposes or observing places of historical value: thanatourism), **3**—walking and contemplating, spending time with children, relaxing (on benches) Photographs: A. Długoński and D. Dushkova.

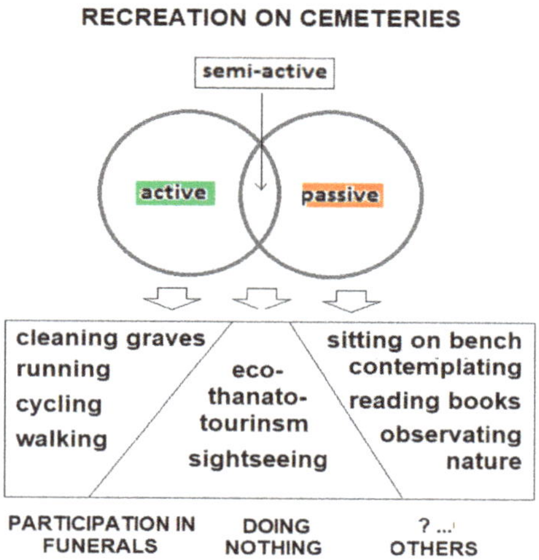

Figure 8. Relationship of overlapping fields of active, semi-active, and passive recreation in cemeteries observed during site observation research in cemeteries in Poland and Germany.

3.3.2. Results from Questionnaire Survey

As already mentioned, considering the perceptions of people, we asked them first about their gender, age, education, and marital status. In the second set of questions, we collected their opinions toward cemetery use and potential as follows:

- Q1: What was the reason for your visit to the cemetery?
- Q2: What is your opinion on the management of cemeteries?
- Q3: What kind of recreational uses do cemeteries provide?
- Q4: Should cemeteries be part of the city's recreational inventory?
- Q5: Are the cemeteries accessible to you?

According to the first question (Q1), most respondents visited the cemetery pragmatically, to place a candle on the grave of a loved one (Poles: 50 visitors, 64.1%; Germans: 26 visitors, 59.1%). Another set of visitors visited for tourist purposes such as walking, contemplating, a family trip, sightseeing (Poles: 17 visitors, 21.8%; Germans: 11 visitors, 25%) or to participate in a funeral, cleaning the graves, observing nature and sculptures, research, etc., (Poles: 11 visitors, 14.1%; Germans: 7 visitors, 15.9%) (Table 2).

Table 2. Reasons for visiting the studied cemeteries in Poland and Germany.

Q1. What was the Reason for Your Visit to the Cemetery?			
		Poles	**Germans**
A	Putting a candle	50 (64.1%)	26 (59.1%)
B	Tourist purposes	17 (21.8%)	11 (25%)
C	Others	11 (14.1%)	7 (15.9%)
	Total	78 (100%)	44 (100%)

Regarding the second question (Q2), most respondents from Poland responded that they lacked knowledge about the management of urban cemeteries (57 visitors, 73.1%). The results presented a completely different situation for Germans, who found that cemeteries were well managed in their opinion (18 visitors, 40.9%). Some Poles also paid attention to lack of organization (9 visitors, 11.5%) and integration with other organizations or communities of interest (5 visitors, 6.4%). Few found that cemeteries were well managed (2 visitors, 2.6%), or, on the contrary, they claimed that these areas were neglected places (3 visitors, 3.8%) and had improper plant selection in their management (2 visitors, 2.6%). For Germans, other replies were distributed as follows: 15 visitors (34.1%) noted that they lacked knowledge about the management of urban cemeteries, 5 visitors (11.3%) thought that cemeteries were neglected places, 4 visitors (9.1%) noted a lack of integration of interests, 1 visitor (2.3%) highlighted a lack of organization, and also 1 visitor (2.3%) noted improper plant selection in cemetery management (Table 3).

Regarding the third question (Q3), less than half of the respondents in Łódź (38 visitors, 48.7%) disagreed with the idea that cemeteries should be used for recreational uses. One third of the respondents from Łódź said that cemeteries have a historical meaning (23 visitors, 29.5%), and that they know some people who use the space for recreational purpose but they personally do not practice it (17 visitors, 21.8%). In Leipzig, 43.1% (19 visitors) claimed that they knew people who practice recreation on cemeteries, but they did not refer to them, while 36.4% of respondents (16 visitors) disagreed that cemeteries should be used for recreation, and only 20.5% (9 visitors) noted that the cemeteries have historical meaning (Table 4).

Table 3. Respondents' opinion on management aspects of selected cemeteries in Łódź and Leipzig.

Q2. What is Your Opinion on Management in Cemeteries?		
	Poles	Germans
A. Lack of knowledge about management of cemeteries	57 (73.1%)	15 (34.1%)
B. Improper plants	2 (2.6%)	1 (2.3%)
C. Lack of organization	9 (11.5%)	1 (2.3%)
D. Well managed	2 (2.6%)	18 (40.9%)
E. Lack of integration of interests	5 (6.4%)	4 (9.1%)
F. Neglected place	3 (3.8%)	5 (11.3%)
Total	78 (100%)	44 (100%)

Table 4. Respondents' opinion on selected cemeteries' usage in Łódź and Leipzig.

Q3.	What Kind of Recreational Uses do Cemeteries Provide?		
		Poles	Germans
A	I know that people use it but not me	17 (21.8%)	19 (43.1%)
B	Historical meaning	23 (29.5%)	9 (20.5%)
C	I disagree that cemeteries should be used for recreation	38 (48.7%)	16 (36.4%)
	Total	78 (100%)	44 (100%)

With regard to the recreational potential of the cemetery (Q4), half of the respondents from Germany (22 visitors, 50%) mentioned that cemeteries should be a place of recreation, mainly because of the lack of such sites in the neighborhood. However, one third of German respondents (16 visitors, 36.4%) replied "no", due to the sacral character of such places, 4 visitors (9.1%) were not sure (they doubted it was a good idea), and 2 visitors (4.5%) would rather deny, but saw some possibilities of the use of such places for walking (especially for people living nearby). In the case of Poles, two thirds of respondents (47 visitors, 60.3%) claimed "no" due to the character of such places "designated only for funerary uses", 16 visitors (20.5%) were not sure (they doubted it was a good idea), 15 visitors (19.2%) would rather deny, but saw some possibilities of the use for walking (especially for people who live nearby). Interestingly, none of the Poles answered "yes" to this question, either because of the occurrence of some recreational places in the cemetery's surrounding area (0 visitors, 0%), or that they did not think that the given cemetery should be relocated or removed from this area to provide space for organizing more activities (0 visitors, 0%) (Table 5).

Table 5. Respondents' opinion on the studied cemeteries as part of the city's recreational inventory of Lodz and Leipzig.

Q4.	Should Cemeteries be Part of the City's Recreational Inventory?		
		Poles	Germans
A	No	47 (60.3%)	16 (36.4%)
B	Doubts	16 (20.5%)	4 (9.1%)
C	May be for walking	15 (19.2%)	2 (4.5%)
D	Not needed anymore	0 (0%)	0 (0%)
E	Yes, due to lacking other recreation sites	0 (0%)	22 (50%)
	Total	78 (100%)	44 (100%)

With regard to accessibility of cemeteries (Q5), most respondents from Łódź (68 visitors, 87.2%) in addition to half of the respondents from Leipzig (22 visitors, 50%) found that cemeteries were accessible. They claimed that actually a lot of people used it ("you can see a lot of cars driving through, also people walking"). Other visitors, both from Leipzig (14 visitors, 31.8%) and Łódź (6 visitors, 7.7%) believed that they had difficulties with finding the entrance to the cemetery, or a place for parking, or how to travel by public transport, or did not expect all the rules for being in the area of cemetery, or were not aware about the opening hours. The rest of the respondents believed that the cemeteries were not accessible to them, as stated by eight visitors (18.2%) in Leipzig, and four (5.1%) in Łódź. Respondents based their opinions on the fact that they did not want to go there because the given cemetery was unpleasant, or a horrible and overwhelming space for them (Table 6).

Table 6. Respondents' opinion on accessibility of the studied cemeteries in Łódź and Leipzig.

Q5.	Are the Cemeteries Accessible to You?		
		Poles	**Germans**
A	Yes	68 (87.2%)	22 (50%)
B	Not sure	6 (7.7%)	14 (31.8%)
C	No	4 (5.1%)	8 (18.2%)
	Total	78 (100%)	44 (100%)

The next part of the survey consisted of a statistical comparison of the results of the response rates between subjects by nationality in the selected five questions (Q1-Q5), of respondents from Poland and Germany. The results of this analysis, with information about $p < 0.05$ (statistically significant value) and $p > 0.05$ (statistically insignificant value), are presented in Appendix A1 (Table A1). Regarding the first question (Q1), which related to the purpose of visiting the cemetery, the chi-square test showed no significant factor. From the analysis, we can conclude that the remaining questions of the questionnaire survey (Q2–Q5) were statistically significant. The chi-square test thus showed statistically significant values in the responses of different nationalities of the respondents, which were related to their opinions on the management, recreational uses of cemeteries, understanding cemeteries as part of the city's recreational inventory, and accessibility of cemeteries in Łódź and Leipzig.

4. Discussion

4.1. *Reflection on Pilot Study*

Although contemporary literature mostly embeds ecosystem services and functions provided by cemeteries such as biodiversity support and regulating ES, it neglects the experience-rich potential of cemeteries as cultural products. However, it has already been confirmed [5,14–16,19] that cemeteries offer both nature-based and cultural activities. It is worth noting that some specific social groups who need passive recreation which cannot be provided by urban parks, gardens, etc. can be perceived as neglected in terms of not being able to find an appropriate place for recreation. Within the urban context, cemeteries can play a key role in contributing to the UGI and delivery of a wide range of ES (supporting, provisioning, regulatory, cultural ES). It is essential to highlight the multiple benefits provided by cemeteries being a part of UGI. On one hand, they are habitats for unique species of vegetation and animals (supporting and provisioning ES); and on the other hand, they present a type of urban green, as one of the important elements of a city's greenery which provides a special type of recreation (cultural ES) [4,16,47]. Cemeteries could once again make a more substantive contribution to delivering a range of regulatory ES that help to mitigate the effects of urban heat island, air and water quality, and loss of biodiversity [47]. Cultural ES provided by urban cemeteries refers to the fact that natural burial landscape might also help in preserving and enhancing the character and cultural identity of the cemetery landscape [15,19,48,58]. Whilst the potential benefits for mental health and well-being that can be derived from urban green spaces are well documented [59,60], only a few research papers focus on the unique opportunity provided by the urban cemetery to deliver cultural ES, especially those supporting human health and well-being [14,16,21–23,55]. It is important to analyze what qualities correspond to societal changes and landscape design history [24,25]. At the same time, the research often relates to burial landscape as nature, architecture and art, while individuals' use of the cemetery as a place for memory and meaning-making remains overlooked. Moreover, it would be of great importance to explore how they do relate to the fact that death becomes an area for negotiation between different cultural, religious and individual opinions or needs, and social norms/rules, especially in relation to the intentions for recreational use [48–56]. Thus, further research is needed in order to highlight certain tendencies in place-making strategies.

4.2. Cemeteries as Places of Multiple Diversity and Mental Barriers

The results indicate that urban cemeteries now witness a variety of usages and are not limited to commemoration practices. However, the social acceptance of nonconventional activities on cemetery sites is still debatable [51,52,54,55]. Further research on this issue should investigate different ways of using cemeteries that are clearly evident to be part of the cemeteries' daily life, to explore different activities, and to understand people's opinions about these activities (as was explored within the pilot study). There is also a need for research to prove the functions of urban cemeteries in comparison with other elements of UGI and their potential for accommodating a variety of functions. In the context of increased urban development, a better understanding of the current role of cemeteries within the urban fabric appears to be highly relevant not only for UGI planners and scientists, but also for the neglected groups of people who need other types of recreation that cannot be provided by crowded public parks or gardens. This includes cultural ES that gives potential benefits to users in terms of mental health and well-being. Additionally, cemeteries also play a role as sacred places with "spatial vessels of civic identity, telling diverse histories of the city, and representing intangible notions of the characteristics of a given place" [19].

The pilot study presented in the article shows that cemeteries might be understood as "slow places" in hectic cities. First of all, they have high potential as refuge areas (e.g., place for retreat) for flora and fauna in terms of biodiversity, and they provide cities with habitat for native wildlife. Secondly, cemeteries are repositories of natural and cultural diversity, and they are also unique places with their own ecological sanctuary and an ecological niche. Therefore, they serve as a "peace of mind" refuge area for humans who visit the graves of relatives; for tourism goals or so called thanatourism (e.g., observing historical values of city and past city's citizens' history) [3]; for spending time with family—relaxing and enjoying urban fauna; and also for those people (e.g., elderly, disabled, depressive, or hypersensitive individuals) who are looking for perfect peace and silence amidst the buzz and noise of a city area. All these factors make those places ecologically, culturally and historically different than other urban green spaces such as parks, pocket parks or forests [4,16,23].

In Łódź city, some activities (cycling and running) are prohibited in cemeteries, whereas in Germany, more physical activities are also not allowed but are acceptable (Südfriedhof). However, physical activities are prohibited in both Jewish cemeteries because of its culture and religion. This difference is not only due to different cultural approaches of Christian and Jewish communities but also because of their nationalities. Germans often treat cemeteries as part of urban green public places (Südfriedhof), compared with Polish people who perceive cemeteries (Ogrodowa Street Cemetery) as sacred/holy places that need to be separated from the city and protected by maintaining silence and using only for the purpose of resting.

As mentioned earlier, cemeteries make a fascinating cultural display for visitors by offering both nature-based and cultural activities [19]. They present one of the important elements of a city's greenery that provide a special type of recreation. This recreation is closer to passive (low-impact) recreation, which differs from jogging, picnic or ball games, but mostly relates to observing nature and exploring the history of the place, contemplation and sightseeing (thanatourism) [3]. On one hand, cemeteries are specific areas with certain rules, but have barriers that enclose visitors to spend more time in these areas. All of the studied cemeteries were available at certain times, which organized and imposed certain activities on the cemetery. Even more difficult were the (Old) Jewish cemeteries in Leipzig and Łódź, which were much more difficult to access than the Christian cemeteries (Südfriedhof and Ogrodowa Street cemeteries). As a rule, their entrance was from a side street and was not exposed. Some, such as Jewish cemeteries, have special rules, usually determined by rules or religions (e. g. obligatory wearing of headgear for male visitors of cemetery). On the other hand, they are places of great potential for a selected group of people (neglected people) who, due to difficult urban conditions or noise pollution in other green areas such as parks, forests and boulevards, need sacred tranquility or silence. However, these are only places for people who want to relax and enjoy this specific green space in peace and silence (Figure 9).

Figure 9. Barriers and restrictions to access the cemeteries in Łódź and Leipzig: Legend: (**A**,**B**)—cemetery regulations in Südfriedhof, (**C**)—side entrance to the premises of the Jewish Cemetery in Łódź, (**D**)—opening hours of the Old Jewish Cemetery in Leipzig, (**E**)—obligatory wearing of headgear by men at Jewish cemeteries with regard to rules and religion. Photographs by: Długoński A., Dushkova D.

4.3. Synthesis of Questionnaires Survey

The questionnaire survey showed some differences in the perception of the cemetery space by Poles and Germans. Poles treat cemeteries as a sphere of the sacred, while Germans more often see these places as recreational and touristic places of high historical and cultural importance. The knowledge about cemetery management and belonging to the green infrastructure system is much broader among Germans. Most of the respondents from Poland (73.1%) had no knowledge about the management of the cemetery, while Germans (40.9%) found that cemeteries were well managed. In addition, there were minority responses both from Germans and Poles that the cemeteries were neglected, with lack of integration of interests or usage of improper plants in land use of the given area. The users of the cemeteries in Łódź indicated that the historical cemeteries, although still in use, were already used as recreational places (19.2%), mainly for walking purposes or watching/observing nature. In Germany, the proportion was 4.5%. In total, 29.5% of Poles and 20.5% of Germans drew attention to the historical value of cemeteries. The proportion of respondents from Łódź who did not see a recreational potential of historical cemeteries was much higher (48.7%), compared with only 36.4% of respondents from Leipzig. Cemetery users in Poland and Germany were also asked if they thought the cemetery area should be integrated into UGI and used for recreational purposes. The majority of the respondents recognized cemeteries as memorial sites and not places for any activity other than walking, contemplating and sightseeing/leisure activities.

4.4. Possibilities for the Future Development and Challenges in Research on Cemeteries

Understanding how the different elements of UGI and especially urban ecosystems presented by cemeteries work in addition to multifunctionality and multiculturality provided by urban ecosystems is an important issue for both science and management/practice of cemeteries. With regard to research on cemeteries, there is acknowledgement that they perform important functions in personal, familial and community life [4,5,16,19]. However, this issue of social function needs to be examined in more detail in the direction of current policies (for maintenance, security and possible re-use) that sustain the nature of cemeteries and contribute to the renewal of cemetery life [47–50]. All this requires a careful understanding of the historical, cultural, religious, legislative, economic, but also ecological and social roles of cemeteries, for a wide range of visitors, and an issue of public acceptance here, plays an important role [51,52,54,55]. Moreover, exploring the visitor (user) perspective can provide first-hand understanding of underlying meanings and uses of cemeteries for different cultural and ethnic groups, for which the cemetery landscape is ever present as the material outcome of sets of interests.

The results of this study can be used as a background and essential element for elaboration of the framework/guidelines to inform urban management and decision makers from the field of urban cemeteries, on how to best integrate the natural environment and human needs, by applying the methodical approach from the concepts of UGI, ES and BCD. For this purpose, the next step will be to broaden the research including more cemeteries in the two cities, to repeat the study and to make the findings more robust, e.g., to uncover other categories of ES provided by cemeteries and identify further trends in cemetery use, by analyzing the preferences among users (local residents and touristic visitors) in terms of the perception, use practices and management of UGI. Based on this fundamental knowledge on the interrelations of ES provided by cemeteries and the identification of particular types of cemetery use practices, we could implement a more in-depth study to determine ecological and cultural properties and performance of ES of urban cemeteries for cities, in addition to exploring how these types are allocated and distributed in urban space, and if the differences within a city are more than those between studied cities. Another direction for future research can be seen in broadening the choice and sample of test sites including other cities and other religions (we had already started with one Orthodox and one Muslim cemetery in Berlin, some years ago, as a test). Using an innovative approach (multi-mixed method of cemeteries' assessment as suggested by research framework, Figure 1) can bring together academia, city governance, and civil society, aiming towards the co-production of usable knowledge in the form of a transferable framework that can guide managers through the processes of evaluating sociocultural and biophysical conditions, determining desired future conditions, and assessing how to progress from the present to desired future conditions through collaborative creation and implementation of a management plan for future cemetery use.

5. Conclusions

Cemeteries are an important element of cultural heritage and landscape of cities, and have a specific biodiversity potential for native species and respective niches.

1. Cemeteries are a space of silence and contemplation in dense cities;
2. Cemeteries have specific ecological potential and deliver a bundle of ecosystem services that differentiates them from other urban green spaces;

3. Cemeteries serve also specific groups of population, in particular, neglected people and 'invisible groups' such as elderly, disabled, depressive, or hypersensitive individuals who:
    a. sometimes do not find space in other types of UGI;
    b. are searching for safe and silent places in dense cities, because they are significantly different from groups that use other green spaces within the city in terms of multifunctionality, and multiculturalism being a characteristic for urban cemeteries;

Biocultural diversity of cemeteries of different religions (Christian, Jewish) differs considerably within one city, in addition to land cover and management.

4. The perception of cemeteries by nationalities varies and depends on the level of secularization of the society. For Poles, a cemetery is mostly a sacrum sphere, for Germans, it is an open place with remembrance of relatives but also some selected recreational and touristic opportunities;
5. Cemeteries are thus far undervalued as public green resources that can perform important functions in the sociocultural life and mental well-being of the general public; however, this issue of social function needs to be examined in more detail in the direction of current policies. Users' perspectives related to cultural practices, management aspects, or usage are of special importance for sustainable living in ever more multicultural societies;
6. As land demand in cities everywhere increases, cemeteries are considered as an open space to accommodate passive recreational activities such as walking, nature observation or sightseeing, and provide an area for visual satisfaction;
7. Further research on urban cemeteries should:
    a. Highlight certain tendencies in place-making strategies based on the fact that the cemetery is an area for negotiation between different cultural, religious and individual opinions or needs;
    b. Specify the role of cemeteries to safeguard local (native) biodiversity, act as specific niches for local ecosystems and thus a local gene pool;
    c. Investigate different ways of using cemeteries which are already a clearly evident part of the cemeteries' daily life, to explore different activities, and to understand people's opinions about these activities;
    d. Determine the desired future conditions of urban cemeteries, and assess how to progress from the present to the desired future conditions through collaborative creation and implementation of a management plan.

Author Contributions: Conceptualization, A.D., D.D. and D.H.; methodology, A.D., D.D. and D.H.; software, A.D.; validation, A.D., D.D. and D.H.; formal analysis, A.D. and D.D.; investigation, A.D. and D.D.; resources, A.D.; data curation, A.D.; writing—original draft preparation, A.D. and DD.; writing—review and editing, A.D., D.D. and D.H.; visualization, A.D. and D.D.; supervision D.H.; project administration, A.D.; funding acquisition, A.D. All authors have read and agreed to the published version of the manuscript.

Funding: This research received no external funding.

Institutional Review Board Statement: Not applicable.

Informed Consent Statement: Informed consent was obtained from all subjects involved in the study.

Data Availability Statement: The database for the questionnaire survey is synthetically discussed in the article. The sources of elaboration on the database were based on the visitors' opinions and ratings related to given case studies from the questionnaire survey during conducted field studies in cemeteries in Łódź and Leipzig. All visitors' opinions from the above-mentioned sources are automatically processed to detect inappropriate content during the database elaboration and conducting of statistical analyses. Hence, the greatest value to the development of the authors' database on investigated areas was the honest and objective content posted.

Acknowledgments: We thank Justyna Marchewka from the Department of Human Biology, Institute of Biological Sciences, Faculty of Biology and Environmental Sciences, and Cardinal Stefan Wyszynski University in Warsaw, for help in statistical analysis. The authors would like to thank the reviewers for all useful and helpful comments on our manuscript.

Conflicts of Interest: The authors declare no conflict of interest.

Appendix A

Table A1. Comparison of response rates between respondents by nationality (chi square test).

Question	Answer	Poles	Germans	Chi Square
Q1	Putting a candle	50 (64.1%)	26 (59.1%)	0.8600 [2]
	Tourist purposes	17 (21.8%)	11 (25%)	
	Others	11 (14.1%)	7 (15.9%)	
Q2	Lack of knowledge about management of cemeteries	57 (73.1%)	15 (34.1%)	0.0000 [1]
	Improper plants	2 (92.6%)	1 (2.3%)	
	Lack of organization	9 (11.5%)	1 (2.3%)	
	Well managed	2 (2.6%)	18 (40.9%)	
	Lack of integration of interests	5 (6.4%)	4 (9.1%)	
	Neglected place	3 (3.8%)	5 (11.3%)	
Q3	I know that people use it but not me	17 (21.8%)	19 (43.1%)	0.0449 [1]
	Historical meaning	23 (29.5%)	9 (20.5%)	
	I disagree that cemeteries should be used for recreation	38 (48.7%)	16 (36.4%)	
Q4	No	47 (60.3%)	16 (36.4%)	0.0000 [1]
	Doubts	16 (20.5%)	4 (9.1%)	
	May be for walking	15 (19.2%)	2 (4.5%)	
	Not needed anymore	0 (0%)	0 (0%)	
	Yes, due to lacking other recreation sites	0 (0%)	22 (50%)	
Q5	Yes	68 (87.2%)	22 (50%)	0.0000 [1]
	Not sure	6 (7.7%)	14 (31.8%)	
	No	4 (5.1%)	8 (18.2%)	

Note: Q1. What was the reason for your visit to the cemetery?; Q2. What is your opinion on management in cemeteries?; Q3. What kind of recreational uses do cemeteries provide?; Q4. Should cemeteries be part of the city's recreational inventory?; Q5. Are the cemeteries accessible to you? [1] If $p < 0.05$, the association is significant (which implies that the analyzed factor influences the phenomenon being investigated). [2] If $p > 0.05$, the association is not significant.

References

1. Andersson, E.; Langemeyer, J.; Borgström, S.; McPhearson, T.; Haase, D.; Kronenberg, J.; Barton, D.N.; Davis, M.; Naumann, S.; Röschel, N.L.; et al. Enabling Green and Blue Infrastructure to Improve Contributions to Human Well-Being and Equity in Urban Systems. *BioScience* **2019**, *69*, 566–574. [CrossRef] [PubMed]
2. Elmqvist, T.; Andersson, E.; McPhearson, T.; Bai, X.; Bettencourt, L.; Brondizio, E.; Colding, J.; Daily, G.; Folke, C.; Grimm, N.; et al. Urbanization in and for the Anthropocene. *npj Urban Sustain.* **2021**, *1*, 6. [CrossRef]
3. Sallay, Á.; Mikházi, Z.; Gecséné Tar, I.; Takács, K. Cemeteries as a Part of Green Infrastructure and Tourism. *Sustainability* **2022**, *14*, 2918. [CrossRef]
4. McClymont, K.; Sinnett, D. Planning Cemeteries: Their Potential Contribution to Green Infrastructure and Ecosystem Services. *Front. Sustain. Cities* **2021**, *3*, 789925. [CrossRef]
5. Nordh, H.; Evensen, K.H. Qualities and functions ascribed to urban cemeteries across the capital cities of Scandinavia. *Urban For. Urban Green.* **2018**, *33*, 80–91. [CrossRef]
6. Kamran, S.; Khan, S.M.; Ahmad, Z.; Ur Rahman, A.; Iqbal, M.; Manan, F. The role of graveyards in species conservation and beta diversity: A vegetation appraisal of sacred habitats from Bannu, Pakistan. *J. Forest. Res.* **2020**, *31*, 1147–1158. [CrossRef]
7. Morelli, F.; Mikula, P.; Benedetti, Y.; Bussière, R.; Tryjanowski, P. Cemeteries support avian diversity likewise urban parks in European cities: Assessing taxonomic, evolutionary and functional diversity. *Urban For. Urban Green.* **2018**, *36*, 90–99. [CrossRef]
8. Kowarik, I.; von der Lippe, M.; Fischer, L.K. The ecological heritage in the Weißensee. Weißensee cemetery: Nature conservation versus monument preservation? *ICOMOS J. Ger. Natl. Comm.* **2011**, *53*, 46–51.

9. Kowarik, I.; Buchholz, S.; von der Lippe, M.; Seitz, B. Biodiversity functions of urban cemeteries: Evidence from one of the largest Jewish cemeteries in Europe. *Urban For. Urban Green.* **2016**, *19*, 68–78. [CrossRef]
10. Tryjanowski, P.; Morelli, F.; Mikula, P.; Krištín, A.; Indykiewicz, P.; Grzywaczewski, G.; Kronenberg, J.; Jerzak, L. Bird diversity in Urban Green space: A large-scale analysis of differences between parks and cemeteries in Central Europe. *Urban For. Urban Green.* **2017**, *27*, 264–271. [CrossRef]
11. Nordh, H.; Evensen, K.; Skår, M. A peaceful place in the city–A qualitative study of restorative components of cemetery. *Landsc. Urban Plan.* **2017**, *167*, 108–117. [CrossRef]
12. de Lacy, P.; Shackleton, C.M. Woody plant species richness, composition and structure in urban sacred sites, Grahamstown, South Africa. *Urban Ecosyst.* **2017**, *20*, 1169–1179. [CrossRef]
13. McClymont, K. That eccentric use of land at the top of the hill': Cemeteries and stories of the city. *Mortality* **2016**, *21*, 378–396. [CrossRef]
14. Skår, M.; Nordh, H.; Swensen, G. Green urban cemeteries: More than just parks. *J. Urban. Int. Res. Placemaking Urban Sustain.* **2018**, *11*, 362–382. [CrossRef]
15. Francis, D. Cemeteries as cultural landscapes. *Morality* **2003**, *8*, 222–227. [CrossRef]
16. Długozima, A.; Kosiacka-Beck, E. How to enhance the environmental values of contemporary cemeteries in an urban context. *Sustainability* **2020**, *12*, 2374. [CrossRef]
17. De Lacy, P.; Shackleton, C. Aesthetic and Spiritual Ecosystem Services Provided by Urban Sacred Sites. *Sustainability* **2017**, *9*, 1628. [CrossRef]
18. Tanaś, S. *Przestrzeń Turystyczna Cmentarzy. Wstęp do Tanatoturystyki. [Tourist Space of Cemeteries. Introduction to Cemeteries Tourism]*; Wydawnictwo Uniwersytetu Łódzkiego: Łódź, Poland, 2008. (In Polish)
19. Pécsek, B. City Cemeteries as Cultural Attractions: Towards an Understanding of Foreign Visitors. *DETUROPE Cent. Eur. J. Reg. Dev. Tour.* **2015**, *7*, 44–61. [CrossRef]
20. Young, C.; Light, D. Interrogating spaces of and for the dead as 'alternative space': Cemeteries, corpses and sites of Dark Tourism. *Int. Rev. Soc. Res.* **2016**, *6*, 61–72. [CrossRef]
21. Evensen, K.H.; Nordh, H.; Skaar, M. Everyday use of urban cemeteries: A Norwegian case study. *Landsc. Urban Plan.* **2017**, *159*, 76–84. [CrossRef]
22. Grabalov, P. Public life among the dead: Jogging in Malmö cemeteries. *Urban For. Urban Green.* **2018**, *33*, 75–79. [CrossRef]
23. Rae, R.A. Cemeteries as public urban green space: Management, funding and form. *Urban For. Urban Green.* **2021**, *61*, 127078. [CrossRef]
24. Elands, B.; Wiersum, F.; Buijs, A.; Vierikko, K. Policy interpretations and manifestations of biocultural diversity in urbanized Europe: Conservation of lived biodiversity. *Biodivers. Conserv.* **2015**, *24*, 3347–3366. [CrossRef]
25. Vierikko, K.; Elands, B.; Niemela, J.; Andersson, E.; Buijs, A.; Fischer, L.K.; Haase, D.; Kabisch, N.; Kowarik, I.; Luz, A.C.; et al. Considering the ways biocultural diversity helps enforce the urban transformation. *Curr. Opin. Environ. Sustain.* **2016**, *22*, 7–12. [CrossRef]
26. Haase, D.; Dushkova, D.; Długoński, A. Ecosystem Services Provisioning of Urban Cemeteries from a Multidimensional Perspective-A Comparative Study in Lodz, Poland and Leipzig, Germany. In Proceedings of the ECOSERV 2018, Poznań, Poland, 17–19 September 2018.
27. Długoński, A. Ecology of Cemeteries. First Results from the German-Polish Survey. In Proceedings of the Seminar of the Landscape Ecology Lab, Berlin, Germany, 2 May 2019.
28. OpenStreetMap. 2020. Available online: https://www.openstreetmap.org/ (accessed on 23 January 2022).
29. Urban Atlas 2012. Available online: https://land.copernicus.eu/local/urban-atlas/urban-atlas-2012 (accessed on 23 January 2022).
30. *The Economics of Ecosystems and Biodiversity: Ecological and Economic Foundations*; Earthscan from Routledge: London, UK; Washington, DC, USA, 2010.
31. Haines-Young, R.; Potschin, M.B. Common International Classification of Ecosystem Services (CICES) V5.1 and Guidance on the Application of the Revised Structure. 2018. Available online: www.cices.eu (accessed on 12 December 2021).
32. Długoński, A. Ecosystem services of recreational parks. *Ecol. Quest.* **2018**, *29*, 113–117.
33. Dushkova, D.; Ignatieva, M.; Konstantinova, A.; Yang, F. Cultural Ecosystem Services of Urban Green Spaces. How and What People Value in Urban Nature? In *Advanced Technologies for Sustainable Development of Urban Green Infrastructure*; Vasenev, V., Dovletyarova, E., Valentini, R., Cheng, Z., Calfapietra, C., Inostroza, L., Leuchner, M., Eds.; Springer: Cham, Switzerland, 2021; pp. 292–318. [CrossRef]
34. Haase, D. Integrating Ecosystem Services, Green Infrastructure and Nature-Based Solutions—New Perspectives in Sustainable Urban Land Management. In *Sustainable Land Management in a European Context. Human-Environment Interactions*; Weith, T., Barkmann, T., Gaasch, N., Rogga, S., Strauß, C., Zscheischler, J., Eds.; Springer: Cham, Switzerland, 2021; Volume 8. [CrossRef]
35. Larondelle, N.; Haase, D. Back to nature! Or not? Urban dwellers and their forest in Berlin. *Urban Ecosyst.* **2017**, *20*, 1069–1079. [CrossRef]
36. Lorance Rall, E.D.; Haase, D. Creative Intervention in a Dynamic City: A Sustainability Assessment of an Interim Use Strategy for Brownfields in Leipzig, Germany. *Landsc. Urban Plan.* **2011**, *100*, 189–201. [CrossRef]
37. Pueffel, C.; Haase, D.; Priess, J.A. Mapping ecosystem services on brownfields in Leipzig, Germany. *Ecosyst. Serv.* **2018**, *30*, 73–85. [CrossRef]

38. Długoński, A.; Dushkova, D. The Hidden Potential of Informal Urban Greenspace: An Example of Two Former Landfills in Post-Socialist Cities (Central Poland). *Sustainability* **2021**, *13*, 3691. [CrossRef]
39. Długoński, A.; Szumański, M. Analysis of green infrastructure in Lodz, Poland. *J. Urban Plan. Dev.* **2014**, *141*, A5014001.
40. Szymańska, D.; Lewandowska, A.; Rogatka, K. Temporal trend of green areas in Poland between 2004 and 2012. *Urban For. Urban Green.* **2015**, *14*, 1009–1016. [CrossRef]
41. Długoński, A. Assumptions for the Analysis Method of Green Infrastructure in Big Cities. In *New Ideas in Planning for Territorial Development Architecture*; Węcłowicz-Bilska, E., Ed.; Tadeusz Kościuszko Cracow University of Technology: Kraków, Poland, 2017; Volume 5, pp. 33–58.
42. Schwarz, N.; Haase, A.; Haase, D.; Kabisch, N.; Kabisch, S.; Liebelt, V.; Rink, D.; Strohbach, M.W.; Welz, J.; Wolff, M. How Are Urban Green Spaces and Residential Development Related? A Synopsis of Multi-Perspective Analyses for Leipzig, Germany. *Land* **2021**, *10*, 630. [CrossRef]
43. Bryman, A. *Social Research Methods*; Oxford University Press: Oxford, UK, 2016.
44. Emerson, R.M.; Fretz, R.I.; Shaw, L.L. *Writing Ethnographic Fieldnotes*, 2nd ed.; University of Chicago Press: Chicago, IL, USA, 2011.
45. Holm, G. Photography as a Research Method. In *The Oxford Handbook of Qualitative Research*, 1st ed.; Leavy, P., Ed.; Oxford University Press: Oxford, Great Britain, 2014. [CrossRef]
46. StatSoft Polska. Statistica 13.0. Available online: https://www.statsoft.pl/statistica_13/ (accessed on 20 March 2022). (In Polish).
47. Clayden, A.; Green, T.; Hockey, J.; Powell, M. Cutting the lawn–Natural burial and its contribution to the delivery of ecosystem services in urban cemeteries. *Urban For. Urban Green.* **2018**, *33*, 99–106. [CrossRef]
48. Johnson, P. The modern cemetery: A design for life. *Soc. Cult. Geogr.* **2008**, *9*, 779–790. [CrossRef]
49. Harvey, T. Sacred spaces, common places: The cemetery in the contemporary American city. *Geogr. Rev.* **2006**, *96*, 295–312. [CrossRef]
50. Quinton, J.M.; Duinker, P.N. Beyond burial: Researching and managing cemeteries as urban green spaces, with examples from Canada. *Environ. Rev.* **2019**, *27*, 2. [CrossRef]
51. Huang, S.L. Intentions for the recreational use of public landscaped cemeteries in Taiwan. *Landsc. Res.* **2010**, *32*, 207–223. [CrossRef]
52. Hussein, J. The cemetery as a garden: Why a new understanding of the word "cemetery" is necessary. *J. Inst. Cemet. Crematorium Manag.* **2006**, *1*, 13–16.
53. Kjøller, C.P. Managing green spaces of the deceased: Characteristics and dynamics of Danish cemetery administrations. *Urban For. Urban Green.* **2012**, *11*, 339–348. [CrossRef]
54. Rugg, J. Defining the place of burial: What makes a cemetery a cemetery? *Mortality* **2000**, *5*, 259–275. [CrossRef]
55. Swensen, G. Between romantic historic landscapes, rational management models and obliterations—urban cemeteries as green memory sites. *Urban For. Urban Green.* **2018**, *33*, 58–65. [CrossRef]
56. Barrett, G.W.; Barrett, T.V. Cemeteries as Repositories of Natural and Cultural Diversity. *Conserv. Biol.* **2001**, *15*, 1820–1824. [CrossRef]
57. Tudor, C.A.; Ioja, I.C.; Hersperger, A.; Patru-Stupariu, I. Is the residential land use incompatible with cemeteries location? Assessing the attitude of urban residents. *Carpathian J. Earth Environ. Sci.* **2013**, *8*, 153–162.
58. Woodthorpe, K. Sustaining the contemporary cemetery: Implementing policy alongside conflicting perspectives and purpose. *Mortality* **2011**, *16*, 259–276. [CrossRef]
59. Tzoulas, K.; Korpela, K.; Venn, S.; Yli-Pelkonen, V.; Kaźmierczak, A.; Niemela, J.; James, P. Promoting ecosystem and human health in urban areas using green infrastructure: A literature review. *Landsc. Urban Plan.* **2007**, *81*, 167–178. [CrossRef]
60. van den Berg, A.E.; van den Berg, M.M.H.E. Health benefits of plants and green space: Establishing the evidence base. *Acta Hortic.* **2014**, *1093*, 19–30. [CrossRef]

land

MDPI

Article

Baroque Gardens in Transylvania: A Historic Overview

Albert Fekete * and Máté Sárospataki

Department of Garden Art and Landscape Design, Institute of Landscape Architecture, Urban Planning and Garden Art, Hungarian University of Agriculture and Life Sciences, 2100 Gödöllő, Hungary; sarospataki.mate@uni-mate.hu
* Correspondence: fekete.albert@uni-mate.hu

Abstract: For over more than 20 years, Transylvanian ensembles, gardens and parks have been investigated, described and analysed by a research group from Hungary, led by Albert Fekete. The goal of this study of Transylvanian ensembles is to get background information, insight for developing a strategy for landscape preservation and development in the long run that comprises the cultural and historical values and the demands from society on what to do with them in the contemporary context. The goal of the article is to give an overview of what is already known and what could be done from the viewpoint of protection, planning and design. The research methods are mixed, but are largely based on the case study approach, supplemented by experimental design, fieldwork and research by design. The conclusion is that, given the state of what is left over from these historical artefacts, restoration in the strict sense will be impossible. This will be a major challenge for landscape architecture to take into account the historical values, integrate them with new functions and use and the recent demands of improving water management, energy transition and the creation of comfort and healthy living environments for people.

Keywords: history of gardens; castle garden; goosefoot avenue; star-shaped garden layout; Austro-Hungarian Empire; Transylvania as part of Romania

Citation: Fekete, A.; Sárospataki, M. Baroque Gardens in Transylvania: A Historic Overview. *Land* **2022**, *11*, 949. https://doi.org/10.3390/land11060949

Academic Editors: Jan K. Kazak, Katarzyna Hodor, Magdalena Wilkosz-Mamcarczyk and Panayiotis Nektarios

Received: 15 March 2022
Accepted: 13 June 2022
Published: 20 June 2022

1. Introduction

The Carpathian Mountains are not only geographically but also culturally a crucial part of Europe's history. The multi-confessional zones along the fault lines of the Orthodox–Catholic (east–west) schism of 1054 and the Protestant–Catholic (north–south) schism of 1517 meet in the Carpatho–Pannonian region [1]. Regarding its geography, the Carpathian Basin is a coherent entity, while it is one of the most characteristic transition areas in Europe in terms of its spatial structure from political, economic and religious aspects [2]. In this landscape and regional context, Hungary and its neighbouring countries are interdependent not only geographically but also historically and culturally. For centuries, Hungary has played a leading role in this coexistence. From the conquest to the present day, the history of Transylvania is linked to the history of Hungary. For centuries, it existed either as part of Hungary or as an independent principality closely linked to Hungary, and in broad cultural and historical terms it can be interpreted as the easternmost bastion of Europe. The national unity, the intertwined history and the context of the single landscape unit also make it clear that the research of Hungarian garden art can and should be carried out comprehensively, covering the entire Carpathian Basin, and that the Transylvanian gardens from the modern period can be interpreted as some of the easternmost sites of the European garden styles from the 17th century onwards. In the history of Transylvanian gardens, the 18th century Baroque gardens that followed the late Renaissance gardens were mainly developed in connection with the residences of the local aristocracy and the ecclesiastical centres of the Roman Catholic Church (bishoprics, major monasteries). Accordingly, the main goal of the present study is to focus on research of the Baroque castle gardens of the region, highlighting their most important compositional aspects. Accordingly, the research summarises, classifies and introduces the types of the Baroque gardens surveyed through specific examples, providing information on their role and significance in Hungarian and European garden art.

2. Historical Background

After the conquest (896), Hungarian history can be divided into several periods and this division is important for the historical, economic and social aspects of the region. Figure 1 shows the most important historical periods of the Kingdom of Hungary and, in this context, of Transylvania. According to this historical division, until the collapse of the Austro-Hungarian Empire the most significant and prosperous period of the kingdom was the third period (Figure 1c), which started after the Mongol Invasion of 1241, and lasted until the Battle of Mohács in 1526. During this period, the medieval Kingdom of Hungary became a regional power; its position in international policy was the most favourable possible and its internal legislative and administrative reforms reflected a fair spirit and a strong sense of justice [3–5].

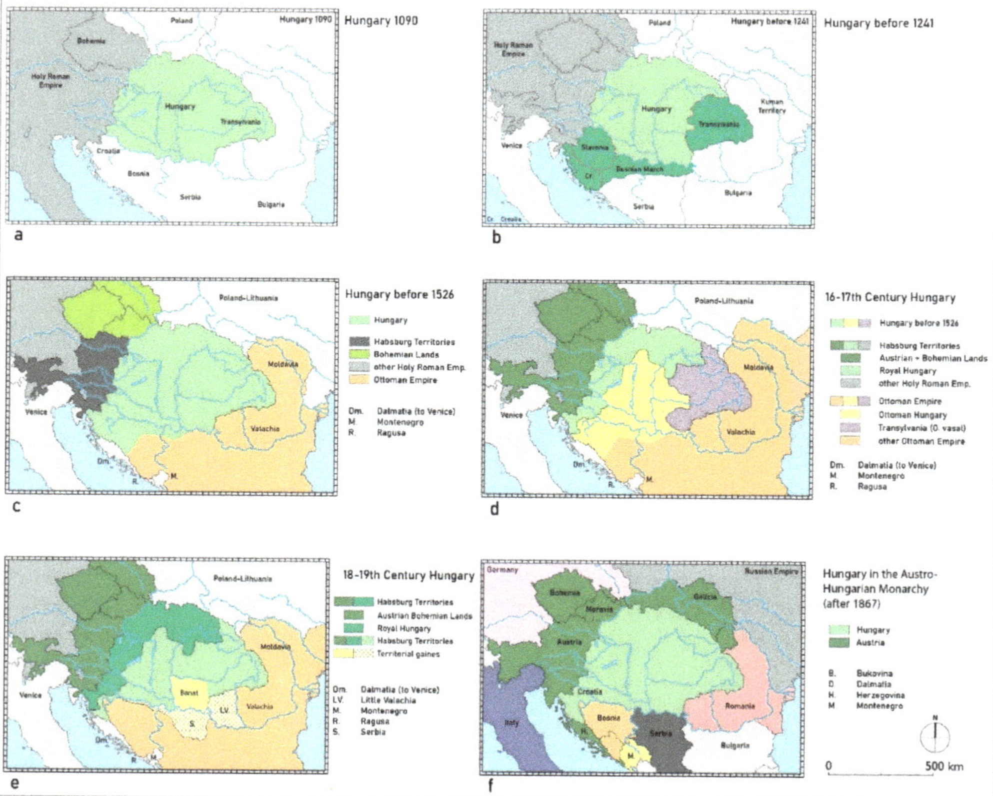

Figure 1. Graphical representation of the major historical periods of the Kingdom of Hungary. Source: prepared by the authors.

The Battle of Mohács, which resulted in an Ottoman victory, caused a major rupture not only in the history of Eastern Europe but also in its culture and landscape history. After the Ottoman rule following the Battle of Mohács, for a hundred and fifty years Western Hungary, Upper Hungary and Transylvania became the last bastions of the spiritual and material heritage and continuity of the European Christian culture in the region.

The fourth stage of the periods shown in Figure 1 includes the "golden age" of the independent Principality of Transylvania, 1613–1683, when Transylvania, although a vassal of the Porte, was economically prosperous, strengthened and flourishing [6–9]. During this period, Transylvanian garden culture in the late Renaissance reached a high level of excellence. At least 60 important Renaissance gardens are authentically mentioned and/or described in archival sources and in some of them it is still possible to detect Renaissance monuments and garden elements of the period [10–15].

The fifth historical period is particularly significant for the present study, as the period of absolutism following the final defeat of the Rákóczi War of Independence in 1711 marked the emergence and completion of the Baroque period in Hungary and Transylvania. From 1712, the supreme governing authority of Transylvania, the administrative and the judicial body of the region was the "Gubernium", directly subordinate to the Viennese court. Under the new balance of forces, Transylvania lost its autonomy and played a strongly subordinate role in the Habsburg Empire, both economically and politically, until the Compromise of 1867. The battles with the Ottomans or the Habsburgs, the political dependency of a once prosperous and powerful Transylvania in the 17th century, the division of the Transylvanian aristocracy and the decline of economic capabilities all contribute to the fact that the 18th century can be considered one of the darkest periods in Transylvanian history [16–19].

3. Materials and Methods

3.1. Timelines of Key European Garden Styles

When examining the characteristics of Transylvanian Baroque garden art, it is not only necessary to know the geographical and spatial location of the area under study as well as its historical background, but it is also important to know the exact chronological boundaries of the subject under study: the Baroque period of garden history. This temporal definition should be done in a European context [20–31], taking into consideration the well-known baroque garden examples and studying their compositional and functional characteristics. In this aspect the gardens from Versailles (FR), Vaux le Vicomte (FR), Montalto (IT), Hampton Court (UK), Het Loo (NL), Frederiksborg (DK), Herrenhausen (D), Nymphenburg (D), Schönbrunn (AU), Peterhof (RU) represent a comprehensive and relevant European pattern.

The development of gardens is always closely linked to the historical and economic development of a country or region. The analysis of the comparative table (Figure 2) shows that, until the national tragedy of Mohács, Hungarian garden art was on a par with European gardens. However, after the Battle of Mohács, during the 150 years of Ottoman occupation, it fell back considerably, and even after the expulsion of the Ottomans it took a century and a half to catch up the cultural delay. In the 19th century it came close to the ideas of Western European garden culture, but in terms of its overall spectrum it was only after the economic boom following the Compromise that it managed to catch up with this time lag that originated from the 16th century history [32,33]. Accordingly, Fatsar links the emergence of Baroque garden art to the appearance of the goosefoot-pattern avenues and the gradual abandonment of the orthogonal layout, which occurred at the turn of the 16th and 17th centuries [34] (Table 1).

In comparison with French or German Baroque, the transition from Renaissance to Baroque in Transylvania is much longer, and the Baroque style spread here about a hundred years later in the middle and second half of the 18th century. By this time, the connection between Western civilisation and Transylvania, which led through Hungary, had opened up. With the spread of the imperial Baroque, several western European architects and sculptors employed initially in Austria were invited to Transylvania by local aristocracy and as a result Transylvanian 18th-century architecture became an integral and representative part of the Baroque art of the monarchy [35,36].

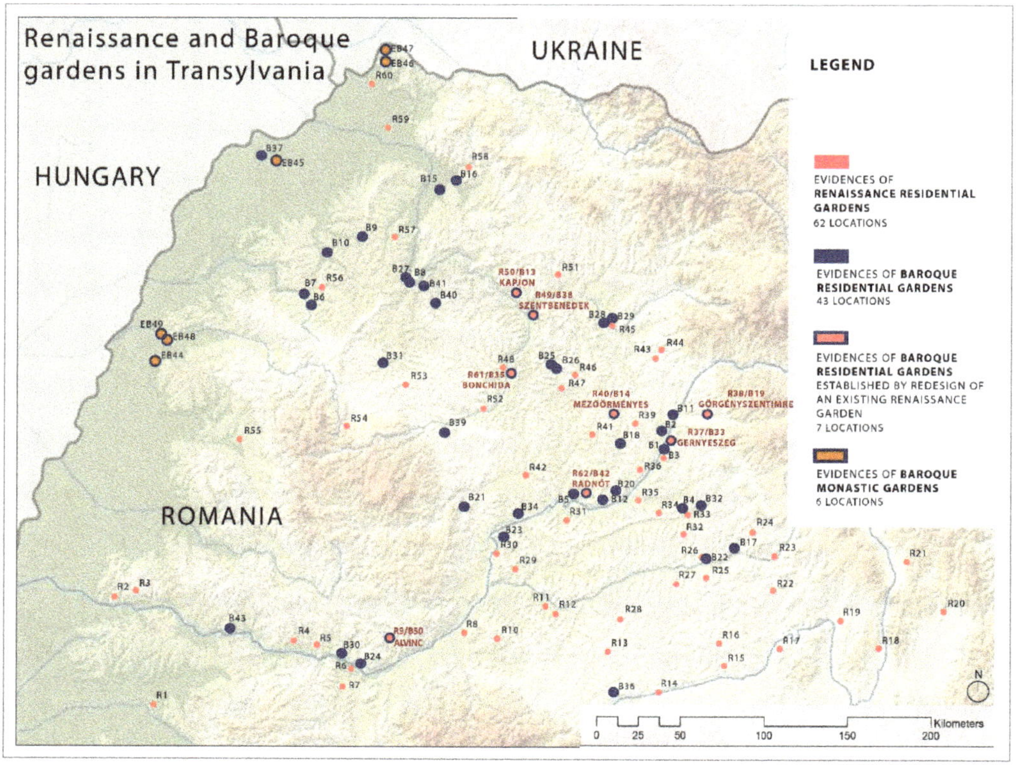

Figure 2. Spatial distribution of Transylvanian Baroque gardens and their relation to the pre-Baroque late-Renaissance Transylvanian garden art; the locations of the Baroque gardens marked on the map can be identified by the numbering assigned to them (B1–B50) using Table 2. Source: prepared by Albert Fekete.

Table 1. Comparative analysis of the periods of major garden styles between Western Europe and Transylvania. In the last column the delayed style periods are marked with dark background. Source: prepared by Albert Fekete based on [14,37–40].

Garden Style		Western Europe	Transylvania	Comments
EARLY RENAISSANCE G		14–15th century	Mathias Corvinus' age (1443–1490)	SIMILAR
LATE RENAISSANCE G		15–16th century	The "Golden Age" of Transylvania (1613–1690)	**CCA 100-150 YEAR DELAY**
BAROQUE G		17–18 century (from 1630s)	2nd part of 18th c.	**CCA 100 YEAR DELAY**
ENGLISH LANDSCAPE G	pictoresque	from 1720s	beg of 19th c. to mid 19th c.	**CONSIDERABLE DELAY**
	gardenesque	from 1830s	mid 19th c.	**SMALL DELAY**
	romantic	from mid 19th c.	mid 19th c. to 2nd part of 19th c.	**ALMOST SIMILAR**
MODERN G	early modern, art nouveau	turn of 19/20th c.	turn of 19/20th c.	SIMILAR
	modern	from 1930 to 1950	from 1930 to 1950	SIMILAR
	postmodern	from 1950 to 1970	from 1950 to 1970	SIMILAR
CONTEMPORARY G		from 1970s	from 1970s	SIMILAR

3.2. *Research Methodology*

Unfortunately, no Transylvanian garden has survived to this day in its original design. In a few places (e.g., Bonchida/Bonţida, Felek/Avrig) the main Baroque compositional elements of the gardens, such as paths, earthworks, alleys, garden structures, etc., can still be seen, but most of the Transylvanian Baroque gardens have been destroyed.

The research is therefore based on archival material and a review of the literature on Transylvanian Baroque garden art. The methodology of the archival research is known.

However, the archival source material on Transylvanian castle gardens is poor. Much of the material has been destroyed, and detailed research of any existing archival material may take a long time due to the lack of proper filing and accessibility of the archives.

The research methodology of Transylvanian baroque gardens was based on the principle that the sites concerned may, and therefore must, be interpreted in context with the related settlements and landscapes as the only way to understand their historic importance and current value [41–43].

For a systematic survey we established the following theoretical framework:

- Identification of all potential sites;
- Thorough garden history research of the sites;
- General landscape assessment of the present conditions of the sites;
- Survey and assessment of the spatial layout and landscape composition.

The goal of the historic research of primary and secondary sources found (archives, library and museum materials, map and postcard collections, thematic bibliography reviews, internet sources, etc.) is to provide a clear idea of the establishment and development of the gardens. It comprises the role the sites play in the landscape and the urban character and layout, and the landscape scale relationships that served as a basis for the establishment of the manor garden and determined the character of the surrounding landscapes to a great extent. The garden history research also deals with the architectural history of the manor house and the family history of the owners.

The site survey precisely records the actual conditions of each garden (sketches, minutes, GPS coordinates, geodetic surveys, plant inventory, digital photographs, etc.) as well as the valuable existing features that are possible to preserve, and thus it serves as a status report and a basis for comparison for conservation strategies and any future restorations. A topographic map (e.g., land registry map) provides the basis for the survey of the general conditions and valuable landscape features.

The garden analysis based on compositional principles helps to define the most important structural types of Transylvanian Baroque gardens, which can serve as a comparison basis for further garden and landscape historical research.

Regarding the literature, we must rely on the works of two of the most dedicated and well-known personalities of Transylvanian architecture and, accordingly, of Baroque garden art, József Biró and Margit B. Nagy.

József Biró examined the palace and the palace garden as a whole in numerous relevant studies [44–46]. In his main work Transylvanian Castles, published in 1943, he devotes a separate chapter to the historical aspects of castle gardens (including the Baroque estates), as well as to the description of their contemporary condition, making it an important source document [47].

Margit B. Nagy primarily researches the architectural heritage of Transylvania, but in her works [11,37,38] she also publishes and processes inventories from the 17th and 18th centuries that contain valuable references to contemporary garden elements and archival source materials presenting them.

Apart from them, several well-known writers in the professional literature and local historians also evoke some aspects of the Transylvanian Baroque garden art [48–54], however, their works are less comprehensive, mostly referring to a single site or detail, and in several cases they refer to the works of Biró or Nagy.

4. Research Results

4.1. Baroque Works and Their Masters

The masters of gardens belong to two groups. The first group is made up of the noble families, i.e., the owners of the palaces, who in many cases are directly involved in the design and transformation of parks, gardens or individual garden elements. The owners are the source of the spirit and cultural content necessary to create the genius loci, the identity of the place and to develop the residential gardens. In many places, the owners have designed the gardens according to their own ideas or have directly influenced their design, so in a sense they can be considered as designers.

The second important group of masters is that of professionals, artists and craftsmen. With the help of archival material and other secondary sources, it has been possible to identify several architects, engineers, stone masons, sculptors, gardeners, hydraulic engineers and other professionals who played an important role in the design, construction and transformation of certain Transylvanian castle gardens or their specific elements (Table 2).

Table 2. List of professionals and estate owners contributed to the design of the significant Baroque gardens in Transylvania. Source: prepared by Albert Fekete based on research of archives and the literature [7,12,44,54,55].

Name of Professional	The Castle Garden Where He Worked	Year of Activity	Owner Family
Blaumann Johann Eberhardt	Bonchida (Bonțida), Bánffy Castle Garden Zsibó (Jibou), Wesselényi Castle Garden Felek (Avrig), Bruckenthal Castle Garden	1776–1783 1780s 1784	Bánffy Wesselényi Bruckenthal
Bode György	Nagykároly (Carei), Károlyi Castle Garden	1790s	Károlyi
Böhm Anton	Görgényszentimre (Gurghiu), Rákóczi-Bornemisza Castle Garden	1790	Bornemisza
Burey Francois	Erdőszentgyörgy (Sângeorgiu de Pădure), Rhédey Castle Garden	from 1800 (36 year long)	Rhédey
Damm Florian	Nagykároly (Carei), Károlyi Castle Garden	1783	Károlyi
Erras Johann Christian	Bonchida (Bonțida), Bánffy Castle Garden	1750s	Bánffy
Gindtner Franz	Hadad (Hodod), Wesselényi Castle Garden	1770	Wesselényi
Hanek Nicolas Philip	Gernyeszeg (Gorneşti), Teleki Castle Garden	1783	Teleki
Hartmann Gottfried	Kolozsvár (Cluj-Napoca), Bánffy Palais	from 1774 to 1783	Bánffy
Hoffmann János	Kendilóna (Luna de Jos), Teleki Castle Garden	1744	Teleki
Jarschel Frantz	Mezőmadaras (Mădăraş), Bethlen Castle Garden	1784	Bethlen
Kopmann Andreas	Száromberke (Dumbrăvioara), Teleki Castle Garden	1782	Teleki
Kováts Sámuel	Zsibó (Jibou), Wesselényi Castle Garden Cege (Ţaga), Wass Ádám Castle Garden Cege (Ţaga), Wass Jenő Castle Garden	turn of 18/19 c.	Wesselényi Wass Wass
Leder Josef	Hadad (Hodod), Wesselényi Castle Garden	1787	Wesselényi
Leerch Ferenc	Görgényszentimre (Gurghiu), Rákóczi-Bornemisza Castle Garden Száromberke (Dumbrăvioara), Teleki Castle Garden Kendilóna (Luna de Sus), Teleki Castle Garden	1782 1783–1784 1780s	Bornemisza Teleki Teleki
Luidor Jean	Koronka (Corunca), Toldalaghi Castle Garden	after 1770	Toldalaghi
Nachtigall János	Bonchida (Bonțida), Bánffy Castle Garden Zsibó (Jibou), Wesselényi Castle Garden	1750s 1760s	Bánffy Wesselényi
Rosenstingl Franz	Nagykároly (Carei), Károlyi Castle Garden	1783	Károlyi
Schuchbauer Antal	Bonchida (Bonțida), Bánffy Castle Garden	1750s	Bánffy
Serbán Sándor	Gernyeszeg (Gorneşti), Teleki Castle Garden	1795	Teleki
Überlacher Anton	Hadad (Hodod), Wesselényi Castle Garden	1790s	Wesselényi
Wrabetz Franz	Zsibó (Jibou), Wesselényi Castle Garden	from 1786 to 1800	Wesselényi
"bécsi Inschíner Kapitán" (*engineer from Vienna*)	Bonchida (Bonțida), Bánffy Castle Garden	1767	Bánffy
"abafáji német kertész" (*German gardener*)	Gernyeszeg (Gorneşti), Teleki Castle Garden	1792	Teleki
"felső-magyarországi kertész" (*gardener from Upper Hungary*)	Gyulafehérvár (Alba Iulia), Princely Castle Garden	1680s	Principality of Transylvania
"wiener Gartner (der zuvor bei Graf Lacy beschaftigt war)" (*gardener from Vienna, working earlier for Count Lacy*)	Felek (Avrig), Bruckenthal Castle Garden	1774	Bruckenthal

4.2. Main Types of Baroque Gardens in Transylvania

Imre Ormos describes the four main types of Baroque gardens, which are defined in the literature on garden history on the basis of the compositional relationship between garden and building [32]. However, the division of the main Baroque garden types is not so obvious. Several interpretations of the main Baroque garden types exist in the literature, which also classify the main Baroque garden types according to a plausible logic based on different classification criteria (e.g., relief, open or closed main axis, main axis perpendicular or parallel to the axis of the palace, etc.) [34,55–64].

In the course of our research on the history of gardens, we have managed to identify a total of 50 sites in Transylvania, where we can assume the presence of Baroque garden art on the basis of mention, detailed description or representation (layout, design, painting, etc.), or can clearly establish it on the basis of surviving garden fragments and garden elements. The majority of the gardens surveyed (forty-three sites) served secular representational purposes associated with the palaces of the aristocracy, while a smaller proportion of the gardens (seven sites) were ecclesiastical (ornamental gardens of religious orders) (Table 3).

Table 3. The list of Transylvanian Baroque gardens surveyed, classified by the type of sources used. Source: prepared by Albert Fekete based on research of archives, the literature and site visit.

Only Mention of Castle Garden (18 Cases)	Detailed Description of Castle Garden (13 Cases)	Detailed Garden Description, Graphic Representation and/or Surviving Garden Features on Site	
		(12 Cases Castle Gardens)	(7 Cases Monastic Gardens)
B1. Sáromberke (Dumbrăvioara)—*Teleki*	**B19. Görgényszentimre** (Gurghiu)—*Rákóczi-Bornemisza*	**B32. Erdőszentgyörgy** (Sângeorgiu de Pădure)—*Rhédey*	**EB44. Püspökfürdő** (Băile 1 Mai) —*Nagyváradi Bishop's possession*
B2. Vajdaszentivány (Voivodeni)—*Zichy*	**B20. Kerelőszentpál** (Chirileu)—*Haller*	**B33. Gernyeszeg** (Gorneşti)—*Teleki*	**EB45. Kaplony** (Căpleni)—*Garden of Franciscan monastery*
B3. Nagyernye (Ernei)—*Bálintitt*	**B.21. Torockósztgyörgy** (Colţeşti)—*Thoroczkai-Rudnyánszki*	**B34. Marosújvár** (Ocna Mureş)—*Teleki-Mikes*	**EB46. Máramarossziget** (Sighetu Marmaţiei)—*Garden of Piarist monastery*
B4. Kelementelke (Călimăneşti)—*Simén*	**B22. Fehéregyháza** (Albeşti)—*Haller*	**B35. Bonchida** (Bonțida) —*Bánffy*	**EB47. Bocskó** (Bocicău)—*Parish garden*
B5. Kutyfalva (Cuci)—*Degenfeld*	**B23. Csombord** (Ciumbrud)—*Kemény*	**B36. Felek** (Avrig)—*Bruckenthal*	**EB48. Nagyvárad** (Oradea)—*Garden of Jesuit monastery*
B6. Szilágybagos (Boghiş)—*Bánffy*	**B24. Dédács** (Simeria)—*Gyulay-Fáy-Ocskay*	**B37. Nagykároly** (Carei)—*Károlyi*	**EB49. Biharpüspöki** (Episcopia Bihor)—*Bishop's pheasantry*
B7. Szilágynagyfalu (Nuşfalău)—*Bánffy*	**B25. Cege** (Ţaga)—*Wass György*	**B38. Szentbenedek** (Mănăstireni)—*Kornis*	**EB50. Alvinc** (Vintul de Jos)—*Garden of Franciscan monastery*
B8. Zsibó (Jibou)—*Béldi*	**B26. Cege** (Ţaga)—*Wass Ádám*	**B39. Magyarfenes** (Vlaha)—Jósika	
B9. Hadad (Hodod)—*Wesselényi*	**B27. Zsibó** (Jibou)—*Wesselényi*	**B40. Csákigorbó** (Gârbău)—*Haller-Jósika*	
B10. Sarmaság (Sărmăşag)—*Kemény*	**B28. Kerlés** (Chiraleş)—*Bethlen*	**B41. Szurdok** (Surduc)—*Jósika*	
B11. Abafája (Apalina)—*Huszár*	**B29. Árokalja** (Arcalia)—*Bethlen*	**B42. Radnót** (Iernut)—*Kornis-Rákóczi-Bethlen*	
B12. Marosugra (Ogra)—*Haller*	**B30. Déva** (Deva)—*Bethlen* (Magna Curia)	**B43. Soborsin** (Săvârşin)—*Nádasdy-Forray*	
B13. Kapjon (Coplean)—*Haller*	**B31. Váralmás** (Almaşu)—*Csáky*		
B14. Mezőörményes (Urmeniş)—*Rákóczi-Bethlen*			
B15. Kővárhosszúfalu (Satulung)—Teleki			
B16. Koltó (Coltău)—*Teleki*			
B17. Fiatfalva (Filiaşi)—*Ugron*			
B18. Mezősámsond (Şincai)—*Bethlen*			

As we observe in Figure 2, many of the gardens are located close to rivers, which serve not just as water supply sources but as important traffic and transportation corridors as well. The spatial distribution of the baroque gardens follows the location of estate centres, related mainly to urban environments or to main communication and service corridors (roads).

The typological analysis of the Transylvanian Baroque gardens is based on their compositional characteristics, with particular emphasis on the path system and plant composition that determine the layout of the garden. On this basis, I have classified the gardens into five main groups.

4.2.1. Gardens with Goosefoot-Pattern Avenues

The goosefoot-pattern avenues are a typical element of Baroque spatial composition and garden layout. It is a combination of avenues running in the extension of the central axis of the main building and in a symmetrical 'V' shape along the axis and terminated by a prominent feature or scene. Although the goosefoot pattern of the avenues can be traced back to the front garden of the Roman Villa, Moltanto at the end of the 16th century, the exemplar of hierarchical organisation of space based on the goosefoot avenues radiating from the main axis of the palace was the gardens of Versailles, which were developed over several decades from the 1660s onwards [34].

However, the goosefoot avenue in France was used not only in Versailles but before in the gardens of Le Notre, starting with Vaux-Le-Vicomte (Richelieu Gardens, 1627). In Versailles it was taken to its maximum expression, and, above all, it was also used for the organisation of urban space.

There are also several earlier examples with radial space compositions outside of France, for instance, Il Tridente di Roma (beginning of the 16th century) or Villa Medicea di Pratolino (Bernardo Buontalenti, 1568). In fact, the radial layout as a mechanism of spatial organisation and landscape definition was used earlier in Spain: Palacio de Aranjuez y las Huertas del Picotajo (Juan bautista de Toledo, 1553). In the Baroque period, this type of layout had a much deeper meaning, as it was usually used to cover the large scales in the forest areas and surrounding territories. It was thus contrasted with the orthogonal layout associated with the palace area. In fact, above both types of layout (radial and orthogonal), a superstructure of compositional axes was superimposed in relation to the axis of the palace. This superstructure dominated the arrangement of the entire layout.

The first simplified Transylvanian version of this classical spatial structure is the Bonchida palace and garden, developed between 1748 and 1752 by Count Dénes Bánffy, which is also the most important Baroque residence in Transylvania [44]. In addition to a pheasantry, the 400-acre estate of Bonchida (Bonţida) also included the most typical Transylvanian Baroque goosefoot-pattern French garden.

From the western facade of the palace, located on the plain of the Small Szamos River, three radial linden alleys extended, each approximately 1000 m long. The starting point of the alleys is the bridgehead built on the main axis of the palace. In addition to the three linden alleys, a fourth shorter horse chestnut alley ran to the north, following the water channel perpendicular to the main axis of the composition [44,47] (Figure 3).

Source: compiled by Albert Fekete, based on Dávid, Gy. [49] and the First Ordnance Survey of Transylvania (1769–1773) [65].

4.2.2. Gardens with Mixed (Orthogonal–Goosefoot) Layout

The most typical Transylvanian example of the layout combining elements of both the orthogonal parterres and the goosefoot-pattern walkways is the Bruckenthal Palace Gardens in Felek (Avrig, RO), which was begun to be built between 1761 and 1764 by General Adolf von Buccow, later governor of Transylvania. After the governor's death in 1764, the estate became the property of Baron Samuel von Bruckenthal, who used it as a summer residence and began to develop the Baroque park of about 6 ha [66]. The ornamental garden, terraced and enclosed by a stone wall because of the considerable differences in levels, is essentially orthogonal in layout, but from the water feature located in the third of the main axis away from the castle, the paths no longer run transversely but five radial walkways start, dividing the southern part of the garden into regular sections (Figure 4).

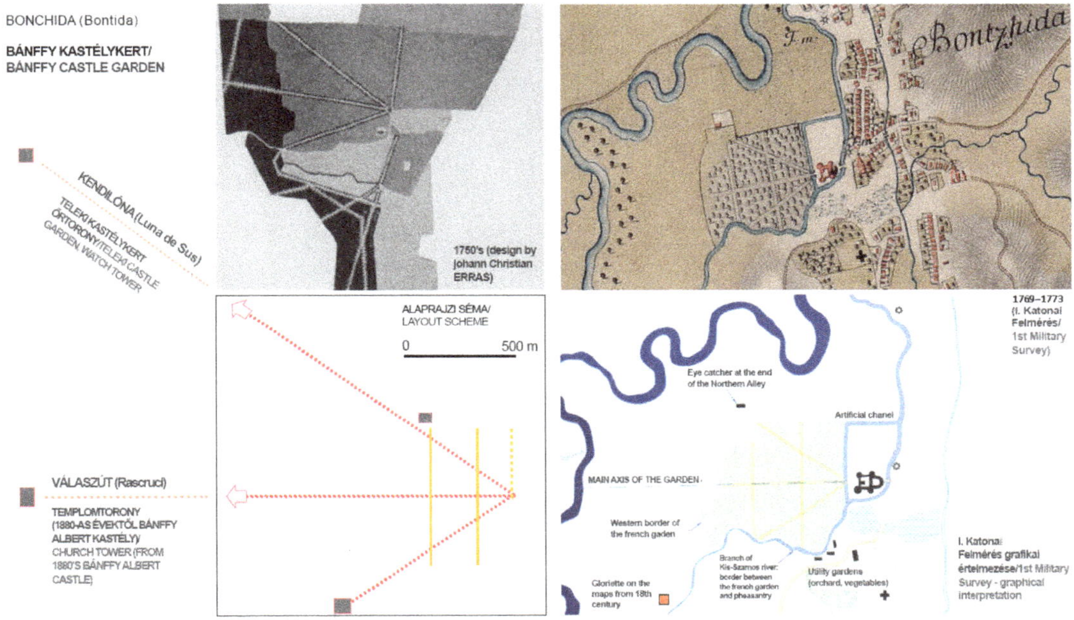

Figure 3. Map-based analysis of the goosefoot-pattern Baroque garden in Bonchida.

4.2.3. Gardens with Orthogonal Layout

The most common layout type of Baroque garden in Transylvania is the orthogonal garden. Its prevalence is also due to its relatively simple transparent layout, similar to a Renaissance parterre garden. It consists mostly of square, or occasionally rectangular or irregular square, compartments, sometimes intersected by diagonal or oblique paths. In the composition of the orthogonal gardens, the structural hierarchy typical of the Baroque (e.g., positioning in relation to the main building, side paths subordinate to the main walk) is only rarely present.

Source: compiled by Albert Fekete based on Feyer (2006), the First Ordnance Survey of Transylvania (1769–1773) [61] and "*Mappa prima der neuen Land- und Post-Strassen von Hermanstat bis Cronstat Mappa secunda*" [67]

The subordinate Baroque composition in the hierarchy of the network of paths or the use of the hedges is clearly visible, for example, in the Baroque garden plan of Franz Anton Hillebrandt from 1783 in Nagykároly (Carei, RO) (Károlyi Castle, Figure 5b). The same Baroque gesture of spatial organisation is much more nuanced in the Baroque garden of the Jesuit monastery in Nagyvárad (Oradea, RO), where only the two ornately designed central embroidered parterres emphasise the composition (Figure 5a).

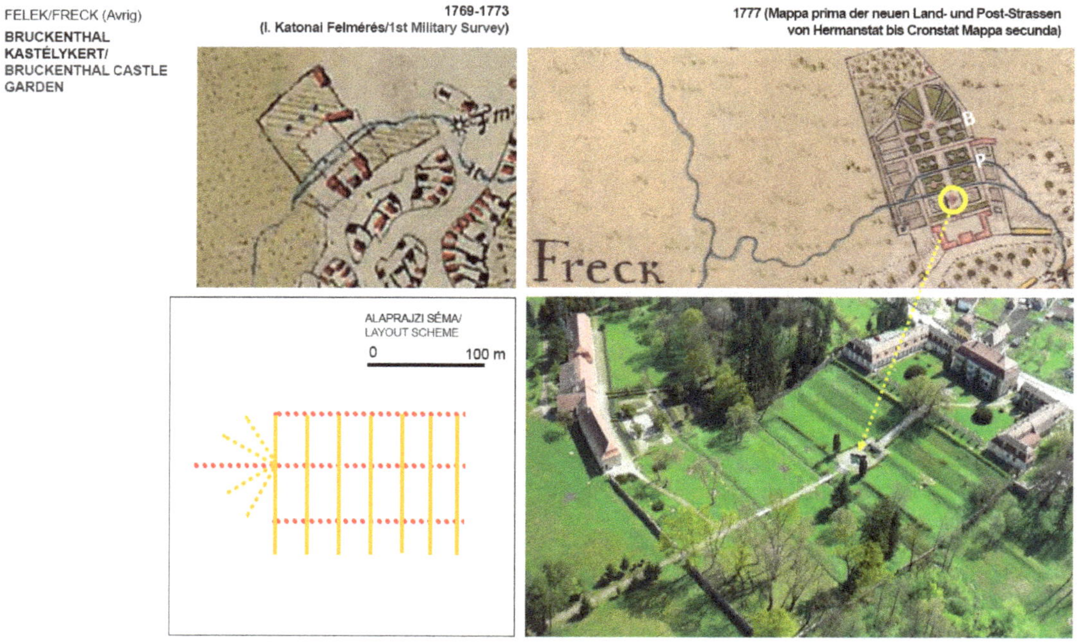

Figure 4. Map-based analysis of the mixed-layout Baroque garden in Felek.

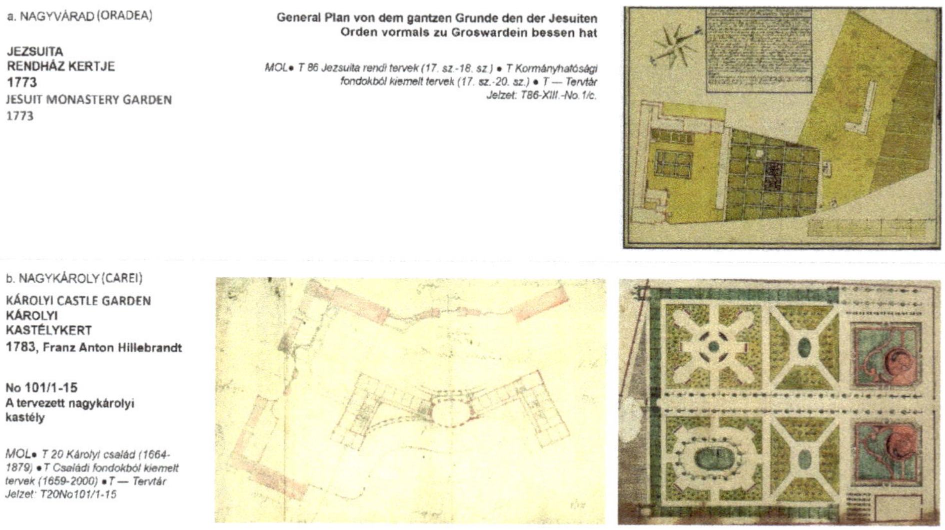

Figure 5. Source: compiled by Albert Fekete. (**a**). The Baroque garden plan of Hillebrandt from 1783 for the Károlyi Castle in Nagykároly (Carei, RO) based on "*A tervezett nagykárolyi kastély*" [68]. (**b**). The Baroque garden plan from 1773 for the Jesuit monastery in Nagyvárad (Oradea, RO) based on "*General Plan von dem gantzen Grunde den der Jesuiten Orden vormals zu Groswardein besessen hat*" [69].

The manuscript map of the Rákóczi-Bornemisza Castle in Radnót (Iernut, RO), which was the seat of the prince in the 17th century, shows two very simple orthogonally divided geometric gardens on the west and south-west sides of the castle in the early 19th century.

Neither of these gardens form a Baroque composition with the four-cornered bastion castle building of Renaissance origin.

An aerial photograph showing the current state of the castle and garden shows the outlines and layout of a garden of a similar geometry and size to the gardens on the manuscript map and the ruins of the Baroque gatehouse (Snake House) built on the main southern axis of the castle (the location of the gate is marked by a red circle).

By superimposing the manuscript map and the aerial photograph, it can be seen that the georeferencing does not support the idea that the orthogonal garden (yellow dotted rectangle) shown in the aerial photograph on the main axis connecting the castle and the gate house would have the same size and location as the old garden depicted on the manuscript map (Figure 6).

RADNÓT (IERNUT)

RÁKÓCZI-BORNEMISZA KASTÉLYKERT

BORNEMISZA CASTLE GARDEN

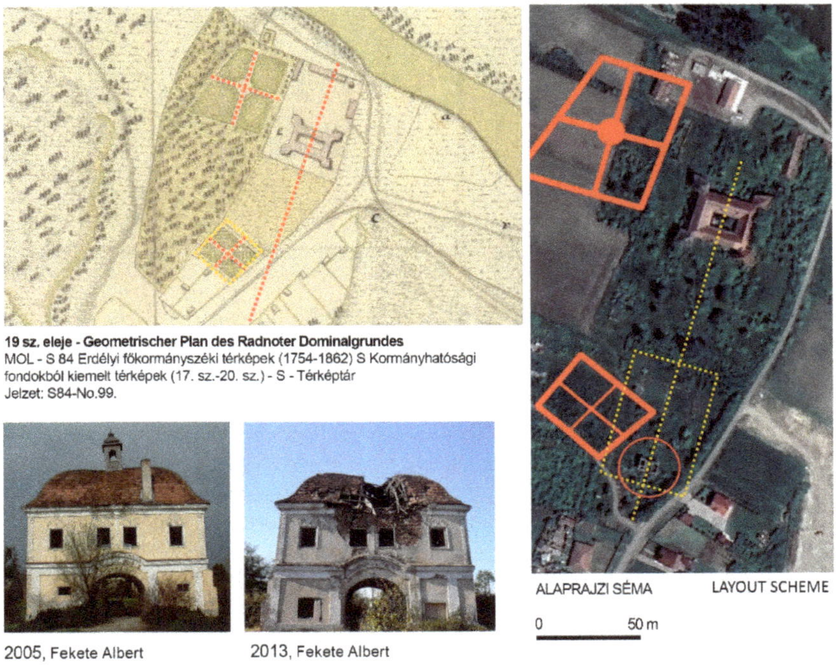

Figure 6. Historic analysis of the surroundings of the princely castle in Radnót (Iernut, RO).

Source: compiled by Albert Fekete based on "*Geometrischer Plan des Radnoter Dominalgrundes*" [70], based on a Google Earth image from 2021 and on own photographs.

4.2.4. Gardens with Radial Walkways (Star-Shaped Layout)

Radial walkways (star-shaped layout) are a typical compositional feature of the Baroque garden layout, with walkways spreading symmetrically around a central feature (structure, fountain, square). The focal point of the walkways may be located both near to the castle or at a greater distance [60–64].

Instances of radial walkways located near the castle are in Nagykároly (Carei, RO; Figure 7b), Soborsin (Săvârşin, RO; Figure 8) or in Alvinc (Vinţul de Jos, RO). Franz Rosenstingl's 1789 plan of the castle garden in Nagykároly shows a square ornamental garden fitted to the building, with a completely symmetrical layout divided into eight sections with a parterre (or radial walkway) in the middle, which is supplemented by further geometric ornamental garden sections along a garden axis parallel to the longitudinal axis of the castle. The plan shows a subtle redesign of the Baroque garden depicted in the first military survey.

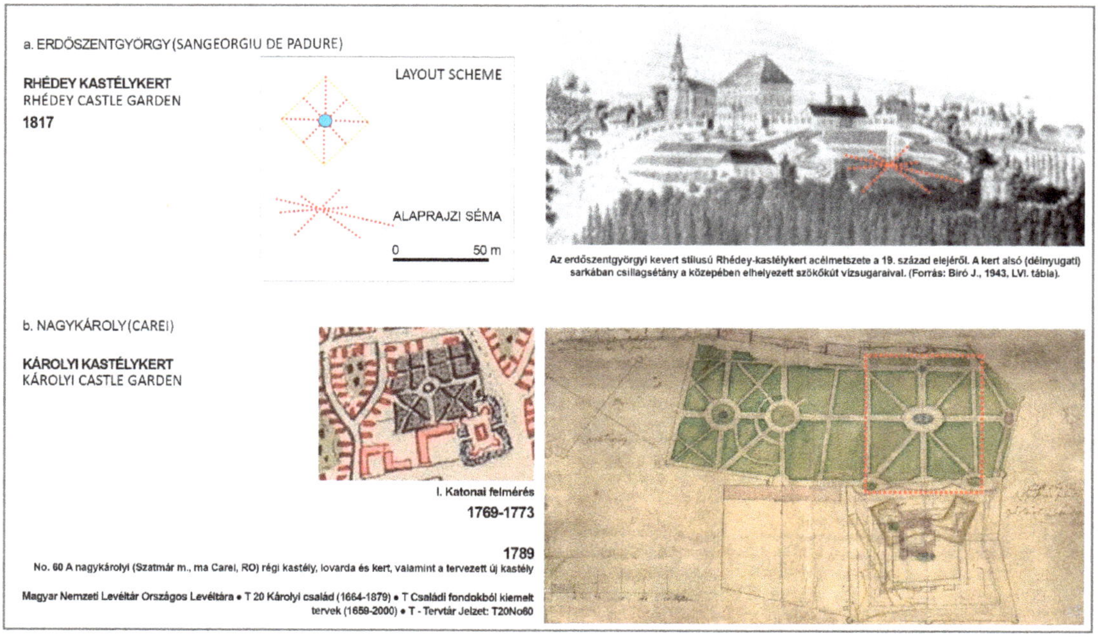

Figure 7. Source: compiled by Albert Fekete. (**a**). Steel engraving of the Rhédey castle garden in Erdőszentgyörgy from 1817 [47]. (**b**). Franz Rosenstingl's 1789 plan of the castle garden in Nagykároly (Carei, RO) based on "A nagykárolyi (Szatmár m., ma Carei, RO) régi kastély, lovarda és kert, valamint a tervezett új kastély" [71].

In the case of the castle garden in Alvinc (Vinţul de Jos, RO), a feature of radially shaped walkways in the ornamental garden is mentioned in an inventory from 1676, that is, from the late Renaissance period in Transylvania: "*The ornamental garden was spread out beyond the moat, east of the bridge. Divided by paths in a radiating shape, the flower and vegetable compartments enclosed a gazebo built over the fishpond, which was the centre of an obviously symmetrical composition.*" [37].

A good example of the radial walkways set further away from the castle building is the garden of the Rhédey Castle in Erdőszentgyörgy (Sângeorgiu de Pădure, RO; Figure 7a). It is special due to its "mixed" character, which supports Anna Zádor's observation that 19th century English gardens in Transylvania "*often retained something of the previous garden design, thus showing a somewhat mixed style, adhering to the old and the traditional*" [72], as happened in many renaissance–baroque, baroque–landscape and landscape–Victorian gardens.

From the castle in Erdőszentgyörgy (Sângeorgiu de Pădure, RO) a staircase runs down the terraced hillside to the lower garden and at the bottom of the hill is the bosque, divided by radial walkways with a fountain in the middle shooting high water jets [47].

Source: compiled by Albert Fekete based on "Situations Plan Der Marosch Fluss Laage bey Valliemare und jenseiths/!/bey Soborsin" [73] and own photographs.

4.2.5. Baroque Gardens of Local Character

We have descriptions and depictions of several gardens in Transylvania that present Baroque gardens of a specific layout. Among these, one of the most characteristic and unique is the design of the Teleki Castle Garden in Gernyeszeg (Gorneşti, RO), dating from the time of Count József Teleki (I) (Figure 9).

Figure 8. The Forray–Nádasdy Castle and its garden in Soborsin (Săvârşin, RO) in 1782.

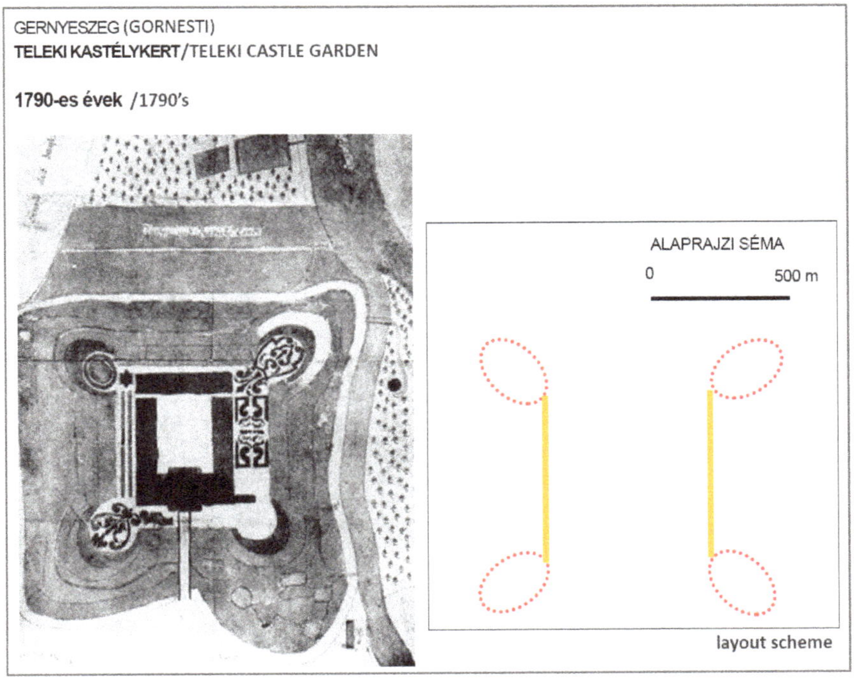

Figure 9. The Baroque garden design for the Teleki Castle Garden in Gernyeszeg (Gornesti, RO) from 1792. Source: compiled by Albert Fekete based on "*A gernyeszegi francia kert tervrajza*" [33].

The count's fascination with gardens is well known. During his youthful travels in Western Europe, he visited the famous gardens of his time and, like all works of art, he showed a keen interest in them. Between 1759 and 1761, he travelled through Austria, Bavaria, Switzerland, the Rhineland principalities, the Netherlands and France. In Germany, he visited the gardens of Rastatt, Karlsruhe, Mannheim and Augustusburg in Brühl, and in the Netherlands he visited the castle and park of ten Bosch, among others [41].

In the 1780s he decided to undertake a major garden renovation in Gernyeszeg (Gorneşti, RO), and thus the plan of 1792 by the "German gardener from Abafája" was born, the design of which was inspired by motifs of local folk decoration. Despite the interesting design elements, the count was not entirely impressed by the design, considering it to be lacking the characteristic elements of the landscape gardens that were then fashionable in Europe. [74].

Following the French Baroque garden pattern, water surfaces (the aquatic element was essential in the French Baroque and was in fact a French addition when the classical gardens were imported to France) were popular elements in some of the Transylvanian Baroque castle gardens and their placement in the geometric Baroque composition also led to specific solutions. In addition to the Haller Castle Garden in Fehéregyháza (Albeşti, RO) and the Bethlen Castle Garden in Kerlés (Chiraleş, RO), there are references in the case of the Thoroczkay–Rudnyánszky Castle Garden in Torockószentgyörgy (Colţeşti, RO) to a large water basin, which could be used as a fish pond, giving the garden a unique character: "*The large canal pond is an old Transylvanian tradition, as the fish pond in the former ornamental garden of the Count's castle in Torockószentgyörgy, for example, is even associated with a legend*" [44].

5. Discussions

In the garden history of Transylvania, the Baroque style was almost one hundred years late in spreading compared with other parts of Europe. In terms of the number of gardens in Transylvania, the Baroque cannot be called the leading garden style in this part of the country. The number of Baroque gardens is much lower than the number of earlier Renaissance or later Landscape Gardens (Figure 10). This is a consequence of the forced political and economic dependence of Transylvania in the 18th century.

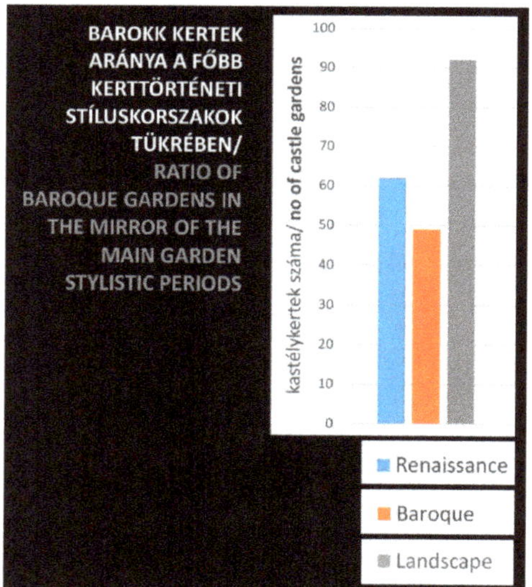

Figure 10. The number of Renaissance, Baroque and Landscape Gardens in Transylvania. Source: compiled by Albert Fekete.

The majority of the Baroque gardens realised in Transylvania were of modest design, reflecting the financial constraints of an economically underdeveloped part of the country.

Nevertheless, many foreign masters (gardeners, sculptors, architects, painters, jacks-of-all-trades) of Austrian, German, French, Walloon, Flemish, Czech and Polish origin worked on the design and construction of Baroque gardens of the 18th century Transylvanian noble residences. Their presence proves that Transylvanian Baroque garden art, although geographically marginalised, did not exist in isolation, but was linked to the centre of ideas through its creators and designers. The owners of the estates were aware of the European examples to follow and their artistic qualities.

Somewhat more representative Baroque gardens were developed at the residences of noble families with close ties to the Viennese court and of the dignitaries in the Roman Catholic Church.

The Transylvanian Baroque garden history is mainly about ornamental gardens, although in many places various kinds of kitchen and vegetable gardens or orchards were also part of the Baroque garden. In terms of their compositional characteristics, Transylvanian Baroque ornamental gardens are largely integrated into the general European system. They followed the models, sometimes as simplified small-scale paraphrases of them, but there are also examples that have specific local characteristics (in ethnographic and topographic terms or in plant use).

Given the state of what is left over from these historical artefacts, restoration in the strict sense is almost impossible in the majority of the locations. Devastation, missing archival sources, changing ownerships and sustainability reasons make the restoration work even harder. During the investigation, the analysis and the fieldwork of the Transylvanian ensembles, we had ample contact with local stakeholders, politicians, owners, NGOs, users and other people related to the Transylvanian ensembles. The core of the problem is concentrated around two poles: one of heritage and cultural meaning, the other on the search for new functions and uses. These two are often contradictory and conflicting; they can be categorised in the polarity between development and conservation. In all landscape architectural projects this contradiction plays a role, but in the case of historical phenomena they are even more pronounced and demand special attention [41–43,75].

This will be a major challenge for landscape architecture to take into account the historical values and to integrate them with new functions and use and the recent demands of improving water management, energy transition and the creation of comfort and healthy living environments for people.

6. Conclusions

Landscape architecture is an applied science. According to the guidelines on the preparation of scientific publications in garden history, the results of our research are based on general preliminary studies of garden and landscape history, the research results and experiences of several decades, the exploration and analysis of authentic historical sources and the site surveys and assessments. The most important findings of the research are the following:

- **Baroque castle gardens are part of the most significant garden heritage assets in Transylvania.** This finding is supported by the history and the relation to the landscape of dozens of European, Hungarian and Transylvanian castle gardens examined during our preliminary studies and earlier research [33–55], and also by the results of our research based on the survey of 50 Transylvanian baroque gardens introduced in detail in the paper. Due to their compositional characteristics and integration with the surrounding landscape, the baroque gardens have a complex importance as heritage assets.
- **The baroque castle, the garden and the surrounding landscape together represent a single artistic and compositional unit.** Neither the castle nor the garden should be interpreted independently. They are a single unit, and all the man-made and natural elements of the garden and often specific elements beyond the garden boundaries, make part of this unit. The castle and the garden together represent a composition

that is an integral part of a complex system developed on artistic, cultural, historical, ecological and economic bases.

- **Due to the scarce sources in garden history, garden restoration is rarely a feasible option in Transylvania.** In Transylvania, the current conditions of Baroque gardens and the availability of historical sources (often still uncatalogued, inaccessible and incomplete) in most of the cases do not allow for a full restoration of the original design. Nevertheless, the gardens still represent a part of the garden heritage that is possible to restore in the most authentic way on the basis of the features preserved and the data available on the actual conditions.

In summary, it can be said that the Transylvanian Baroque garden is considered an integral part of Hungarian garden history, although its instances are of lower quality than the Hungarian and European averages. Their importance in the field of garden history is due to the fact that Transylvania is the easternmost location of the typical garden styles of early modern Europe, and the Baroque castles and gardens created in Transylvania, mainly on the model of Western Europe, represent some of the easternmost examples of 18th century European garden design.

Author Contributions: Conceptualisation, A.F. and M.S.; methodology, A.F.; software, M.S.; validation, A.F. and M.S.; formal analysis, A.F.; investigation, A.F. and M.S.; resources, A.F.; data curation, M.S.; writing—original draft preparation, A.F., writing—review and editing, A.F. and M.S.; visualisation, A.F.; supervision, M.S.; project administration, A.F. All authors have read and agreed to the published version of the manuscript.

Funding: This research received no external funding. The publication was supported by the Hungarian University of Agriculture and Life Sciences (MATE).

Data Availability Statement: Not applicable.

Conflicts of Interest: The authors declare no conflict of interest.

References

1. Madeley, J.T.S. A framework for the comparative analysis of Church–State relations in Europe. *West Eur. Politics* **2003**, *26*, 23–50. [CrossRef]
2. Kocsis, K. A Kárpát-medence történeti vallásföldrajza (10–21. század). In *Tanulmánygyűjtemény*; Gál, A., Frisnyák, S., Kókai, S., Eds.; Nyíregyházi Egyetem Turizmus és Földrajztudományi Intézete és a szerencsi Bocskai István Katolikus Gimnázium: Nyíregyháza, Hungary; Szerencs, Hungary, 2016; Volume 1, pp. 9–30.
3. Fine, J.V.A. *The Late Medieval Balkans: A Critical Survey from the Late Twelfth Century to the Ottoman Conquest*; The University of Michigan Press: Ann arbor, MI, USA, 1994; Available online: https://www.press.umich.edu/7807 (accessed on 13 January 2022).
4. Hóman, B. *Magyar Középkor*; Attraktor: Máriabesnyő, Hungary, 2003.
5. Sedlar, J.W. *East and Central Europe in the Middle Ages, 1000–1500*; University of Washington Press: Seattle, WA, USA, 1994.
6. Köpeczi, B. (Ed.) *Erdély Története I–III*; Akadémiai Kiadó: Budapest, Hungary, 1988.
7. Csetri, E. *Bethlen Gábor Életútja*; Kriterion: Bukarest, Romania, 1992.
8. Kovács Kiss, G.Y. (Ed.) *Studies in the History of Early Modern Transylvania*; Atlantic Research and Publications, Inc.: Highland Lakes, NJ, USA; Columbia University Press: New York, NY, USA, 2011; 616p.
9. Sugar, P.F. *Southeastern Europe under Ottoman Rule, 1354–1804*; University of Washington Press: Seattle, WA, USA, 1996; pp. 150–154.
10. Herczeg, Á. *Szép és Kies Kertek*; Pro-Print Kiadó (n.y): Csíkszereda, Romania, 2008.
11. Margit, B.N. *Várak, Kastélyok, Udvarházak, Ahogy a Régiek Látták*; Kriterion: Bukarest, Romania, 1973.
12. Stirling, J. *Magyar Reneszánsz Kertművészet a XVI–XVII. Században*; Enciklopédia Kiadó: Budapest, Hungary, 1996.
13. Kovács, A. *Késő Reneszánsz Építészet Erdélyben: 1541–1720*; Teleki László Alapítvány–Polis Könyvkiadó: Budapest, Hungary; Kolozsvár, Romania, 2003.
14. Fekete, A. *Az Erdélyi Kertművészet-Maros Menti Kastélykertek*; Művelődés Műhely: Kolozsvár, Romania, 2007.
15. Fekete, A. Late Renaissance Garden Art in the Carpathian Basin. *Landsc. Environ.* **2020**, *14*, 10–28. [CrossRef]
16. Gyémánt, L. The Habsburgs' Policy: Absolutism, Reformism and Liberalism. In *The History of Transylvania (from 1711 to 1918)*; Pop, I.A., Nagler, T., Magyari, A., Eds.; Center for Transylvanian Studies, Romanian Academy: Cluj Napoca, Romania, 2010; Volume 1, pp. 215–241.
17. Macartney, C.A. Hungary. In *The European Nobility in the Eighteenth Century; Studies of the European Nobilities of the Major European States in the Pre-Reform Era*; Goodwin, A., Ed.; A & C Black: London, UK, 1953; p. 133.
18. Meszlenyi, A. *A Jozefinizmus Kora Magyarországon: 1780–1846*; Stephaneum: Budapest, Hungary, 1934; pp. 15–30.
19. Szijártó, I.M. *Dieta. A Magyar Rendek és az Országgyűlés 1708–1792*; Osiris: Budapest, Hungary, 2005; pp. 178, 194.

20. Gothein, M.L. *Geschichte der Gartenkunst—(Band 2): Von der Renaissance in Frankreich bis zur Gegenwart*; Diederichs: Jena, Germany, 1914.
21. Cormier, C.M. Seventeenth-Century Garden Theory in France: Its Culmination at Vaux-le-Vicomte. Master's Thesis, Department of Art History, McGill University, Montreal, QC, Canada, March 1992.
22. Mukerji, C. The Political Mobilization of Nature in Seventeenth-Century French Formal Gardens. *Theory Soc.* **1994**, *23*, 651–677. [CrossRef]
23. Weiss, A. *Mirrors of Infinity: The French Formal Garden and 17th-Century Metaphysics*; Princeton Architectural Press: New York, NY, USA, 1995.
24. Steenbergen, C.; Reh, W. *Architecture and Landscape: The Design Experiment of the Great European Gardens and Landscapes*; Prestel: München, Germany, 1996.
25. Baridon, M. The scientific imagination and the Baroque garden. *Stud. Hist. Gard. Des. Landsc.* **1998**, *18*, 5–19. [CrossRef]
26. Morgan, L. The Baroque garden. In *The Baroque: Style in the Age of Magnificence, 1620–1800*; Snodin, M., Llewellyn, N., Eds.; Victoria and Albert Museum: London, UK, 2009; pp. 276–277.
27. Dobrilovič, M. Vegetation elements in Baroque Gardens (The influence of foreign plants on the Baroque Programme). *Acta Hortic.* **2010**, *881*, 899–904. [CrossRef]
28. Turner, T. *European Gardens—History, Philosophy and Design*; Routledge: London, UK, 2011.
29. Vermes, G. "Extra Hungariam non est Vita" and the Baroque in 18th and Early 19th Century Hungary. *Hung. Stud. Rev.* **2012**, *39*, 21–34.
30. Nurme, S.; Nutt, N.; Hiob, M.; Hess, D.B. Talking ruins: The legacy of baroque garden design in Manor Parks of Estonia. In *Landscape Archaeology between Art and Science: From a Multi- to an Interdisciplinary Approach*; Kluiving, S.J., Guttmann-Bond, E.B., Eds.; Series: Landscape and Heritage Research; Amsterdam University Press: Amsterdam, The Netherlands, 2012; pp. 115–126. Available online: https://www.jstor.org/stable/j.ctt6wp79m (accessed on 28 January 2022).
31. Zgarbová, M. Philosophical and socio-philosophical base of historical garden type development—From Continental French Formal Garden of the 17th century to Insular English Landscape Garden of the 18th century. *Eur. Sci. J.* **2014**, *2*, 381–389.
32. Ormos, I. *A Kerttervezés Története és Gyakorlata*; Mezőgazdasági Kiadó: Budapest, Hungary, 1967.
33. Rapaics, R. *Magyar Kertek*; Királyi Magyar Egyetemi Nyomda: Budapest, Hungary, 1940.
34. Fatsar, K. *Magyarországi Barokk Kertművészet*; Helikon Kiadó: Budapest, Hungary, 2008.
35. Genthon, I. *Erdély Művészete*; Hungária R.T.: Budapest, Hungary, 1936.
36. Kovács, A. Az alvinci kastély és leltárai. In *Erdély XVII–XVIII. Századi Építészetének Forrásaiból*; Kovács, Z.S., Ed.; Scientia Kiadó: Kolozsvár, Romania, 2004; pp. 15, 63.
37. Margit, B.N. *Reneszánsz és Barokk Erdélyben, Művészettörténeti Tanulmányok*; Kriterion: Bukarest, Romania, 1970.
38. Margit, B.N. *Stílusok, Művek, Mesterek*; Kriterion: Bukarest, Romania, 1977.
39. Kőváry, L. *Erdély Építészeti Emlékei: Erdély Régiségei*; Stein: Kolozsvár, Hungary, 1866.
40. Kelemen, L. *Művészettörténeti Tanulmányok*; Kriterion: Bukarest, Romania, 1977; Volume 1.
41. Fekete, A. *Transylvanian Garden History. Castle Gardens along the Maros River*; Művelődés Műhely: Kolozsvár, Romania, 2007.
42. Fekete, A. *Az Erdélyi Kertművészet-Szamos Menti Kastélykertek*; Művelődés Műhely: Kolozsvár, Romania, 2012.
43. Fekete, A. *Komponált Látványok. Történeti Kertek és Tájak Helyreállítása*; SZIE: Gödöllő, Hungary, 2019.
44. Biró, J. A bonczhidai Bánffy-kastély. *Erdélyi Múz. Egyes. Kiadása* **1935**, *75*, 5–36. Available online: https://eda.eme.ro/handle/10598/8768 (accessed on 26 February 2022).
45. Biró, J. *A Gernyeszegi Teleki-Kastély*; Sárkány Nyomda Rt.: Budapest, Hungary, 1938.
46. Biró, J. *A Zsibói Kastély*; Franklin Társulat: Budapest, Hungary, 1942.
47. Biró, J. *Erdélyi Kastélyok*; Új Idők Irodalmi Intézet: Budapest, Hungary, 1943.
48. Toroczkai, W.E. *Régi Kert s Míesei*; Táltos Kiadó: Budapest, Hungary, 1917.
49. Gyula, G.D. *A Bonchidai Bánffy-Kastély*; Polis: Kolozsvár, Romania, 2011.
50. Sisa, J.; Wiebenson, D. (Eds.) *The Architecture of Historic Hungary*; The MIT Press: Cambridge, MA, USA; London, UK, 1998.
51. Zsolt, K. (Ed.) *Erdély XVII–XVIII. Századi Építészetének Forrásaiból*; Scientia Kiadó: Kolozsvár, Romania, 2004.
52. Zoltán, B.; Zsolt, O. *Isten Segedelmével Udvaromat Megépítettem. Történelmi Családok Kastélyai Erdélyben*; Gutenberg Kiadóvállalat: Budapest, Hungary, 2011.
53. Balogh, J. *Kolozsvári Kőfaragó Műhelyek XVI. Század*; MTA Művészettörténeti Kutató Csoportja: Budapest, Hungary, 1985; Available online: http://adatbank.transindex.ro/html/cim_pdf923.pdf (accessed on 6 February 2022).
54. Sas, P. Egy erdélyi művészettörténész munkássága leveleinek tükrében (Biró József levelei Lyka Károlyhoz). *Ars Hung.* **1995**, *23*, 95–104.
55. De Bay, P.; Bolton, J. *Gartenkunst im Spiegel der Jahrhunderten*; Collection Rolf Heyne: München, Germany, 2000.
56. Adams, W.H. *The French Garden 1500–1800*; George Braziller: London, UK, 1979.
57. Mader, G.; Neubert-Mader, L. *The English Formal Garden Five Centuries of Design*; Aurum: London, UK, 1997.
58. Conan, M. (Ed.) *Baroque Garden Cultures: Emulation, Sublimation, Subversion*; Dunbarton Oaks: Washington, DC, USA, 2005.
59. Magni, A. Principles of the French Garden in the Czech Garden Design. *Acta Univ. Agric. Silvic. Mendel. Brun.* **2018**, *66*, 1171–1181. [CrossRef]
60. Fowler, P. World Heritage Cultural Landscapes, 1992–2002: A Review and Prospect. In Proceedings of the Cultural Landscapes: The Challenges of Conservation, Ferrara, Italy, 11–12 November 2002.

61. Benevolo, L. *La Cattura Dell'Infinito*; Laterza: Roma, Italy; Bari, Italy, 1991.
62. Lazzaro, C. *The Italian Renaissance Garden from the Conventions of Planting, Design, and Ornament to the Grand Gardens of Sixteenth-Century Central Italy*; Yale University Press: New Haven, CT, USA, 1990.
63. Hansmann, W. *Gartenkunst der Renaissance und des Barock*; DuMont: Köln, Germany, 1983.
64. Hazlehurst, F.H. *Gardens of Illusion. The Genius of André Le Notre*; Vanderbilt University Press: Nashville, TN, USA, 1980.
65. First Ordnance Survey of Transylvania. Available online: https://maps.arcanum.com/hu (accessed on 25 February 2022).
66. Feyer, C.; Hermann, C. *Parksanierungskonzept Historsiche Parkanlage Sommerresidenz "Samuel von Bruckenthal" Freck/Avrig, Rumanien, Siebenbürgen*; Nationales Denkmalschutzamt: Sibiu, Romania, 2006.
67. "Mappa prima der Neuen Land- und Post-Strassen von Hermanstat bis Cronstat Mappa Secunda" Magyar Nemzeti Levéltár Országos Levéltára—S 84 Erdélyi Főkormányszéki Térképek (1754–1862)—S Kormányhatósági Fondokból Kiemelt Térképek (17. sz.–20. sz.)—S–Térképtár. Index No: S84-No.24/1-6. Available online: https://maps.hungaricana.hu/hu/search/results/?list=eyJmaWx0ZXJzIjogeyJGT1JSIjogWyJNT0wiXSwgIlBJQyI6IFsiWUVTIl19LCAicXVlcnkiOiAiRkVKWj0oXCJTIDg0IEVyZFx1MDBlÖWx5aSBmXHUwMTUxa29ybVx1MDBlMW55c3pcdTAwZTlraSB0XHUwMGU5cmtcdTAwZTlwZWsgKDE3NTQtMTg2MilcIikifQ&per_page=50&page=1 (accessed on 25 February 2022).
68. "A Tervezett Nagykárolyi Kastély" Magyar Nemzeti Levéltár Országos Levéltára—T 20 Károlyi Család (1664–1879)—T Családi Fondokból Kiemelt Tervek (1659–2000)—T–Tervtár; Index No.: T20No101/1-15. Available online: https://maps.hungaricana.hu/hu/search/results/?list=eyJmaWx0ZXJzIjogeyJTT1VSQ0UiOiBbIkxUX01PTCJdfSwgInF1ZXJ5IjogIkZFSlo9KFwiVCAyMCBLXHUwMGUxcm9seWkgY3NhbFx1MDBlMWQgKDE2NjQtMTg3OSlcIikifQ&per_page=100&page=2 (accessed on 26 February 2022).
69. "General Plan von dem Gantzen Grunde den der Jesuiten Orden Vormals zu Groswardein Besessen hat" Magyar Nemzeti Levéltár Országos Levéltára—T 86 Jezsuita Rendi Tervek (17. sz.–18. sz.)—T Kormányhatósági Fondokból Kiemelt Tervek (17 sz.–20. sz.)—T–Tervtár—Index No.: T86-XIII.-No.1/c. Available online: https://maps.hungaricana.hu/hu/search/results/?list=eyJmaWx0ZXJzIjogeyJGSUVMRFMiOiBbIkNJTVxcVFJHWSJdLCAiUElDIjogWyJZRVMiXX0sICJxdWVyeSI6ICJuYWd5dlx1MDBlMXJhZCJ9&per_page=100 (accessed on 26 February 2022).
70. "Geometrischer Plan des Radnoter Dominalgrundes" Magyar Nemzeti Levéltár Országos Levéltára—S 84 Erdélyi főkormányszéki térképek (1754–1862)—S Kormányhatósági fondokból kiemelt térképek (17. sz.–20. sz.)—S–Térképtár—Index No.: S84-No.99. Available online: https://maps.hungaricana.hu/hu/search/results/?list=eyJmaWx0ZXJzIjogeyJEQVRBQkFTRSI6IFsiTUFQUyJdLCAiRklFTERTIjogWyJISUVSXFxUSVRMRSJdLCAiUlRJUCI6IFsiVEVSS0VQIl0sICJTT1VSQ0UiOiBbIkxUX01PTCJdfSwgInF1ZXJ5IjogIkZFSlo9KFwiUyA4NCBcdTIwMTMgRXJkXHUwMGU5bHlpIGZcdTAxNTFrb3JtXHUwMGUxbnlzelx1MDBlOWtpIHRcdTAwZTlya1x1MDBlOXBla1wiKSJ9&per_page=100 (accessed on 27 February 2022).
71. A Nagykárolyi (Szatmár m., Carei, RO) Régi Kastély, Lovarda és Kert, Valamint a Tervezett új Kastély Magyar Nemzeti Levéltár Országos Levéltára—T 20 Károlyi Család (1664–1879)—T Családi Fondokból Kiemelt Tervek (1659–2000)—T–Tervtár—Index No.: T20No60. Available online: https://maps.hungaricana.hu/hu/search/results/?list=eyJmaWx0ZXJzIjogeyJGSUVMRFMiOiBbIkhJRVJcXFRJVExFIl0sICJQSUMiOiBbIllFUyJdfSwgInF1ZXJ5IjogIkZFSlo9KFwiVCAyMCBLXHUwMGUxcm9seWkgY3NhbFx1MDBlMWQgKDE2NjQtMTg3OSlcIikiLCAic29ydCI6ICJSRUNOOVU0ifQ&per_page=50&page=2 (accessed on 27 February 2022).
72. Zádor, A. Az angolkert Magyarországon. In *Régi Magyar Kertek*; Géczi, J., Stirling, J., Eds.; Vár ucca Tizenhét: Veszprém, Hungary, 1999.
73. "Situations Plan der Marosch Fluss Laage Bey Valliemare und Jenseithts /!/ Bey Soborsin" Magyar Nemzeti Levéltár Országos Levéltára—S 11 Kamarai Térképek (1747–1882) Magyar Nemzeti Levéltár Országos Levéltára S Kormányhatósági Fondokból Kiemelt Térképek (17. sz.–20. sz.)—S–Térképtár—Index No.: S11-No.1145. Available online: https://maps.hungaricana.hu/hu/search/results/?list=eyJmaWx0ZXJzIjogeyJHRU8iOiBbIllFUyJdLCAiUElDIjogWyJZRVMiXX0sICJxdWVyeSI6ICJGRUpaPShcIlMgXHUyMDEzIFRcdTAwZTlya1x1MDBlOXB0XHUwMGUxclwiKSJ9&page=14&per_page=100 (accessed on 28 February 2022).
74. Ecsedy, A. Huszadik Századi Szerzemények a Magyarországi Kertek Barokk Szoboranyagában: Schmidt Miksa Kertdekorációs Tevékenysége a "gernyeszegi Sorozat" Kapcsán. Ph.D. Thesis, Hungarian University of Agriculture and Life Sciences, Budapest, Hungary, 2003.
75. Fekete, A. Konzepte für Neugestaltung/Concepts Pour Réamenagement. *Anthos Z. Freiraumgestalt. Grün Landsch.* **2016**, *13*, 17–20.

land

Article

Water Dams of the Krakow Fortress: Potential of a Vanishing Heritage

Wojciech Korbel [1], Filip Suchoń [1,*] and Marta Łapuszek [2]

1 Faculty of Architecture, Cracow University of Technology, 31-155 Kraków, Poland; wojciech.korbel@pk.edu.pl
2 Faculty of Environmental and Power Engineering, Cracow University of Technology, 31-155 Kraków, Poland; marta.lapuszek@pk.edu.pl
* Correspondence: filip.suchon@pk.edu.pl

Abstract: Cultural heritage conservation is a constant process of preserving the valuable historical legacy and transferring it to future generations. The ability to adapt the matter under conservation to changing needs and environmental conditions is an essential element of this process. In this context, climate change and its consequences are a growing challenge, requiring innovative and often simultaneous efforts. This study was conducted in response to the discovery of previously unknown documents on nineteenth-century impoundment structures of the Krakow Fortress's defensive system. At present, the facilities are almost entirely ruined, yet the need to restore and preserve the memory of their culturally valuable legacy merits investigation. The conditions and requirements of the management of Krakow's changing hydrological environment became a vital component of this study. The uncovered archival documents were subjected to historical-interpretative analysis. Virtual modeling contributed to identifying the original scope of the dams' impact. Analysis of the city's spatial planning documents pointed to their contemporary potential. The entirety of the material collected aided in determining the framework in which protective measures targeting this dying heritage are currently possible. This study features a proposal for a new form of recreating the structures under investigation by assigning them a range of possible simultaneous uses. Thus, the presented research proposal is a form of concern for preserving this historical legacy and an attempt at rising to contemporary challenges posed by an intensively changing environment.

Keywords: water dams; Krakow fortress; climate change; environmental management; cultural heritage conservation

Citation: Korbel, W.; Suchoń, F.; Łapuszek, M. Water Dams of the Krakow Fortress: Potential of a Vanishing Heritage. *Land* **2021**, *10*, 1273. https://doi.org/10.3390/land10111273

Academic Editors: Magdalena Wilkosz-Mamcarczyk and Jan K. Kazak

Received: 22 October 2021
Accepted: 17 November 2021
Published: 20 November 2021

1. Introduction

Military engineering structures have always been an essential element of cultural heritage and have had a key impact on the development of civilization. The technical means they featured by shaping defensibility often significantly affected the natural and cultural landscape of the area they protected [1].

Along with technological advancement, changing needs, and intensive urbanization, many such structures were (and still are) altered in numerous ways, and many have irreversibly disappeared [2,3].

Therefore, the issue at stake is the ability to protect this heritage, especially where its value is undeniable. Equally essential is the ability to transfer this knowledge to successive generations while understanding dynamically changing conditions and accounting for environmental changes and sustainable development [4]. It can be observed in recent years that due to climate change the frequency and magnitude of extreme climate events such as heavy rains, flooding, and drought is increasing throughout Europe (Linnerooth-Bayer, Dubel and Sendzimir, 2015) [5]. Different risk management strategies are currently enforced in Central Europe, dealing with challenges of potential resilience-building measures and protection of cultural heritage against floods or fire (Cacciotti, Kaiser, and Sardella, 2021) [6].

The concept of heritage itself has evolved. It began with UNESCO's 1962 recommendation, raising the need to protect landscapes, natural and cultural environments that form a natural, harmonious whole. In the 1972 UNESCO Convention on the Protection of Heritage, "cultural heritage" appears, encompassing monuments and sites of outstanding universal value (Vecco, 2010) [7].

There has been some critical discussion of value typologies for heritage conservation and management in light of a review of published literature on heritage values (Fredheim and Khalaf, 2016). Considering the specificity of typologies derived from a particular context, a broad framework for assessing and communicating this meaning is proposed. This approach prevents the false division and separation of heritage into cultural and natural or tangible and intangible [8].

Another nuance is that the future of heritage is also loaded with our contemporary perceptions of the past. Given the relationship between attachment to a physical place, a sense of belonging, and the resulting importance to community life, heritage places suffer if the community does not feel connected to them. This results in neglect and vandalism (Spennemann 2011) [9].

This study was based on a separate query (performed by one of the authors) of archival documents concerning the original functioning of the Krakow Fortress. This extensive and culturally valuable complex that surrounds the city of Krakow, Poland, was built in the nineteenth century on the order of Emperor Franz Joseph I and is an element of numerous academic studies [10–12].

The original investigation conducted in the years 2014–2020 focused on the linkages between the city and its fortifications [3,13], urban structure revitalization, heritage site revitalization, and the role of modern-period fortifications in urban space, which accurately aligned itself with the scope of this study.

As a result, archival photographic and cartographic documentation was collected, presenting three poorly known water damming structures that had once been an element of the fortress's defensive system. Their general location has been shown on a plan of contemporary Krakow (Figure 1).

Figure 1. Diagrammatic plan of the Krakow Fortress with the locations of previously unknown impoundment structures marked in dark brown, accompanied by the boundaries of defensive areas I through VIII.

This study was based on documentation that includes the historical maps of Krakow's fortification system (drawn to a scale of 1:10,000, sheets: no. 16 "Rakowice," no. 20

"Olszanica," no. 33 "Opatkowice") [14], extension plans from the First World War (drawn to a scale of 1:10,000), and unpublished archival photographs from 1915.

In the literature on the fortress, most studies have a cross-sectional and synthetic character and refer to its history as one structure. The use of water obstacles in Krakow's defensive system has not been previously investigated. Thus, the archival materials found have supplemented the knowledge resulting mainly from the war manuals and historical standards, including an Austrian instruction for engineering and sapper forces [15,16], an Austrian permanent fortifications manual [17], and a Polish field fortifications manual [18].

These materials also provided a basis for evaluating the contemporary management of the city's aquatic environment in the context of current challenges and needs, including heritage conservation. The fortification elements studied had long lost their defensive significance, and despite the hypothetical protection afforded to historical monuments, became utterly forgotten. However, it appears that they can and should find a new application under current environmental conditions, restoring the invaluable memory of their operation in the cityscape. Based on the collected archival material and recent planning documents, this study reports on an investigation of solutions that can be best suited to preserve the disappearing heritage while also serving the goals of sustainable development.

2. Materials and Methods

An archival query was performed in search of source documentation. The study was divided into stages, and began with historical-interpretative research. The use of archival materials, iconographic sources (primarily photographs), and field research demonstrated the link between the structures identified during the query with their spatial context and relevant past events.

The authors also performed a simplified military site analysis. With this tool, hidden under the acronym KOCOA (meaning Key Terrain Observation and Fields of Fire, Cover, and Concealment, Obstacles, Avenues of Approach), it is possible to read the broad background of design and construction decisions [19]. The results of this analysis are included in the illustrations of the existing condition, provided further in the text. The location of the forts, the range of fire and mutual fire support, the blind spots, the course of wire obstacles, the relief of the terrain making observation difficult, and the expected attack directions were read from staff maps. All these elements, when superimposed, showed a rather broad approach to be dangerous for the defenders in the area of the Wilga, Dłubnia, and Rudawa river valleys.

The investigation of the context included an inspection of selected sites and the collection of photographic documentation. A literature review was performed during the comparative analysis of damming structures to other Austrian fortifications across Europe. Furthermore, spatial and environmental policy documents for Krakow were analyzed, especially in their relationships with the areas under study and the role they assigned to them.

Initial virtual modeling was applied in the study. Sections of the site in the vicinity of the former dams were modeled by CAD software. Since relief and land use have changed over the centuries, contours from historical maps were used as input data. An attempt was made to present the damming result in its original state and in a form that can currently be reconstructed. The work was carried out in three stages.

2.1. Analysis of Documentation Content—The Past Significance of the Structures

During the first stage, the archival documentation was analyzed. The technical details of the damming structures were excluded from the study as a separate problem related to heritage conservation rather than landscape planning and the environmental impact of fortifications.

During the investigation, an attempt was made to estimate the area and depth of the flooding area that resulted from impoundment during the period when Krakow's fortifications had been in active use. This analysis also concerned the potential consequences of these effects. Analytical work was based on similar studies by other teams of researchers. In the Polish literature, similar research was performed by a group of academics from Wrocław [20,21]. The subject of this study covered light fortifications erected at the turn of the twentieth century and linked with Wrocław's contemporaneous fluvial network. This research involved the modelling of historical flows and recreating the scope of flooding on the Ślęza and Widawa rivers caused by impounding the water with military dams. Based on the data collected, it also became possible to build a hydraulic model of water damming by historical weirs of the Wrocław Fortress [20].

The publications presented above became an inspiration for this study, as they covered:

- Identification of impoundment functioning and impact in the original period;
- Investigation of the obtained information's significance about state of the art.

2.2. *Analysis of the Existing State and Applicable Formal and Legal Documents—Heritage in the Present Day*

The insight gathered during the first stage was confronted with the current functioning of the structures or their remains. The second stage of the study began with a site visit. The investigation also focused on the role and possible significance of the facilities under study in Krakow's applicable spatial development policy documents.

The investigation included:

- A survey of the current state of the structures and an identification of changes in local development (relative to the original form);
- An assessment of the significance of the structures in currently applicable frameworks of local management and development policies and reference to the threat posed by climate change.

2.3. *Synthesis of the Information Collected in the Context of the Potential for Contemporary Conservation and Use of Disappearing Cultural Heritage*

During the final stage, the information collected during the previous stages became a basis for an in-depth discussion of the contemporary potential for preserving the heritage under study against Krakow's dynamic development background. Thus, key elements of phase three were:

- An identification of possible weaknesses in the policy formulated by the municipality (especially in the light of the archival materials uncovered);
- A search for directions of integrating efforts towards sustainable development, including the conservation of a disappearing cultural heritage and its incorporation into contemporary forms of use and countering climate change.

3. Results and Discussion

3.1. *Results of the Analysis of the Collected Documentation and Its Application Potential*

3.1.1. General Overview of the Three Artificial Dams

Nineteenth-century defensive doctrines assumed organizing most fortresses as ring-type structures, with an external ring of forwarding forts located in the foreground and an internal ring of continuous fortifications in direct proximity to the city they defended. Considerable tracts of land were used for defensive purposes and occupied by defensive structures (divided into covers, emplacements and obstacles) [22]. Construction bans were enacted around cities to leave a free field of fire and facilitate observation. Furthermore, military doctrine assumed supplementing the permanent fortification strip (forts and batteries) with field fortifications, necessary during a state of immediate threat and mobilization [17,22] (Figure 2).

Figure 2. Fragment of the original plans from 1915 concerning field fortifications, an obstacle course, and a water damming on Dłubnia. From the authors' private collection.

Adaptive means and actions mainly involved "clearing the foreground" which entailed clearing forests, demolishing buildings, land leveling, etc. A constant readiness to quickly destroy bridge crossings was maintained. The potential for using water bodies and waterways for local impoundment was noted, as it would flood specific areas and make them impassable. Sapper instructions [16] highlighted the potential use of wide riverbeds and other terrain fortifications as obstacles. A watercourse could be turned into an obstacle by building hydraulic structures that formed artificial wetlands, blockages (weirs), or flooding. Depending on local conditions, they were either makeshift or comprehensive structures.

In the Krakow Fortress, the first and second defensive circuits were formed by the cities of Krakow and Podgórze, respectively, separated by the Vistula River. The borders of the remaining defensive circuits included water divisions. In the north, the third circuit stretched between the Vistula and Rudawa Rivers, the fourth from the Rudawa to the Białucha, the fifth from the Białucha to the Dłubnia, and the sixth from the Dłubnia to the Vistula. In the south, the fortification ring ran from east to west, towards the Vistula and was divided by the Wilga River into two defensive circuits [23]. With defensibility in mind, at the junction of the circuits mentioned earlier (Wilga, Rudawa, and Dłubnia), hydraulic structures were designed and built in 1914, enabling the creation of artificial inundation areas parallel to the frontline. The quality and effectiveness of these obstacles were defined using "storm resistance" ratings (originally called *Sturmfreiheit* in German). The mild or heavy flooding and wetlands created with them were to become a significant terrain obstacle [16].

Design operations preceded the process of creating such obstacles. The borders of future inundation areas were delineated, and the potential associated costs were assessed. The evaluation covered the profile of a river or channel, soil characteristics, bank incline parameters, water velocities, and information about pre-existing structures (bridges, levees, sluices, etc.). The sites for the newly designed crossings were also marked in the field, while accounting for the apparent optimization of means and resources [18].

Similar operations concerned the construction of the three dams on Dłubnia, Wilga, and Rudawa.

The Dłubnia River was impounded by improvising a blockage of the clearance under a bridge, with the concrete abutments simultaneously constituting the abutments of the weir (initially named the *Dłubnia Stauanlage*). These abutments featured vertical niches visible in the figure below, used to fasten a flat barrier (Figure 3). The dam was located along a road that connected Mogiła with Krakow. The nearest known and surviving fortification is Fort 49a Dłubnia, situated around 1 km to the north.

The Wilga's waters were dammed using a particular complex of buildings and devices. Its course could be closed using the fortress's valve weir and a second levee across the river valley (originally named *Wilga Schleuße, Staudamm a.d. Wilga*). They formed an extensive hydraulic structure dedicated strictly to military needs (Figure 4). Relatively near the weir, at around 0.5 km to the west, is the surviving Fort 52a Jugowice-Łapianka.

Water impoundment in the Rudawa river valley was performed between two waterways, namely between the Rudawa and its millrace, the Młynówka. It was located around 0.5 km to the south of Fort 41a Mydlniki. The impoundment was possible by building an earthen levee across the valley along with the channel connecting both waterways (originally named *Rudawa Inundationsdamm*). The inundation channel along with an embankment between the Rudawa and Rudawa's millrace were built in an agricultural area. An archival photograph shows a fragment of the structure with the completed earthworks and the reinforcement of the levee bank using a layer of stone. The bed of the channel had a circular cross-section, typical for structures from the beginning of the twentieth century. The photograph also shows the reinforcement of the bank zone with a system of fascines, further strengthened by fences from wooden posts and wicker. From the side where attacks were to take place, a strip of obstacles from wire on barrier rods was built (Figure 5).

Figure 3. View of a bridge on the Dłubnia, archival photo from 1933. Note the massive concrete abutments that were also used by the weir. From the collection of the Polish National Digital Archives, signature no. 3/1/0/8/3240.

Figure 4. View of the damming wall of the *Wilga-Schleuße* fortress weir, seen from the side of the inundation area. Archival photograph from 1915. From the author's collection.

Figure 5. View of the levee and inundation channel from the side of the Rudawa millrace. Archival photograph from 1915, from the author's collection.

3.1.2. Overview of the Size and Depth of the Inundation Areas under Study

The material collected allowed for an in-depth analysis of the size of potential inundation areas. The literature (Biesiekierski, 1922) [18] points to the application of needle dams [24] with a maximum restriction level set at around 1.8 m. By analogy, this level was also assumed for the structures under analysis.

Archival maps drawn to a scale of 1:10,000 from 1899–1902 and 1914–1915 were used, as they contained a sufficient amount of information to complete the investigation. The authors determined the flood zones based on the analysis of the topographic layout of the area. The contour lines existing on the historical map were taken into account. The assumed maximum possible level of water accumulation (approx. 1.8 m) and the contour lines indicated the range of potential inundation zones.

In the case of archival maps from the end of the nineteenth and the beginning of the twentieth century, the investigation encountered obstacles in identifying permanent points (objects) that could be referenced to contemporary landmarks. In most cases, only the forts that have survived to the present day provided an anchor. The historical maps were overlaid on the existing maps of spatial development plans and the current topographic maps to identify the relation between the historical flood zones and the recent development of the analyzed areas.

Throughout the investigation concerning the Dłubnia weir, it was possible to determine the old inundation area as potentially being up to 62.2 ha (Figure 6). The inundation surrounding Fort Dłubnia from the west strengthened the designated division between the defensive districts.

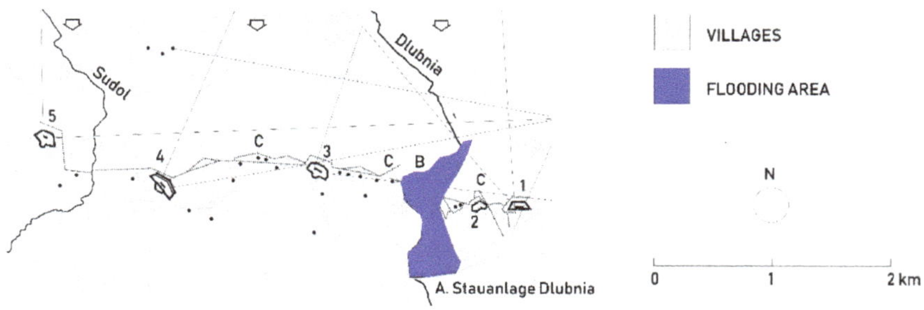

Figure 6. Reconstruction of the Dłubnia inundation area for the state from 1915. The watercourse was blocked by an improvised blockade under a bridge clearance. The water flowing downstream (due south) spilled successively into the river valley. Markings: A. location of the levee, B. inundation area, C. wire obstacle strip; 1. Fort Dłubnia, 2. Earthwork 1/Dłubnia, 3. Fort Mistrzejowice, 4. Fort Batowice, and 5. Fort Sudół. The expected avenues of attack are marked with arrows.

As a result of the impoundment of the Wilga with a fortress levee, an inundation was produced that was confined by a railway embankment from the west and the natural terrain layout from the east (Figure 7). The estimated total area of the historical inundation is around 29.7 ha.

Fort Mydlniki and its foreground were protected using a levee that closed a historical inundation and its accompanying parallel channel. The inundation area closed off by the levee, estimated based on historical maps and calculations, was approximately 17.9 ha (Figure 8).

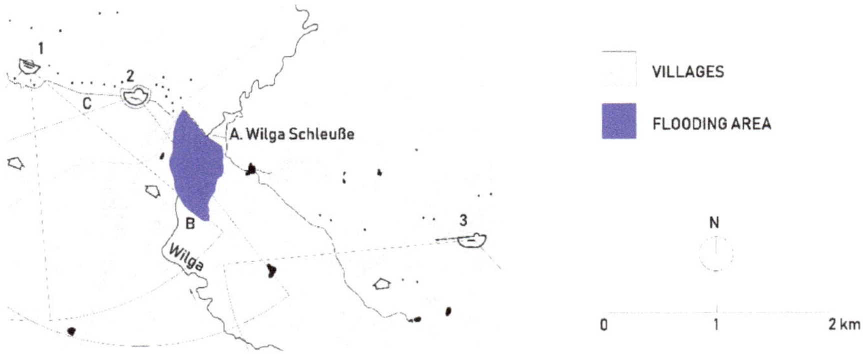

Figure 7. Reconstruction of the Wilga inundation area for the state from 1915. The watercourse was blocked by the fortress weir. The water flowing downstream (due north) spilled successively into the river valley. Markings: A. location of the levee, B. inundation area, C. wire obstacle strip; 1. Fort Borek, 2. Fort Jugowice (a.k.a. Łapianka), and 3. Fort Wróblowice. The expected avenues of attack are marked with arrows.

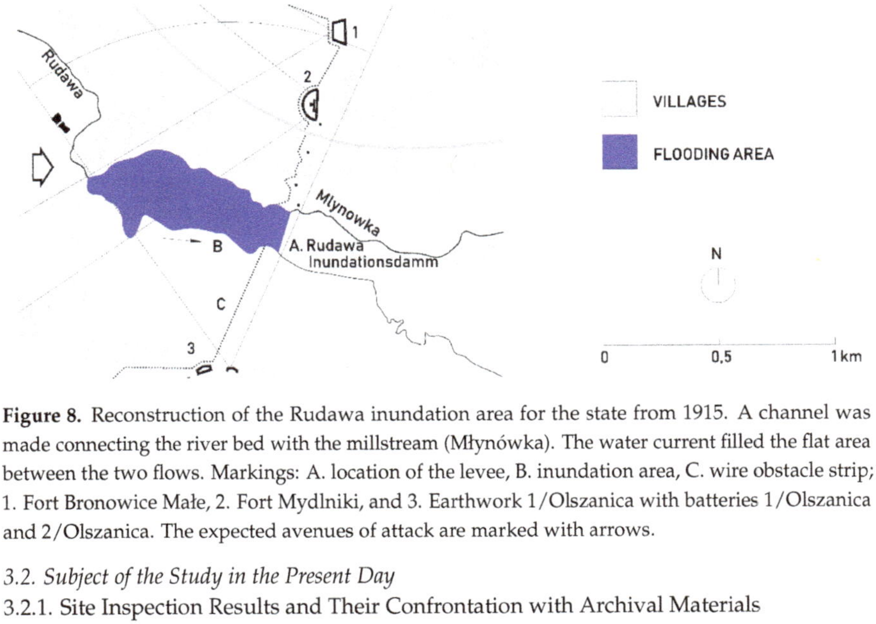

Figure 8. Reconstruction of the Rudawa inundation area for the state from 1915. A channel was made connecting the river bed with the millstream (Młynówka). The water current filled the flat area between the two flows. Markings: A. location of the levee, B. inundation area, C. wire obstacle strip; 1. Fort Bronowice Małe, 2. Fort Mydlniki, and 3. Earthwork 1/Olszanica with batteries 1/Olszanica and 2/Olszanica. The expected avenues of attack are marked with arrows.

3.2. *Subject of the Study in the Present Day*

3.2.1. Site Inspection Results and Their Confrontation with Archival Materials

The investigation was located in the inundations in the contemporary structure of Krakow. A detailed site inspection was performed to search for any remains.

In the Dłubnia River area, due to the remodeling of the transport system, the construction of a new bridge crossing (Figure 9), and dynamic urbanization, no remains of historical hydraulic structures have survived.

Figure 9. Current state of preservation of the Dłubnia inundation area, 2021.

In the Wilga riverbed, one can find the remains of its weir, in the form of a broken-up concrete slab and the fragment of the northern abutment of an earthen levee. The structural elements of the levee are visible, namely, the spaces between the stone walls filled with non-reinforced concrete and large-fraction lime aggregate. The area of the levee and weir complex is currently located in direct proximity to Krakow's expressway bypass. A part of the complex abuts the roadway, while another is an overgrown area covered with wild plants (Figure 10).

Figure 10. Current state of preservation of the remains of the fortress weir on the Wilga River, 2021.

In the Rudawa valley, urbanization processes and development expansion did not transform the area under study to a significant degree. Yet, agriculture and melioration work erased the course of the channel and the levee. Only a meadow is now present, slowly being overgrown with high greenery (Figure 11). The area is used to a minimal degree, mostly for farming and partially for strolling through non-landscaped greenery. The only site fully developed for recreation is a footpath that abuts the millrace bed. The technical infrastructure in the area is also not without significance: it consists of two underground Ø1200 water supply collector pipes and a high-voltage overhead power line.

Figure 11. Current state of preservation of the Rudawa impoundment area, 2021.

3.2.2. Analysis of the Municipality's Management Policies for the Areas under Study

The investigation of spatial development and environmental protection policy documents for Krakow first focused on the Spatial Development Conditions and Directions Study [25]. The analysis also covered other, more detailed documents on this subject, i.e., local spatial development plans drafted based on said Study [26–29] and relevant sectoral documents concerning development management.

In terms of the desired spatial structure [25] (Map K1: Spatial structure—development directions and precepts) the areas under study were assigned in both the study and applicable local plans for non-landscaped greenery conservation and design. The presence of various forms of surface waters and landscaped greenery, including parks, green squares, river parks and buffer greenery was also permitted [25] (Volume III—Guidelines for Local Plans, p. 10). Furthermore, in the Rudawa River area, a land reserve (crossing a green area) was created for the construction of a road that would connect Krakow's downtown area with the Balice airport.

The study identifies a range of climate determinants, including potential threats such as "floods and inundations, draughts, cold and heat waves, mass soil movements" [25] (Volume I Development determinants, p. 74). In terms of proposed development rules and directions, all three locations were included in a landscape preservation and shaping zone, but not in the fortified landscape conservation zone [25] (Map K2: cultural environment—development and protection directions and precepts). Consequently, local development plans were also found not to highlight the historical significance of the locations in question. They were all included in the wildlife system structure as elements of river parks and cross-ventilation ecological corridors [25] (Map K3: natural environment—conservation and development conditions and precepts).

Another crucial document for this analysis is the Framework Conservation and Revitalization Program for the Historical-Landscape Complex of the Krakow Fortress [30].

The declared objectives of the program are to place fortification structures and their surrounding "fortified" landscape under legal protection, which includes the provision of local spatial development plans. The program assumes the need to revitalize these structures via their adaptive reuse as public or commercial facilities. The most valuable sites should be used as museums and exhibition spaces. The others had a range of research and academic, cultural, educational, recreational, and tourism-related uses assigned to them. The program identified a need for the general accessibility of restored fortifications by regulating legal and ownership-related matters. The structures of the fortress in possession of the Krakow municipality should be combined into a uniform historical-landscape complex, in which fortress greenery is to form a part of park areas. This document is a general collection of guidelines for the protection of fortification elements but does not mention water impoundment structures.

The document titled Development and Management Directions for Green Areas in Krakow for the years 2017–2030 is another source important to this study [31].

It asserts the necessity to form sequences of green areas that would act as ecological corridors. It indicates a need to integrate the scattered greenery structure into a continuous system, namely, the linking of existing and new public green areas via pedestrian and bicycle paths and green space sequences, to improve circulation potential and develop recreation. The document assumes a need to provide at least 10 m^2 of landscaped recreational green areas per citizen. It also points to a need to support the conservation of cultural landscape zones by managing existing green space and establishing new green places.

The 2030 Krakow Climate Change Adaptation Plan [32] is an especially crucial document to this investigation. It concerns the consequences of ongoing climate change that are compounded by spatial development. It points to frequent torrential rain, inundation, flooding, and long periods with high temperatures, as well as intensive expansion of development and the growing number of vehicles. These phenomena result in a decrease in biologically active areas and water assets. The document specifies necessary preventative measures, including "informational and educational actions," "organizational actions," and "engineering actions" [20] (p. 6). The document provides a detailed list of necessary forms of activity and their associated capital expenditure. The highest priority was given to countering the consequences of floods, heatwaves, and worsening air quality via the development of broadly understood blue and green infrastructure.

The 2020–2030 Environment Protection Program Draft for the City of Krakow [33] is the last major document. Until September 2021, it had not been completed, yet its draft included crucial proposals for regulation intended to shape a socially acceptable city development strategy focused on environmental protection. In parallel, the program should facilitate socio-economic development. It includes a range of long-term goals, including the protection of Krakow's wildlife and landscape assets, increasing the share of green areas, a rational water and waste management policy, improvements to air quality, a minimization of flood risk, and ecological education for residents. The draft comprehensively corresponds with the subject of the city's spatial development and references other development strategy documents.

3.3. Analysis of the Collected Documentation in the Context of Modern Needs and Possibilities of Protecting the Heritage under Investigation

3.3.1. Analysis Conclusions

The analysis of archival materials presented in this study and their comparison with contemporary development strategy documents for Krakow indicated a need to take parallel action. The primary measures to be featured in these actions include:

a. The need for greater protection of the Krakow Fortress complex, including measures intended to conserve and restore the memory of buildings which are the most endangered and at risk of being forgotten.

This is the most obvious need, tied to raising public awareness about Krakow's cultural identity. The materials present new elements of Krakow's fortifications that were not previously covered by development plan provisions. The need for their greater protection nevertheless has a basis in the city's general development strategies. It is also reflected in the Polish Act on Monument Protection and Preservation [34].

b. The need to develop the city's green spaces to strengthen recreational and educational functions.

In this context, establishing a waterfront park system is an excellent opportunity to create valuable leisure spaces that would be in a state of symbiosis with the river. The necessary directions of action highlighted in the strategy documents examined are confirmed in the literature. The cases of London, Toronto, or New York demonstrate how establishing waterfront park systems facilitates ecological education and even the promotion of the city [35].

c. The need to find and design small-scale water retention sites in the spatial management system with the intent to make better use of dwindling water assets and limit the risk of local inundation.

The intensive urbanization observed over the past decades, in the absence of an effective planning system, increased Krakow's vulnerability to floods. Creating water-absorbing areas is therefore a priority that is extensively noted in the city's development strategies. Furthermore, these measures should account for the proper distribution of retention areas and thus improve their effectiveness [36], reduce the intensity of water surges [37,38], and enhance biodiversity in floodplains [39,40].

d. The need to use and create a reserve of undeveloped land for creating cross-ventilation corridors, especially along waterways.

River valleys are areas of intensive air circulation. Ensuring they are corridors free from development is thus a priority for the correct cross ventilation of the city. This need is highlighted both in the analyzed strategy document and in many academic studies [41].

e. The need for integrated action.

Listed last, the need for integrated action deserves a special mention. This integration, often articulated in the analyzed strategy documents, supports parallel action intended to diversify the funding of advanced and costly projects. This direction finds extensive application in numerous European and global means of formulating local development policies [42,43]. The literature highlights the significance of ensuring the spatial continuity of systems, which enhances recreational potential, improves the quality of the natural environment, and the effectiveness of valley retention [44,45]. It also points to cases of comprehensive attempts at solving hydrotechnical problems, matters of shaping landscape composition and creating recreational spaces for the city [46,47] in the form of "green chains" abutting rivers.

3.3.2. Proposal of Development Action for the Subject of the Study

The need for integration listed above resulted in an analysis of possible reconstructive action in the three spaces under study. The analysis covered possible forms of action, including the capacity to form new impoundments, their potential location under contemporary conditions, possible size, and significance.

Impoundment of the Dłubnia:

The confrontation of the original inundation area with the current land development showed that contemporary recreation of the historical state is not possible. A renewed attempt at impoundment would negatively affect the structure of a roadway embankment and would reduce the safety of Okulickiego Street, which runs to its south. However, such action is partially in alignment with the city's policy, in the form of a waterfront park that would stretch along the bed of the Dłubnia, covering undeveloped space. This is an area with a length of around 500 m and a width that ranges between 80 and 230 m, occupying an area of 10.8 ha. This area could become the site of a so-called rain garden with a system of scattered, small reservoirs (ponds) which would infiltrate stormwater directly into the soil. An example of the possible layout of such ponds shall be presented in a reconstruction displayed in Figure 12. It was assumed that their combined area would not exceed half of the available site. By increasing small-scale water retention, this concept would provide a leisure space, develop biodiversity, and protect the surrounding space. This proposal should be supplemented by a properly developed information and education system (including historical photographs and a mockup of a reconstruction of the damming structure), which would propagate knowledge about the no-longer-existing fortress's elements.

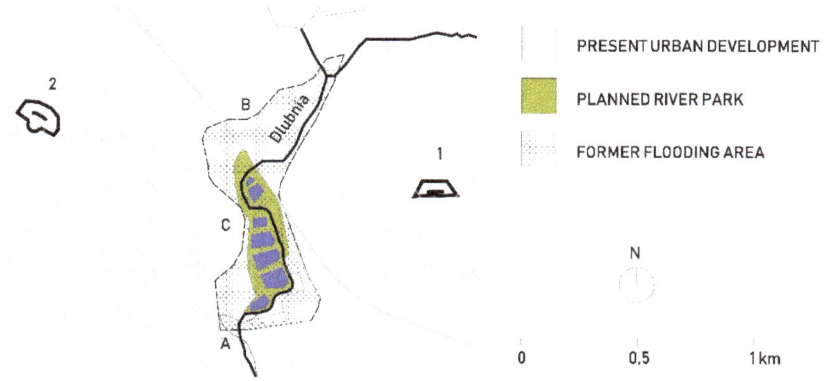

Figure 12. Proposal of a new inundation zone (as a part of a waterfront park) of the Dłubnia River. Markings: A. original location of the impoundment weir, B. border of the historical inundation area; 1. Fort Dłubnia, and 2. Fort Mistrzejowice.

Impoundment of the Wilga:

The historical inundation area of the Wilga River determined in this study was also found to be impossible to recreate. The original impoundment significantly extends beyond present-day undeveloped areas. Photographic documentation allows for the precise determination of the architectural form and dimensions of the sluice itself. One possible action is to rebuild it and delineate a small inundation area that would be safe for the present-day surroundings, occupying around 3.8 ha. This proposal would be a reference to the historical impoundment, creating attractive conditions for recreation, including fishing. The inundation area could be supplemented by a waterfront park, located along the left bank of the Wilga and the highway, namely, at the site in which a larger inundation area would potentially threaten the road's structural integrity. Such a park would occupy around 4.9 ha of land, and no more than half of the space would consist of water reservoirs laid out in a "chain" pattern (Figure 13). The recreated sluice, with displays showcasing its history, would be the main element of the complex, in compliance with the provisions of the Fortress Revitalization Program. Encompassing the nearby Borek, Jugowice and Wróblowice forts, all restored in 2021, the resultant complex would be highly valuable in terms of education and recreation.

The Rudawa River inundation area:

The former inundation area of the Rudawa is currently difficult to recreate. The primary hindrance here is underground municipal infrastructure that perpendicularly crosses the area, thereby limiting potential interventions. The land reserve stipulated in the area's local spatial development plan, intended for road construction, is an obstacle here. Another difficulty is the inability to precisely identify the channel's location. There are no clear traces of its presence in the area, and the applicable land development plan does not feature an obligation to reconstruct it. Despite this, the need for non-landscaped greenery which predominates in planning guidelines, can be used as a pretext for the location of a new hydrotechnical structure, and the channel identified in this study appears to be the most attractive element for reconstruction. Placed as it had been originally, perpendicular to the bed of the Rudawa, it would act as a retention reservoir that would allow water infiltration into the soil, with the possibility of draining its excess into the river [48]. The estimated area of the channel, delimited by the reserve of the planned road, is proposed to be around 0.3 ha. The channel would be primarily an earthen structure, linked with nearby transport and circulation routes, especially pedestrian ones. It would also be an excellent example for implementing similar solutions used for periodical retention in contemporary development. Analogously to other sites, it would require proper infrastructure and annotation, providing historical knowledge (Figure 14).

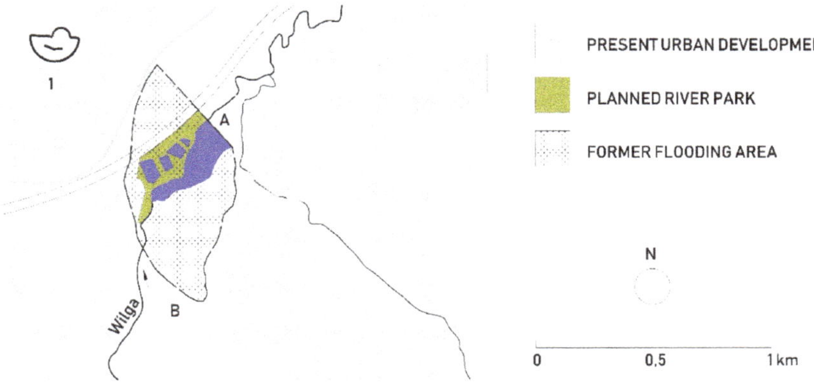

Figure 13. Proposal for a new inundation zone for the Wilga River. Markings: A. location of the weir and levee, B. border of the historical inundation area; 1. Fort Jugowice (a.k.a. Łapianka, currently a Scout Movement Museum).

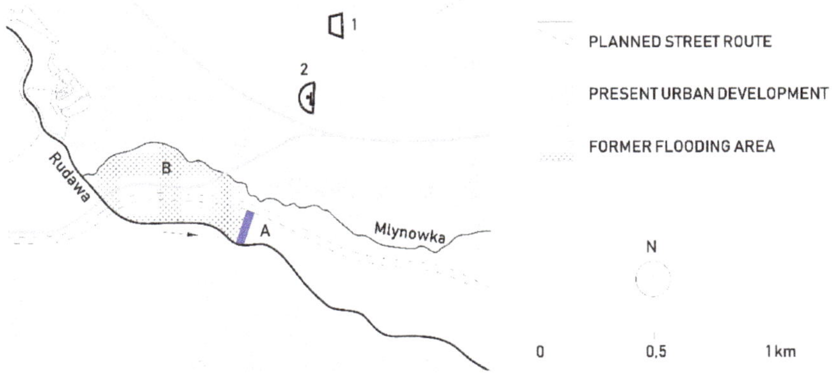

Figure 14. Proposal of an infiltration channel for the Rudawa River. Markings: A. location of the historical levee and irrigation channel—proposed site for the channel and infiltration reservoir, B. border of the historical inundation area; 1. Fort Bronowice Małe, and 2. Fort Mydlniki.

As a part of the investigation, an attempt was made to determine the effectiveness of the proposals presented by simulating their operation based on area estimates (Table 1).

The analysis covered the infiltration potential of the proposed solutions, as defined in three working models, while noting that the infiltration would be performed by the proposed reservoirs. The effectiveness of the total operation of the reservoirs in the waterfront parks of the Dłubnia and the Wilga was calculated, assuming maximum use. Analogous calculations were performed for the Rudawa channel. The total area of the reservoirs was set at a level of 40% of the area of the accompanying waterfront park. Each reservoir (pond) would receive surface runoff, retain it, and allow it to be absorbed into the soil. Any excess runoff would be redirected to the adjacent river [49,50]. Filtration speed was assumed to have the conditions of a unidimensional set flow through the soil, either partially or fully saturated with water, as found in Darcy [47]. It was also assumed that soil humidity increases along with depth, until the groundwater table is reached [51]. The initial groundwater level was assumed in concordance with the real-world conditions present in the areas under study [52]. The infiltration volumes for the proposed reservoir sizes, listed in $m^3 \cdot s^{-1}$, are presented in Figure 15.

Table 1. Estimation of the area size of historical and proposed inundation areas.

River	Historical Inundation Area (ha)	Presently Available Space for Inundation Area Reconstruction (ha)	Proposed Form of Use	Proposed Form of Use Area [ha]/Infiltration Area (ha)
Wilga	29.7	8.7	Reservoir	3.8/—
			Waterfront park with infiltration reservoirs	4.9/2.0
Dłubnia	65.2	10.8	Waterfront park with infiltration reservoirs	10.8/4.3
Rudawa	17.9	0.3	Infiltration channel	0.3/0.3

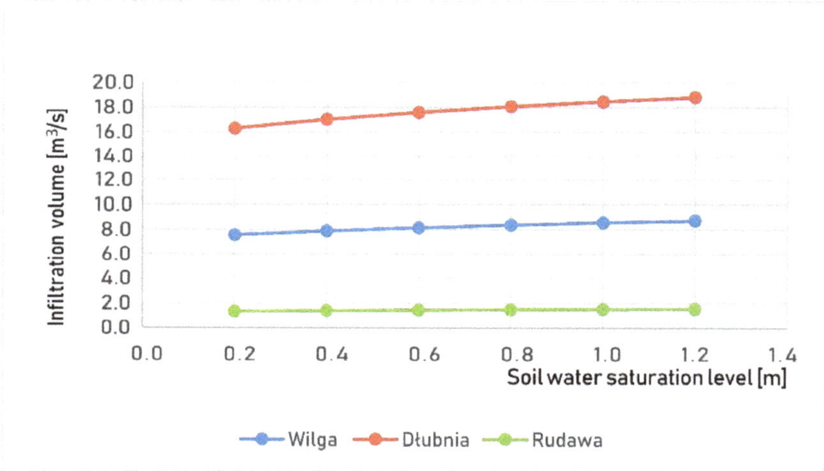

Figure 15. Infiltration volume changes for the proposed infiltration reservoirs located in the Dłubnia and the Wilga waterfront parks and the Rudawa channel.

Furthermore, assuming an average depth for the infiltration reservoirs of only 1 m, their retention capacity would amount to 3000 m^3 in the case of the Rudawa, 19,700 m^3 in the case of the Dłubnia, and 43,000 m^3 for the Wilga. These are volumes that can retain runoff from a fifteen-minute bout of torrential rain on a fully paved surface with an area of 30.8 ha. The areas were identified from which runoff towards the receiving waters is possible in terms of their topographic configuration. The retention capacity of the adopted solutions was estimated, The literature shows that the most significant influence on the value of the runoff coefficient is the type of catchment area coverage [53,54]. Therefore, to determine the average value of the runoff coefficient, the authors analyzed the land use of these areas (impervious roofs, roadways with asphalt surface, clinker surfaces, prefabricated concrete substrates, gravel paths, green spaces, etc.).

The analysis of the effectiveness of the proposed solutions demonstrates their significant potential. The final size selection of reservoirs requires detailed hydrological, hydraulic and topographic data. However, the findings presented prove the feasibility of simultaneous actions in which retention and infiltration significantly diminish the consequences of climate change, contributing to the conservation of cultural heritage. The authors primarily considered the risks associated with water management and extreme hydrometeorological events: heavy, violent rains, rising rivers, and periods of drought in the absence of adequate drainage and water accumulation.

4. Summary & Conclusions

The fortress structures under study, previously unknown in the literature and the city's strategic documents, presented against the background of Krakow's spatial development needs and dynamically changing climate conditions, were used to perform an in-depth analysis of their contemporary potential.

The investigation revealed potential for the protection of a dying heritage in the form of simultaneous efforts aligned with a wider spatial management and development conducive to recreation and even the promotion of the city. This protection, more directly than before, would also address climate threats and challenges, while creating the necessary blue–green infrastructure.

The investigation produced three proposals of design solutions that combine the following goals:

- A presentation of knowledge about disappearing elements of the Krakow Fortress to a wide audience;
- The revitalization of selected spaces, in full compliance with the applicable city development framework policy and higher-order legal acts;
- The creation of new, attractive park spaces;
- The implementation of solutions that would counteract the growing threats related specifically with sudden water surges or long-term water shortages;
- The implementation of measures in a holistic, integrated system;
- The implementation of activities towards diversifying funding sources (during construction and later operation) due to significance to cultural heritage, flood protection, establishing ecological corridors, developing tourism, recreation, and active social education.

The findings presented in this paper are a comprehensive proposal for taking action that would restore the memory of a widely unknown and disappearing cultural heritage and lay the groundwork for the implementation of the wider policy of sustainable development. Moreover, the concept of heritage, identification as part of the history of a place and its reception by inhabitants of Krakow, appears as a future research field.

Author Contributions: Conceptualization, methodology: F.S., W.K., validation, M.Ł., F.S., and W.K.; formal analysis: W.K.; investigation and resources: F.S. and M.Ł.; data curation, M.Ł. and F.S.; writing—original draft preparation, M.Ł., F.S., and W.K.; writing—review and editing, W.K.; visualization, F.S.; supervision, W.K.; project administration, W.K.; funding acquisition, W.K. All authors have read and agreed to the published version of the manuscript.

Funding: This publication received financial support from the Polish Ministry of Science and Higher Education under subsidy for science for the FA CUT in 2021.

Institutional Review Board Statement: Not applicable.

Informed Consent Statement: Not applicable.

Data Availability Statement: The input data in this research can be found within the article or accessed freely from online sources.

Conflicts of Interest: The authors declare no conflict of interest. The funders had no role in the design of the study; in the collection, analyses, or interpretation of data; in the writing of the manuscript, or in the decision to publish the results.

References

1. Bogdanowski, J. *Architektura Obronna w Krajobrazie Polski: Od Biskupina do Westerplatte*; PWN: Warszawa, Poland; Kraków, Poland, 1996; ISBN 83-01-12223-4. (In Polish)
2. Brzoskwinia, W.; Janczykowski, J. *Zabytki Fortyfikacji Twierdzy Kraków: Ochrona i Konserwacja w Latach 1991–1998*; seria II, t. 1; Atlas Twierdzy Kraków; Agencja Wydawnicza "ZEBRA": Kraków, Poland, 1998; ISBN 83-87219-18-5. (In Polish)
3. Suchoń, F. "The iron ring". The fortification system as the constituting factor in the hybrid structure of the city-fortress. In *Hybrid Urban Structures*; PK: Kraków, Poland, 2016.
4. Myczkowski, Z.; Wielgus, K.; Środulska-Wielgus, J.; Rymsza-Mazur, W.; Chajdys, K. *Rejestracja Krajobrazów Warownych Jako Podstawa Rewitalizacji Dawnych Twierdz*; Zarząd Główny Stowarzyszenia Konserwatorów Zabytków: Kraków, Poland, 2008. (In Polish)
5. Linnerooth-Bayer, J.; Dubel, A.; Sendzimir, J.; Hochrainer-Stigler, S. Challenges for mainstreaming climate change into EU flood and drought policy: Water retention measures in the Warta River Basin, Poland. *Reg. Env. Chang.* **2015**, *15*, 1011–1023. [CrossRef]
6. Cacciotti, R.; Kaiser, A.; Sardella, A.; De Nuntiis, P.; Drdacky, M.; Hanus, C.; Bonazza, A. Climate change-induced disasters and cultural heritage: Optimizing management strategies in Central Europe. *Clim. Risk Manag.* **2021**, *32*, 100301. [CrossRef]
7. Vecco, M. A definition of cultural heritage: From the tangible to the intangible. *J. Cult. Herit.* **2010**, *11*, 321–324. [CrossRef]
8. Fredheim, L.H.; Khalaf, M. The significance of values: Heritage value typologies re-examined. *Int. J. Herit. Stud.* **2016**, *22*, 466–481. [CrossRef]
9. Spennemann, D.H.R. Beyond "Preserving the Past for the Future": Contemporary Relevance and Historic Preservation. *CRM J. Herit. Steward.* **2011**, *8*, 7–22.
10. Środulska-Wielgus, J. *Rola Turystyki Kulturowej w Obronie i Udostępnieniu Krajobrazu Warownego*; PK: Kraków, Poland, 2016. (In Polish)
11. Środulska-Wielgus, J. *Zieleń Twierdzy Kraków*; seria II, t. 4; Atlas Twierdzy Kraków; Urząd Miasta Krakowa. Wydział Kultury i Dziedzictwa Narodowego. Oddział Ochrony Zabytków: Kraków, Poland, 2005. (In Polish)
12. Myczkowski, Z.; Środulska-Wielgus, J.; Staniewska, A.; Wielgus, K.; Rymsza-Mazur, W. *O przyszłość Twierdzy Kraków: Opracowanie Studialne Dla Programu Planu Ochrony Zespołu Historyczno-Krajobrazowego Twierdzy Kraków*; CZUWAJMY: Kraków, Poland, 2010. (In Polish)
13. Suchoń, F. 20th century fortress gates in Kraków as spatial marks of the city limits. In *Tradition and Heritage in the Contemporary Image of the City: Monograph. Vol. 3, Practice and Process*; Wydawnictwo PK: Kraków, Poland, 2014.
14. Festungs-Umgebungs-Plan von Krakau, Special-Übersichts-Plan. K.u.k. Militär-Comité 1899. Hadtörténeti Intézet és Múzeum Budapest. Available online: https://maps.hungaricana.hu/en/HTITerkeptar/35477/view (accessed on 6 July 2021).
15. Erd-, Bekleidungs- und Zimmermannsarbeiten: Befestigung. In *Feld-Taschenbuch fur Officiere des Geniestabes und der Pionnier-Truppe*; L.W. Seidel & Sohn: Wien, Austria, 1902. (In German)
16. Theil, A: Feld- und beständige Befestigung. In *Technischer Unterricht für die k. u. k. Pionnier-Truppe*; k.k. Hof- und Staatsdruckerei: Wien, Austria, 1898. (In German)
17. Brunner, M., Jr. *Die beständige Befestigung. Für die k. u k. Militärbildungsanstalten und zum Selbstunterrichte für Offiziere aller Waffen*; L.W. Seidel & Sohn: Wien, Austria, 1909. (In German)
18. Biesiekierski, K. *Fortyfikacja Polowa*; Główna Księgarnia Wojskowa: Warszawa, Poland, 1922. (In Polish)
19. Spennemann, D.H. Using KOCOA Military Terrain Analysis for the Assessment of Twentieth Century Battlefield Landscapes. *Heritage* **2020**, *3*, 753–781. [CrossRef]
20. Czamara, W.; Olearczyk, D.; Stodolak, J.; Pardela, Ł. Możliwości wykorzystania historycznych danych topograficznych, hydrologicznychi hydraulicznych w procesie modelowania przepływu na przykładzie rzeki Ślęzy i Widawy. (Possibility of Historicaal Topographical, Hydrological and Hydraulic Data Use in Flow Modelling on Example of Ślęza and Widawa Rivers). *Zesz. Naukowe. Inżynieria Sr. Uniw./Zielonogórski* **2014**, *154*, 13–21. (In Polish)
21. Stodolak, R.; Olearczyk, D.; Pardela, Ł.; Piecuch, T.; Tymiński, T. Budowle hydrotechniczne na potrzeby obronne Twierdzy Wrocław przełomu XIX i XX w. In *Wybrane Zagadnienia Dokumentowania i Ochrony Zabytków Hydrotechniki*; Kałuża, T., Wosiewicz, B., Eds.; Bogucki Wydawnictwo Naukowe: Poznań, Poland, 2017; pp. 65–77. (In Polish)
22. Wilson, E.E.B.H. Some Recent Tendencies in Field Engineering. *Prof. Mem. Corps Eng. United States Army Eng. Dep. Large* **1912**, *4*, 638–668.
23. Bogdanowski, J. *Warownie i Zieleń Twierdzy Kraków*; Wydawnictwo Literackie: Kraków, Poland, 1979. (In Polish)
24. Kolouszek, S.; Pardela, Ł. Twierdza Wrocław jako unikalny zabytek, Achitectura Militaris, (Fortress Wroclaw—unique heritage of Architectura Militaris). *Zesz. Nauk. WSOWL* **2013**, *2*, 68–82. (In Polish)
25. Kraków City Office. Uchwała Nr CXII/1700/14 Rady Miasta Krakówa z Dnia 9 Lipca 2014 r. w Sprawie Uchwalenia Zmiany Studium Uwarunkowań i Kierunków Zagospodarowania Przestrzennego Miasta Krakowa—Dokument Ujednolicony 2014. 2014. Available online: https://www.bip.krakow.pl/?id=48 (accessed on 2 August 2021).
26. Kraków City Office. Uchwała Nr LXXII/1048/13 Rady Miasta Krakowa z Dnia 24 Kwietnia 2013 r. w Sprawie Uchwalenia Miejscowego Planu Zagospodarowania Przestrzennego Obszaru "Bieńczyce—Park Rzeczny Dłubni". 2013. Available online: https://www.bip.krakow.pl/?dok_id=56021 (accessed on 6 July 2021).

27. Kraków City Office. Uchwała Nr LVIII/776/12 Rady Miasta Krakowa z Dnia 10 Października 2012 r. w Sprawie Uchwalenia Miejscowego Planu Zagospodarowania Przestrzennego Obszaru "Młynówka Królewska—Zygmunta Starego". 2012. Available online: https://www.bip.krakow.pl/?dok_id=53047 (accessed on 6 July 2021).
28. Kraków City Office. Uchwała Nr CVII/2738/18 Rady Miasta Krakowa z Dnia 4 Lipca 2018 r. w Sprawie Uchwalenia Miejscowego Planu Zagospodarowania Przestrzennego Obszaru "Dolina Rudawy". 2018. Available online: https://www.bip.krakow.pl/?dok_id=100574 (accessed on 6 July 2021).
29. Kraków City Office. Uchwała Nr XII/130/11 Rady Miasta Krakowa z Dnia 13 Kwietnia 2011 r. w Sprawie Uchwalenia Miejscowego Planu Zagospodarowania Przestrzennego Obszaru "Swoszowice Uzdrowisko" z późn. Zmianami. 2011. Available online: https://www.bip.krakow.pl/?dok_id=45260 (accessed on 6 July 2021).
30. Kraków City Office. Uchwała Nr CXIX/1294/06 Rady Miasta Krakowa z Dnia 25 Października 2006 r. w Sprawie Przyjęcia Dokumentu: "Ramowy Program Ochrony i Rewitalizacji Zespołu Historyczno—Krajobrazowego Twierdzy Kraków. 2006. Available online: https://www.bip.krakow.pl/_inc/rada/uchwaly/show_pdf.php?id=31042 (accessed on 7 July 2021).
31. Kraków City Office. Kierunki Rozwoju i Zarządzania Terenami Zielonymi w Krakowie na lata 2019-2030", Dokument Przyjęty Zarządzeniem Prezydenta Miasta Krakowa nr 2282/2019 z Dnia 09 Września 2019 r. 2019. Available online: https://www.bip.krakow.pl/?dok_id=115410 (accessed on 6 July 2021).
32. Kraków City Office. Plan Adaptacji Krakowa do Zmian Klimatu do Roku 2030, Dokument Powstały w Ramach Projektu Ministerstwa Środowiska. Opracowanie Planów Adaptacji do Zmian Klimatu Miast Powyżej 100 tys. Mieszkańców. 2019. Available online: https://ngo.krakow.pl/zalacznik/319113 (accessed on 6 July 2021).
33. Kraków City Office. Projekt Programu Ochrony Środowiska dla Miasta Krakowa na lata 2020–2030. 2021. Available online: https://www.bip.krakow.pl/?dok_id=139074 (accessed on 6 July 2021).
34. Art. 6. Ustawy z Dnia 23 Lipca 2003 r. o Ochronie Zabytków i Opiece nad Zabytkam. Dz. U. 2003 Nr 162 poz. 1568 ze Zmianami. 2003. Available online: https://isap.sejm.gov.pl/isap.nsf/DocDetails.xsp?id=WDU20031621568 (accessed on 6 July 2021).
35. Gordon, D.L.A. Planning, Design and Managing Change in Urban Waterfront Redevelopment. *Town Plan. Rev.* **1996**, *67*, 261–290. [CrossRef]
36. Dunglas, J.; Chastan, B.; Lang, M.; Poulard, C.; Royet, P. *Le Ralentissement Dynamique Pour la Prevention des Inondations. Guide des Aménagements Associant L'épandage des Crues Dans le Lit Majeur et Leur Écrętement Dans de Petits Ouvrages*; Irstea: Rennes, France, 2004.
37. Bai, T.; Mayer, A.L.; Shuster, W.D.; Tian, G. The Hydrologic Role of Urban Green Space in Mitigating Flooding (Luohe, China). *Sustainability* **2018**, *10*, 3584. [CrossRef] [PubMed]
38. Xu, D.; Ouyang, Z.; Wu, T.; Han, B. Dynamic Trends of Urban Flooding Mitigation Services in Shenzhen, China. *Sustainability* **2020**, *12*, 4799. [CrossRef]
39. Amoros, C.; Bornette, G. Connectivity and biocomplexity in waterbodies of riverine floodplains. *Freshw. Biol.* **2002**, *47*, 761–776. [CrossRef]
40. Naiman, R.J.; Décamps, H.; Pollock, M. The role of riparian corridors in maintaining regional biodiversity. *Ecol. Appl.* **1993**, *3*, 209–212. [CrossRef] [PubMed]
41. Szmelter, A.; Zdunek-Wielgoaska, J. Pre-war Inspirations in Shaping Green Spaces in Post-war Warsaw. *IOP Conf. Ser. Mater. Sci. Eng.* **2020**, *960*, 042003. [CrossRef]
42. Buurman, J.; Babovic, V. Adaptation Pathways and Real Options Analysis: An approach to deep uncertainty in climate change adaptation policies. *Policy Soc.* **2016**, *35*, 137–150. [CrossRef]
43. Hammond, M.; Chen, A.S.; Batica, J.; Butler, D.; Djordjević, S.; Gourbesville, P.; Manojlović, N.; Mark, O.; Veerbeek, W. A new flood risk assessment framework for evaluating the effectiveness of policies to improve urban flood resilience. *Urban Water J.* **2018**, *15*, 427–436. [CrossRef]
44. Januchta-Szostak, A. Multifunctional Riverside Buffer Parks—The Research on Nature-Urban Revitalisation of River Valleys. *J. Sustain. Arcitecture Civ. Eng.* **2013**, *4*, 42–50.
45. Januchta-Szostak, A. The Water Impact on the Quality of Urban Space. Research on urban river spaces revitalisation processes in the northern Poland. In *Landscape Legacy. Landscape Architecture and Planning Between Art and Science*; Carsjens, G.J., Ed.; Wageningen University: Wageningen, The Netherlands, 2010.
46. Zachariasz, A. *Woda Jako Element Kompozycji Parków Publicznych, (Water as a Composition Element of Public Parks), Woda w Przestrzeni Przyrodniczej i Kulturowej*; Komisja Krajobrazu Kulturowego PTG, Oddział Katowicki PTG: Sosnowiec, Poland, 2003; pp. 325–336. (In Polish)
47. Zachariasz, A. *Architekt Krajobrazu Frederick Law Olmsted Senior Oraz Amerykańskie Parki i Systemy Parków Miejskich, (Landscape Architect Frederick Law Olmsted Senior and American Parks and City Pak Systems)*; Teka Komisji Urbanistyki i Architektury, T.35; Polska Akademia Nauk Oddział Kraków: Kraków, Poland, 2004; pp. 153–166. (In Polish)
48. Zelenakova, M.; Diaconu, D.C.; Haarstad, K. Urban water retention measures. *Struct. Phys. Asp. Constr. Eng. Procedia Eng.* **2017**, *190*, 419–426.
49. Edel, R. *Odwodnienie Dróg*; Wydawnictwa Komunikacji i Łączności WKŁ: Warszawa, Poland, 2000. (In Polish)
50. Geiger, W.; Dreseitl, H. *Nowe Sposoby Odprowadzania wód Deszczowych. Poradnik Retencjonowania i Infiltracji wód Deszczowych do Gruntu na Terenie Zabudowanym*; Oficyna Wydawnicza Projprzem-EKO: Bydgoszcz, Poland, 1999. (In Polish)

51. Zevenbergen, C.; Cashman, A.; Evelpidou, N.; Pasche, E.; Garvin, S.; Ashley, R. *Urban Flood Management*; CRC Press: London, UK, 2011.
52. Pociask-Karteczka, J. Wody podziemne. In *Opracowanie Ekofizjograficzne Miasta Krakowa do Zmiany Studium Uwarunkowań i Kierunków Zagospodarowania Przestrzennego*; Degórska, B., Ed.; Urząd Miasta Krakowa: Kraków, Poland, 2010. (In Polish)
53. Suligowski, Z. *Infrastruktura Kanalizacyjna w Gospodarce Komunalnej*; Wydawnictwo, P.G., Ed.; Politechnika Gdańska: Gdańsk, Poland, 2006. (In Polish)
54. Lallam, F.; Megnounif, A.; Ghenim, A.N. Estimating the runoff coefficient using the analytic hierarchy process. *J. Water Land Dev.* **2018**, *38*, 67–74. [CrossRef]

land

Article

The Problem of Densification of Large-Panel Housing Estates upon the Example of Cracow

Eliza Szczerek

Faculty of Architecture, Cracow University of Technology, 31-155 Cracow, Poland; eszczerek@pk.edu.pl

Abstract: The paper focuses on the phenomenon of intense, uncontrolled densification of large-panel housing estates in Poland. Despite the fact that such housing estates as a legacy of the Modernist concept of segregation of functions are often burdened with problems, they still have considerable potential, which results predominantly from their urban advantages, such as functional and spatial logic, large amounts of open public space, and abundance of greenery. Unfortunately, this potential is being destroyed by introducing new buildings, ignoring the existing urban layout of the housing estate along with its original compositional assumptions. This type of densification results from—without limitations—the pressure exerted by developers in the free-market economy, and it often leads to problems such as the devastation of urban layouts of these housing estates, breaking the continuity of public spaces, appropriation of green areas, strengthening of monofunctionality, etc. This problem is becoming noticeable in the scientific debate, although it is still difficult to obtain reliable data illustrating the densifications of such housing estates. The goal of this paper is to present the scales and character of such densifications of the large-panel housing estates, which pose a threat of devastation of their urban layouts often considered as urban heritage. The paper proposes a method of a quantitative analysis of the housing estates with reference to the increase in the built-up area and a qualitative analysis of the character of development with reference to its distribution. This method comprises a sequence of subsequent steps with relevant criteria. In the results, it demonstrates the scale of the problem, which in many cases is already big and still growing. The resultant threat of devastation of the urban layout and its consequences are presented upon selected examples of housing estates in Cracow, Poland. This paper is a voice in a discussion devoted to the current status, but most of all to the future of large-panel housing estates, particularly in terms of their protection as valuable achievements of urban planning of the second half of the 20th century, and to stopping unfavorable tendencies of urban destruction.

Keywords: urban densification; prefabricated housing estates; spatial devastation; protection of urban layout; Cracow

Citation: Szczerek, E. The Problem of Densification of Large-Panel Housing Estates upon the Example of Cracow. *Land* **2021**, *10*, 1359. https://doi.org/10.3390/land10121359

Academic Editors: Jan K. Kazak, Katarzyna Hodor and Magdalena Wilkosz-Mamcarczyk

Received: 2 November 2021
Accepted: 6 December 2021
Published: 9 December 2021

1. Introduction

The modernist urban thought breaking with the hitherto understanding of urbanity held in the spirit of the doctrine contained in the Athens Charter [1] changed the course and direction of the development of many cities. Years of looking for a new formula of housing contributed—among others—to the development of housing estate structures, which—in contrast to the conventional urban tissue based on the pattern of a street and a square—promoted detached apartments blocks, which were to secure optimal ventilation, insolation, and access to green areas.

The second part of the 20th century saw more and more new housing estates built in Europe. It was a period of economic and demographic development, which accompanied the reconstruction of towns and cities after the war. A crucial role in this respect was played by the industrialization of construction technologies. Advancement in prefabrication processes accelerated the process of building housing estates considerably. Their most prominent development took place in countries of Eastern Europe, where the centrally controlled housing market, an element of the planning strategy of the communist regime, fostered the emergence of large-panel housing estates. Cheap and effective solutions in the housing sector were promoted and implemented with great elan and on a large scale.

In Poland, housing estate structures, most commonly executed in the 1970s, have become inherent elements of landscapes of numerous towns and cities. Their scale is quite significant—in 2016, as much as 40% of urban households were situated in apartment buildings from 1956–1989, mainly in housing estates built in industrialized technologies [2]. It is estimated that in 2020, 30% of residents of Poland (ca. 12 million people) lived in large-panel blocks [3].

Prefabricated housing estates, which constitute a legacy of the modernist segregation of functions, are a difficult heritage in terms of their architecture and urban planning. They are often burdened with a number of spatial problems, such as monofunctionality, monotony of development, repetitiveness of forms, low-quality architecture of apartments buildings, amorphous and undefined spaces, or spatial degradation and isolation [4–8], as well as social ones [9–11] including stigmatization, negative image [12,13], and safety issues [14]. On top of that, there are technical problems associated with the low quality of materials used [15] and poor energy performance [16], although the latter has been dealt with to a great extent by broadly implemented insulation and window replacement projects.

More and more often, large-panel housing estates in Poland are recognized as crisis areas [17]. Nevertheless, the threat of their physical and social degradation is not as big as in similar housing estates in Western Europe, which is predominantly connected with a gap in the housing market as well as a demand for relatively inexpensive apartments [18]. Furthermore, Poland still demonstrates a considerable housing deficit of ca. 640,000 apartments [19]. In relation to the housing estates in Western Europe, it seems that the estates in post-socialist cities will undergo changes according to their own scenarios [20].

Yet, despite many negative phenomena, large-panel housing estates still have considerable potential, resulting predominantly from their urban values and public open spaces associated with them [21]. Frequently, they are characterized by a functional and spatial logic as well as abundance of green areas, which are strongly related to the residential development [22–24]. Time and again, the overall urban layout of a housing estate corresponds to local conditions, making use of such natural advantages as the lay of the land or the vicinity of valuable landscapes [8]. Thus, many of them constitute valuable testimonies to urban planning achievements of Late Modernism, which require adequate protection [3,25].

Against the background of many contemporary housing estates, large-panel housing estates create a housing environment of a better quality and constitute a favorable living environment [26,27]. In many contemporary housing estates, we have to deal with spatial and social problems [28–30]. Small distances between buildings—although still consistent with legal regulations, 'peering into windows', or fencing, the latter being a hallmark of social exclusion, are their standard elements. There is not enough park and public greenery, sports and recreation grounds, sufficient services, education, trade, etc. Consequently, more and more frequently, the contemporary multi-family architecture is referred to as 'housing pathology' [31,32].

In the context of large-panel housing estates, they may pose a particular risk due to two aspects. Firstly, individual buildings or clusters thereof sprawl within the perimeter of housing estates, appropriating valuable open space and destroying their urban composition. Secondly, in many newly erected contemporary developments with multi-family buildings located in the direct vicinity of large-panel housing estates, residents use the existing infrastructure of large-panel housing estates, comprising e.g., schools, healthcare facilities, as well as greenery, as there are no legal regulations that would impose the obligation to build such facilities, as it used to be the case before the planning system reform [29]. Thus, besides certain services in ground floors, such housing estates, oriented toward the highest profits possible generated from residential spaces, often constitute single-use formations, taking a lot from large-panel housing estates but not giving much in return.

The aforementioned excessive and chaotic densification of large-panel housing estates with new development constitutes a common and significant problem, which may lead to the gradual devastation of such housing estates [33]. It happens that new buildings are placed in the areas, which originally were designed as recreational and leisure zones for residents [34]. Since the new buildings most of all are residential ones, monofunctionality is strengthened even more. New buildings are often fenced to be separated from the remaining parts of housing estates, which additionally disturbs the continuity of public spaces of a housing estate and is unfavorable for establishing appropriate relations in the social dimension. This overlaps with the problem of an increased demand for parking spaces, which also take over the space of housing estates considerably [35–39].

The first tendencies of densification of housing estates in Poland started to emerge as early as in the 1980s. They were caused by—among other factors—policies of housing cooperatives, which recognized selling parts of their territories as more economical than maintaining them in an undeveloped state [40]. Intensification in introducing new buildings can be observed since the 1990s, which was supported by the political transformation and new free market mechanisms it entailed. This new situation coincided with rather inefficient planning mechanisms regulating the spatial development of housing estates. However, to a certain extent, they did ensure the protection of unoccupied areas allocated to the greenery of housing estates under the general masterplan from 1994 covering the entire territory of Poland. The situation changed considerably when this plan was repealed in 2003 and no new one was resolved to replace it. This results in allocating these grounds to buildings, in compliance with the law [41]. Intensified and aggressive pressure exerted by developers has far-reaching spatial consequences, which are expressed predominantly in the aforementioned densification of housing estates. It is particularly unfavorable to them, since their urban layout, composition principles, and green areas are decisive for their character and identity as well as for the way they function.

The process of forming as well as functioning of large-panel housing estates in the communist socio-economic system is generally well known and present in the scientific literature. However, the situation of their spatial change after the political transformation are still not sufficiently recognized and described. The problem of densification of large-panel housing estates is becoming noticeable in the scientific debate; however, there are still not enough sources of data referring to the problem, especially to its scale. These processes of densification intensify gradually, and they affect different housing estates to different extents. Even though noticed by urban planners as well as residents, they have yet to be analyzed and provided with comprehensive databases, which would allow developing a methodological foundation for the assessment of the existing or potential threats, as well as formal and legal solutions that could be undertaken to prevent them.

Therefore, the goal of this paper is to present scales and character of such densifications of large-panel housing estates, which pose a threat of devastation of their urban layouts often considered as urban heritage [2,3,25]. Conclusions formulated in this paper may constitute a source of information on tendencies of spatial changes happening in such housing estates as well as provide inspiration to undertake systematic measures toward the rationalization of these processes.

This paper fills a gap in the state of the art concerning the problem of the threat of devastation of the urban layouts of large-panel housing estates in Poland, which are also understood as the structures of urban heritage, caused by chaotic densification, after the introduction of a free-market economy as a result of the transformation of the political system.

The article is divided into five parts. Following the Introduction, Section 2 describes the study area—Cracow large-panel housing estates—and introduces methods comprising a sequence of subsequent steps with relevant criteria, as well as provides datasets sources. Section 3 presents the result of the quantitative analysis related to the scale of densifications as well as a qualitative analysis related to their spatial character. Additionally, an in-depth analysis of selected examples is presented to support the research findings. In Section 4, the author discusses the results in a broader context, referring to selected aspects. Section 5 concludes with a summary of the main results of this research and highlights future research directions.

2. Materials and Methods

2.1. Study Area—Cracow Large-Panel Housing Estates

The study area comprised large-panel housing estates in Cracow (Figure 1)—the main city of the Province of Małopolskie. As the second largest city of Poland, besides the local specificity characteristic for each settlement market with a specific individuality, it demonstrates many tendencies and mechanisms characteristic of big cities. Thus, it illustrates the problems that concern large-panel housing estates in other Polish cities. In addition, a significant number of prefabricated housing estates in Cracow allows for a broader study of the problem.

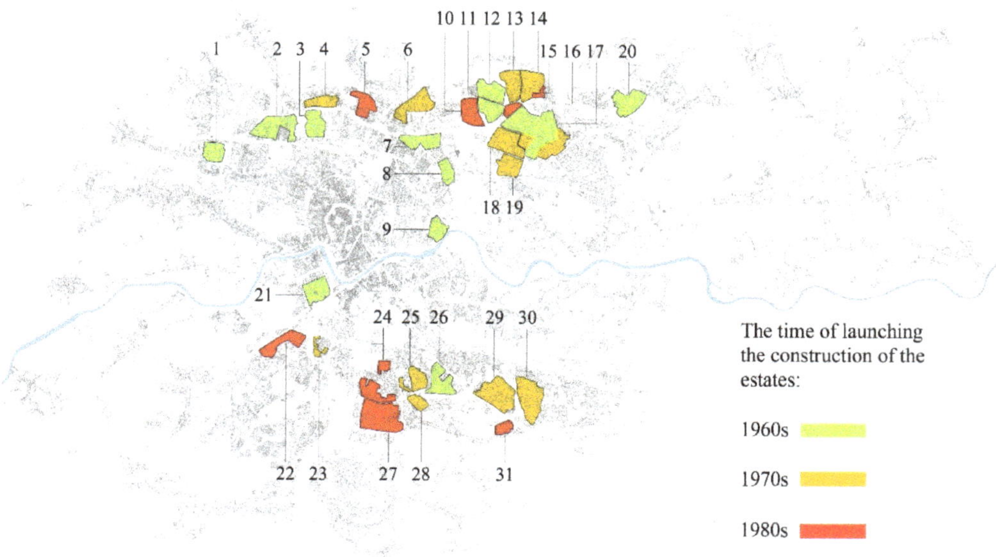

1 – Widok
2 – Azory
3 – Krowodrza Górka
4 – Prądnik Biały Zachód
5 – Prądnik Biały Wschód
6 – Prądnik Czerwony
7 – Olsza II
8 – Ugorek
9 – Dąbie
10 – Oświecenia
11 – Tysiąclecia
12 – Złotego Wieku
13 – Bohaterów Września
14 – Piastów
15 – Mistrzejowice Nowe
16 – Kombatantów
17 – The group of Bieńczyckie housing estates
18 – Dywizjonu 303
19 – II Pułku Lotniczego
20 – Wzgórza Krzesławickie
21 – Podwawelskie
22 – Ruczaj-Zaborze
23 – Cegielniana
24 – Wola Duchacka Zachód
25 – Wola Duchacka Wschód
26 – Na Kozłówce
27 – Kurdwanów
28 – Piaski Nowe
29 – Prokocim
30 – Bieżanów Nowy
31 – Rżąka

Figure 1. Arrangement of large-panel housing estates in the structure of Cracow.

The beginnings of large-panel housing estates in Cracow are associated with the need to provide a large number of apartments, most of all in connection with its constantly growing population after the war. Cracow avoided the need to rebuild its urban structure, including housing resources, as practically, it had not suffered any considerable damage during the war. Housing estates, erected over the span of three decades, were located on the northern and southern side of the city.

The early 1960s saw the emergence of—among others—housing estates in the northern part of the city, such as Azory and Olsza II, as well as housing estates associated with the extension of the metallurgical plant built in the 1950s, i.e., the Wzgórza Krzesławickie housing estate and shortly after that the first from the group of Bieńczyce housing estates. The late 1960s and early 1970s also marked the beginnings of the construction of such housing estates as Krowodrza, Widok, Dąbie, and Prądnik Czerwony. The extension of the development of Nowa Huta was also continued, with a group of four Mistrzejowice housing estates erected there: Tysiąclecia, Złotego Wieku, Bohaterów Września, and Piastów [42]. At the same time, the Podwawelskie and Na Kozłówce housing estates were emerging on the southern bank of the Vistula.

The emerging new developments affected the gradual change of the spatial layout of Cracow from a bipolar to a band-shaped one. The 1970s is a period when distances between one's workplace and residence increased considerably, and the existing tramway no longer sufficed, as it was overloaded. It was a time when a network of bus connections got extended. The year 1973 saw another broadening of the city's administrative limits aimed to mitigate the deficit of construction lands [43]. The subsequent development of the city was based on a plan adopted in 1977, which had a significant impact upon the process of forming housing estates. The plan assumed two parallel bands of development coexisting together, with the historic city center as their keystone [44,45]. In addition to the housing estates whose construction had already begun in the southern band, i.e., Prokocim Nowy and Piaski Nowe, toward the end of the 1970s, the construction of the next housing estates was commenced: Bieżanów Nowy and Wola Duchacka Wschód. In the northern bank, after moving the airport from Czyżyny to Balice, the development of these areas started in 1978, with the Dywizjonu 303 housing estate built to the north from the runway and the II Pułku Lotniczego housing estate to the south from it.

In the 1980s, the construction of this housing estate was continued, but also new ones started to emerge. The Kombatantów housing estate was built in the norther band, which was followed later by the Oświecenia housing estate. Wola Duchacka Zachód, Kurdwanów, and Ruczaj-Zaborze developments were erected in the southern band.

Sporadically, the construction of housing estates continued to the early 1990s, with the Rżąka housing estate as an example. In the ornamentation of its architectural details, it makes a reference to the aesthetics of Post Modernism, which were for the first time applied in Cracow in industrialized residential developments [29]. In the early 1990s, the construction of the Mistrzejowice Nowe housing estate was continued as well.

Housing estates built with the application of industrialized technologies are a significant element in the urban structure of Cracow in terms of territory. Housing developments from the 1960s and early 1970s constitute ca. 25% of all apartments located in the city. The highest number of apartments, ca. 35%, are located in housing estates built in the 1970s and early 1980s [44].

The greatest accumulation of housing estates is in the eastern part of the city, particularly in the northern band, where the extension of the metallurgical plant in Nowa Huta mentioned above was an important stimulus for their coming into being. Even though these housing estates were built in different times and most often are independent from each other in administrative terms, they form a vast complex of housing estates due to their often direct proximity. It has its effect on the identity of this part of the city, which to a great extent is identified with this type of architecture.

Housing estates located in the southern part of the city due to the considerable area they occupy strongly define the character of this part of Cracow as well. However, in contrast to housing estates in Nowa Huta, they do not often border on one another. A feature that can be recognized as characteristic is the fact that they appear as islands among areas of one-family houses, which divide housing estates into separate clusters. Such a situation can be observed in Wola Duchacka, which is a housing estate that is formally divided into an eastern and western part, with the latter additionally consisting of two 'islands' of large-panel buildings. Furthermore, similarly to the Cegielniana and

Na Kozłówce housing estates, the territory occupied by large-panel buildings assumes a shape resembling the letter 'U', surrounding one-family houses. In the northern part, a similar situation can be seen in the Azory housing estates, which are visibly interspersed with one-family houses.

In Cracow, many housing estates from the second half of the 20th century are recognized as areas requiring many issues to be resolved [44] and they have been covered by the 'Program of Multi-family Housing Development Rehabilitation in the Area of Cracow Municipal Commune' [46]. As in many other Polish towns and cities, currently, one of the biggest problems of Cracow is the chaotic densification of such housing estates with new buildings. The way the lands are administered is not beneficial for the spatial situation of housing estates in Cracow. Most lands, even though they are the property of the Municipality of Cracow, are lent for use to numerous entities, and consequently, there is no direct control over them, which in turn makes it harder to make any comprehensive decisions. Furthermore, due to more and more frequent cases of buying them up by different entities, this situation may worsen still [41], since—as it has been mentioned before—their densification is often chaotic, with no respect for the entire context and no well-considered, holistic approach. One of the consequences of densification of housing estates is their fragmentation, which is caused by fencing off areas and by the loss of spatial cohesion due to taking green areas over by cars.

2.2. Methods

The base for pursuing the research goal referring to the scale and the character of urban densifications of the housing estates from the second half of the 20th century is provided by the author's literature studies [8,20,33,34,38,39,47] and scientific experience in this subject matter. This paper presents a more detailed insight into threats to large-panel housing estates, basing on numerical and qualitative data, extending the knowledge in this field. It fills the gap in the existing studies focusing on this problem, particularly in the context of devastation of valuable urban layouts of such housing estates, which are recognized as heritage.

Currently, it is difficult to obtain data illustrating the problem of densification of such housing estates. Different methods are characteristic for the discipline of science: Architecture and Urban Planning have been applied in an attempt to characterize this problem upon the example of Cracow [48]. They are presented in the following steps:

- Step 1. Indication of the analyzed housing estates within the city limits (Figure 1).

  Qualitative criteria: Construction technology, time and place of construction, genesis, urban compositional layout.

  Comment: The analysis covers housing estates built in industrialized technologies (initially in the large-block technology and then in large-panel technologies, which were used on a large scale across European countries, especially in Eastern Europe). Therefore, for the sake of clarity and simplification of deliberations in this paper, they are predominantly referred to as large-panel housing estates. Sometimes, they are also referred to as prefabricated housing estates or estates built in industrialized technologies. Ultimately, all these phrases refer to the same kind of housing estates and are used interchangeably.

  The article refers to housing estates that were built on a large scale in Cracow in the second half of the 20th century. The time adopted is connected with the political system—the time of the communist rule, which had a significant impact on the process of designing such housing estates.

  Such housing estates constitute a legacy of the modernist segregation of functions and urban composition, as they were built in the spirit of modernist ideas promoted by the Athens Charter [1].

- Step 2. Defining the areas of the analyzed housing estates (Figure 2).

  Quantitative criterion: Housing estate's boundary.

Comment: Areas of selected housing estates were defined following Lynch's 'district' category [49], which is understood as a homogenous area with a uniform character in which one has a feeling of being inside or outside of this area.

- Step 3. Characterization of densifications of the analyzed housing estates.

  Quantitative criterion: Increase in the built-up area [%] (Table 1, Figure 3).
  Qualitative criterion: Characteristics of the way new buildings are introduced:

- Densification within the perimeter of the housing estate (Figure 4);
- Densifications on the outskirts of the housing estate—shifting its edge and increasing the original area of the housing estate (Figure 5).

Comment: The increase in the built-up area in the housing estates was defined quantitatively, comparing figures from the mid-1990s and 2021. This time cut-off point is connected with a radical transformation of the political, economic, and social system in Poland, which had a significant impact on such housing estates. The mid-1990s is a transitional period between the already collapsed centralized national economy during the communist time and a newly emerging free market system of a democratic reality. The centrally controlled housing market within the communist regime promoted the implementation of such big spatial and functional urban projects, including large-panel housing estates. With the political transformation and the free market economy, a completely new mechanism in the housing market and urban planning appeared. Even though 1989 was a breakthrough year for the social and political system in Poland, the processes associated with designing such housing estates were sometimes continued in the early 1990s. That is why 1995 was selected as a turning point in the analysis performed, where figures from the mid-1990s and 2021 were compared.

The percentage of the increase in the built-up area was determined by comparing data from a vector map obtained from the City Hall Cracow and other maps available online within the scheme of the Municipal Spatial Information System for Cracow [50] with data from satellite images from 1995.

An analysis was performed covering the densification of central areas of housing estates as well as their peripheral zones, causing disturbances of spatial edges.

- Step 4. Reference to the urban heritage structures and local spatial development plans in the context of the housing estates (Table 2).

  Quantitative criterion (a): Recognition of the estates as urban heritage structures.
  Quantitative criterion (b): Presence of local spatial development plans:

- For the entire territory of the housing estate;
- For a part of the territory of the housing estate or its edge (adjacent areas).

Comment: Housing estates as urban heritage structures were indicated on the basis of the results of a research project [2]. An analysis of coverage of large-panel housing estates with local spatial development plans was performed on the basis of information available in the Public Information Bulletin of the City of Cracow [51].

- Step 5. Indication of monofunctional clusters of apartment blocks in the vicinity of large-panel housing estates. (Figures 6 and 7, Table 3).

Quantitative criterion: Presence of such clusters in areas adjacent to the large-panel housing estates.

Comment: Such complexes of monofunctional residential areas may have a negative impact on the functioning of the housing estates (described in more detail in the Introduction).

- Step 6. An in-depth description of selected examples of housing estates with a valuable urban composition (Figures 8–11).

Qualitative criterion: High spatial devastation of the urban layout of the housing estate, which is caused by intensive densification with new buildings.

Comment: This is to illustrate more vividly the threat of devastation of their urban layouts.

Results of the analyses are presented in a form of a systematic overview of changes within housing estate structures and in areas adjacent to them caused by their densification as well as diagrams illustrating this problem quantitatively.

3. Results

While analyzing spatial metamorphoses within the perimeter of large-panel housing estates in Cracow, it was noticed that their densifications assume different scales and characters and most importantly affect nearly all the housing estates. It is particularly striking in the case of urban layouts of such housing estates with high compositional values, which have been recognized as structures of urban heritage. New buildings were not observed only in one housing estate (Figure 2, Table 1).

Table 1. Densification values for large-panel housing estates in Cracow (%).

	Housing Estates in the Northern Part of the City	Buildings Erected before 1995	New Buildings Erected between 1995 and 2021
1	Widok	90.1%	9.9%
2	Azory	89.1%	10.9%
3	Krowodrza Górka	77.4%	22.6%
4	Prądnik Biały Zachód	89.8%	10.2%
5	Prądnik Biały Wschód	93.8%	6.2%
6	Prądnik Czerwony	86.4%	13.6%
7	Olsza II	93.8%	6.2%
8	Ugorek	97.3%	2.7%
9	Dąbie	80.2%	19.8%
10	Oświecenia	69.4%	30.6%
11	Tysiąclecia	96.6%	3.4%
12	Złotego Wieku	91.9%	8.1%
13	Bohaterów Września	71.8%	28.2%
14	Piastów	89.1%	10.9%
15	Mistrzejowice Nowe	94.5%	5.5%
16	Kombatantów	98.3%	1.7%
17	The Group Of Bieńczyckie Housing Estates	90.5%	9.5%
18	Dywizjonu 303	65.8%	34.2%
19	II Pułku Lotniczego	80.8%	19.2%
20	Wzgórza Krzesławickie	94.1%	5.9%
	Housing estates in the southern part of the city		
21	Podwawelskie	79.0%	21.0%
22	Ruczaj-Zaborze	91.9%	8.1%
23	Cegielniana	100.0%	0.0%
24	Wola Duchacka Zachód	90.3%	9.7%
25	Wola Duchacka Wschód	79.2%	20.8%
26	Na Kozłówce	84.2%	15.8%
27	Kurdwanów	70.0%	30.0%
28	Piaski Nowe	97.0%	3.0%
29	Prokocim	90.0%	10.0%
30	Bieżanów Nowy	74.8%	25.2%
31	Rżąka	85.6%	14.4%

3.1. *Quantitative Characteristics of Densifications*

The quantitative analysis of the increase in built-up areas in the housing estates covered their territories, taking into account their possible extensions if their edges were shifted due to adding new buildings (situations referred to in Section 3.2 as A and B). Quantitative results are presented below (Table 1, Figure 3). The table contains values illustrating the development of the housing estates with reference to the area increased by newly erected buildings: the first column refers to buildings from before 1995, the second refers to buildings erected after 1995 to date (2021).

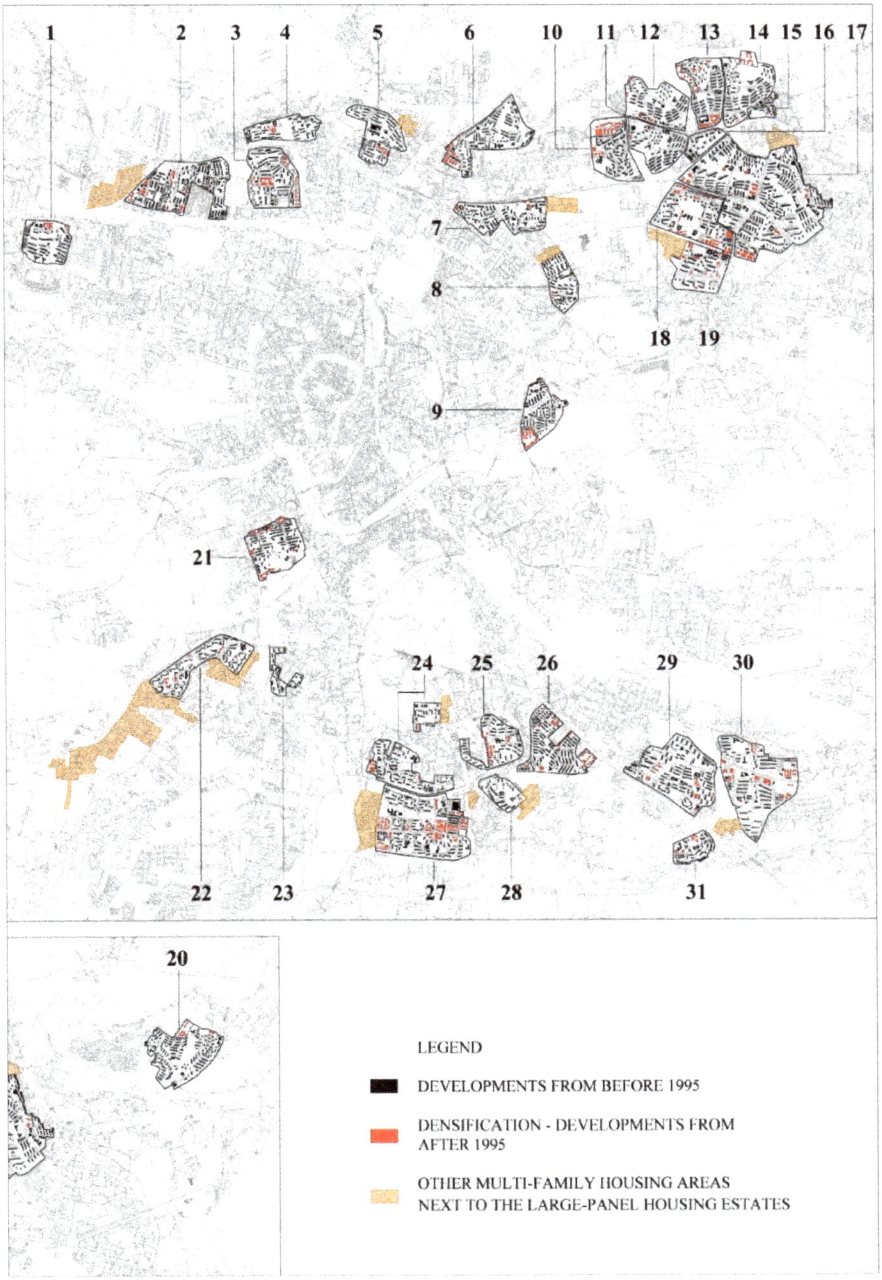

Figure 2. Map of a part of Cracow with areas of the analyzed housing estates marked (black border). 1—Widok, 2—Azory, 3—Krowodrza Górka, 4—Prądnik Biały Zachód, 5—Prądnik Biały Wschód, 6—Prądnik Czerwony, 7—Olsza II, 8—Ugorek, 9—Dąbie, 10—Oświecenia, 11—Tysiąclecia, 12—Złotego Wieku, 13—Bohaterów Września, 14—Piastów, 15—Mistrzejowice Nowe, 16—Kombatantów, 17—The group of Bieńczyckie housing estates, 18—Dywizjonu 303, 19—II Pułku Lotniczego, 20—Wzgórza Krzesławickie, 21—Podwawelskie, 22—Ruczaj-Zaborze, 23—Cegielniana, 24—Wola Duchacka Zachód, 25—Wola Duchacka Wschód, 26—Na Kozłówce, 27—Kurdwanów, 28—Piaski Nowe, 29—Prokocim, 30—Bieżanów Nowy, 31—Rżąka.

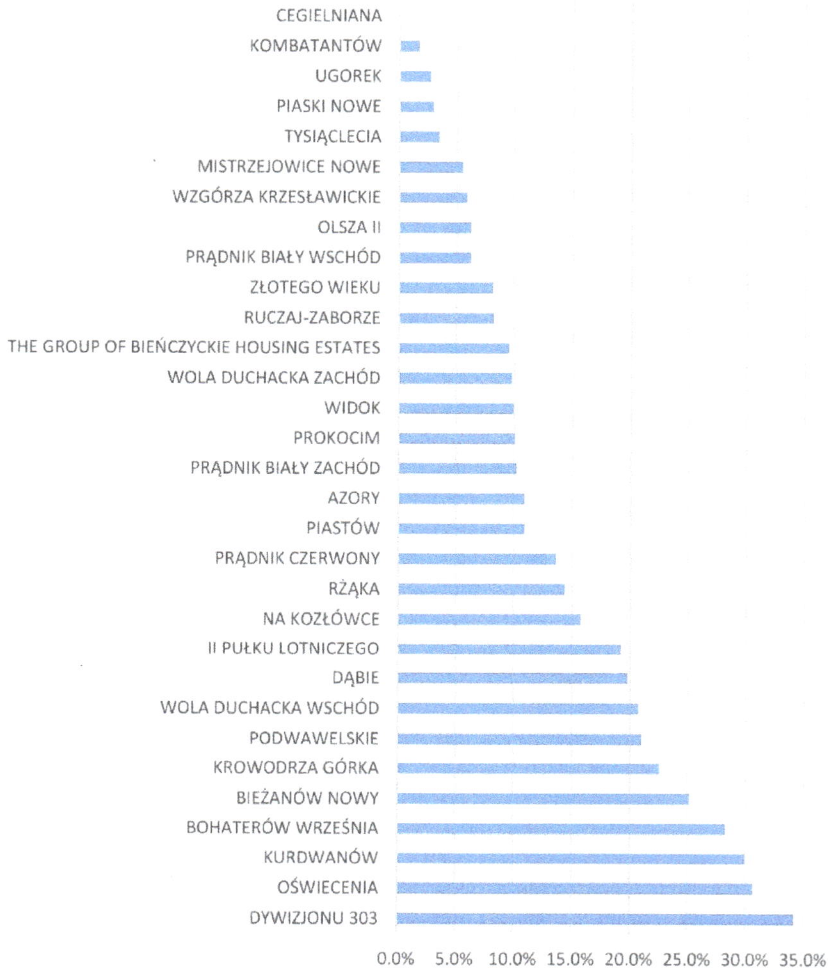

Figure 3. Diagram illustrating a relative development increase in large-panel housing estates in Cracow.

An analysis of the increase in the built-up area demonstrates that densifying housing estates with new buildings is more intensive in the southern part of the city, and on average, it reaches 18.2%. The average increase in built-up areas in the northern housing estates is 13.8%. The total average for all the housing estates is 15.4%.

The biggest number of housing estates—14—increased their built-up area within the range of 1 to 10%. Eight housing estates were in the range 10–20%. In six housing estates, the increase is at the level of over 30%. The housing estates that stand out due to the largest increase in the built-up area, i.e., 30% or more, are Kurdwanów, Dywizjonu 303, and Oświecenia housing estates.

One of the methods of controlled development of housing estates can be local spatial development plans. Some of the housing estates in Cracow have such plans drawn up, covering their entire perimeter or a part thereof. It was also verified whether similar plans cover adjacent areas, which might be significant particularly for the protection of green areas around the housing estates (Table 2). It was presented in the context of the information which estates are recognized as urban heritage structures [2] (Table 2).

Table 2. List of housing estates recognized as urban heritage structures with their entire territories or parts thereof covered with local spatial development plans.

	Housing Estates in the Northern Part of the City	Housing Estates Recognized as Urban Heritage Structures [2]	Local Spatial Development Plan for the Entire Territory of the Housing Estate	Local Spatial Development Plan for a Part of the Territory of the Housing Estate or Its Edge (Adjacent Areas)
1	Widok	X	X	
2	Azory		X	
3	Krowodrza Górka	X		X
4	Prądnik Biały Zachód			X
5	Prądnik Biały Wschód	X		X
6	Prądnik Czerwony		X	
7	Olsza II	X	X	
8	Ugorek	X	X	
9	Dąbie	X	X	
10	Oświecenia	X	X	
11	Tysiąclecia	X	X	
12	Złotego Wieku	X	X	
13	Bohaterów Września	X		X
14	Piastów	X		X
15	Mistrzejowice Nowe			
16	Kombatantów		X	
17	The Group Of Bieńczyckie Housing Estates	X	X	
18	Dywizjonu 303	X	X	
19	II Pułku Lotniczego	X	X	
20	Wzgórza Krzesławickie	X		X
	Housing estates in the southern part of the city			
21	Podwawelskie	X		X
22	Ruczaj-Zaborze	X		
23	Cegielniana			
24	Wola Duchacka Zachód	X		X
25	Wola Duchacka Wschód			
26	Na Kozłówce	X		X
27	Kurdwanów	X	X	
28	Piaski Nowe	X		X
29	Prokocim	X		X
30	Bieżanów Nowy	X		X
31	Rżąka			X

Today, i.e., in 2021, 14 out of the 31 analyzed housing estates or their complexes (11 out of the 23 housing estates recognized as urban heritage structures) are covered with local spatial development plans, whereas seven of them have been in force since 2020. Four housing estates are not covered with overall local spatial development plans at all, and 13 housing developments are only in selected parts or on their edges.

3.2. Qualitative Approach—Characteristics of Spatial Metamorphoses of the Housing Estates

The analysis of densifications of large-panel housing estates allows defining certain trends in the way new buildings are introduced. These trends are characterized below.

A. New development is introduced as single buildings or clusters of buildings in unoccupied spaces between apartments buildings. It happens both in central areas of housing estates and in their peripheral zones (Figure 4).

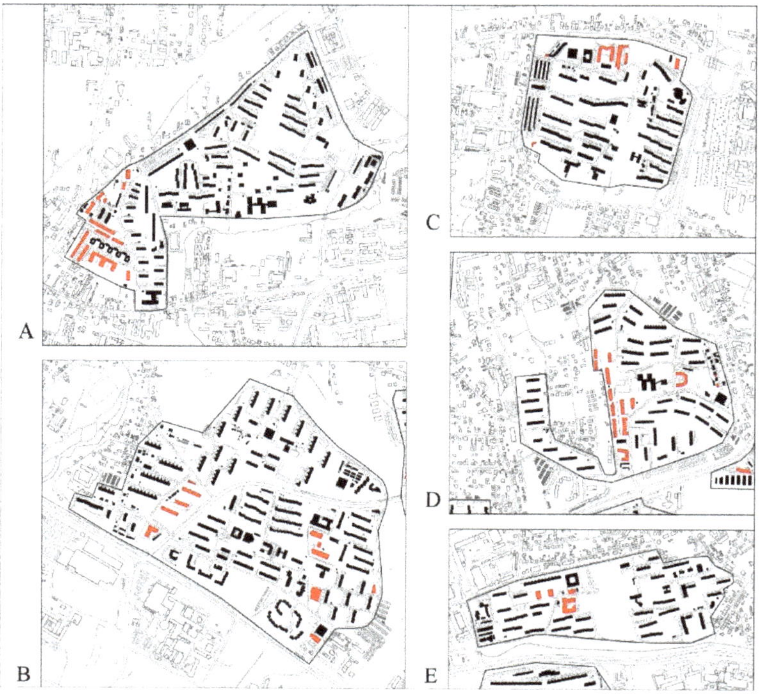

Figure 4. Densification with single or clusters of buildings inside housing estates—selected examples: (**A**)—Prądnik Czerwony (6), (**B**)—Prokocim (29), (**C**)—Widok (1), (**D**)—Wola Duchacka Wschód (25), (**E**)—Prądnik Biały Zachód (4). Color legend: black—developments from before 1995; red—densification—developments from after 1995.

B. In situations where new buildings are introduced in the peripheral zones of a housing estate, such buildings tend to push beyond its perimeter, or to push its edge outwards, to be more precise, thus pushing outwards the boundary of the housing estate, informally increasing its territory (Figure 5). It often entails taking over areas that used to constitute green areas: a buffer zone of the housing estate. In densifications of this type, one still can speak of a certain affiliation of new buildings to the structure of the housing estate, at least in terms of belonging to a certain 'district', which is understood in Lynch's categories as an area that is as homogenous as possible, in which one has a feeling of being inside or outside of this area [49]. The affiliation of new buildings to a large-panel housing estate may be also strengthened by a distinct spatial edge located behind it, e.g., a green belt, a river, or a street, particularly a major one.

C. Furthermore, there are situations where a complex of multi-family buildings has been built in the direct vicinity of a housing estate (and frequently, its construction is still in progress), with the dominating residential function and very few accompanying functions. If the scale, character of buildings, and their location toward the existing large-panel housing estate provide them with features that make them stand out, this study does not recognize them as an extension of the housing estate but as a separate structure. Nevertheless, due to the potential influence of such structures on large-panel housing estates mentioned in the introduction, they are included in this paper as the neighborhood of housing estates (Figure 6).

The nature of structural metamorphoses within housing estates and in areas adjacent to them for all the housing states in Cracow according to the tendencies described above is presented in Table 3.

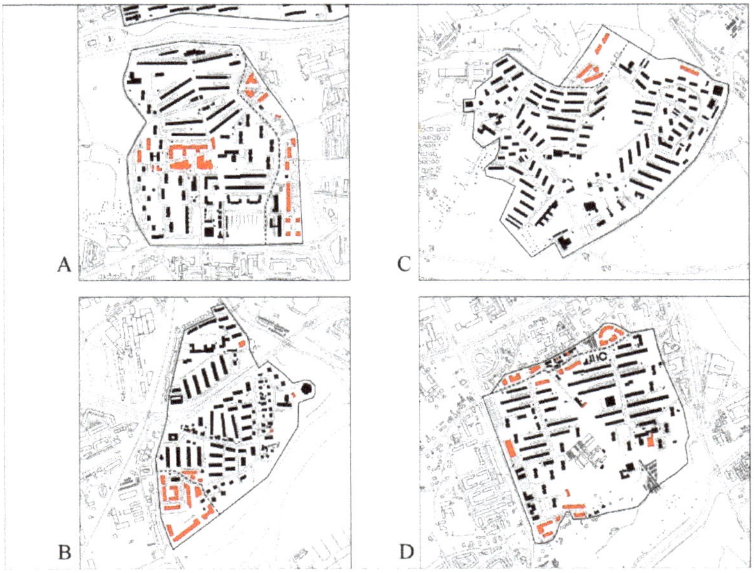

Figure 5. Densification with single buildings or clusters of new buildings on the edges of the housing estates, shifting their boundaries (the extended area is marked with a dotted line)—selected examples of housing estates: (**A**)—Krowodrza Górka (3), (**B**)—Dąbie (9), (**C**)—Wzgórza Krzesławickie (20), (**D**)—Podwawelskie (21). Color legend: black—developments from before 1995; red—densification—developments from after 1995.

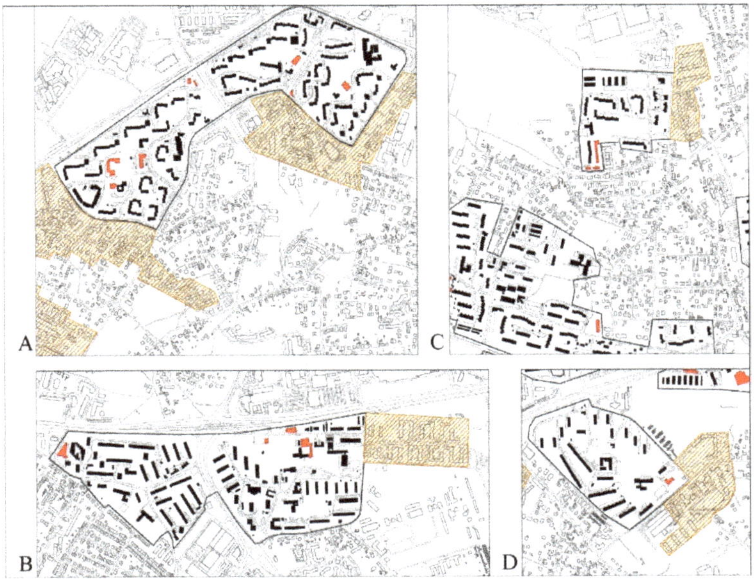

Figure 6. Neighborhood of clusters of multi-family buildings with the dominating residential function—examples in the context of selected housing estates: (**A**)—Ruczaj (22), (**B**)—Olsza II (7), (**C**)—Wola Duchacka Zachód—northern part (4), (**D**)—Piaski Nowe (28). Color legend: black—developments from before 1995; red—densification—developments from after 1995; orange—other multi-family housing areas next to the large-panel housing estates.

Table 3. Characteristics of the way new buildings are introduced within the perimeter and in areas adjacent to large-panel housing estates.

	Housing Estates in the Northern Part of the City	**A. Densifications within the Perimeter of the Housing Estate**	**B. Densifications on the Outskirts of the Housing Estate—Shifting Its Edge and Increasing the Original Area of the Housing Estate**	**C. Proximity to an Area of a Multi-Family Complex with the Dominating Residential Function**
1	Widok	X		
2	Azory	X		X
3	Krowodrza Górka	X	X	
4	Prądnik Biały Zachód	X		
5	Prądnik Biały Wschód	X		X
6	Prądnik Czerwony	X		
7	Olsza II	X		X
8	Ugorek	X		X
9	Dąbie	X	X	
10	Oświecenia	X	X	
11	Tysiąclecia	X		
12	Złotego Wieku	X		
13	Bohaterów Września	X	X	
14	Piastów	X	X	
15	Mistrzejowice Nowe	X		
16	Kombatantów	X		
17	The Group Of Bieńczyckie Housing Estates	X	X	X
18	Dywizjonu 303	X	X	X
19	II Pułku Lotniczego	X		X
20	Wzgórza Krzesławickie	X	X	
	Housing estates in the southern part of the city			
21	Podwawelskie	X	X	
22	Ruczaj-Zaborze	X		X
23	Cegielniana			
24	Wola Duchacka Zachód	X	X	X
25	Wola Duchacka Wschód	X		
26	Na Kozłówce	X	X	
27	Kurdwanów	X	X	X
28	Piaski Nowe	X		X
29	Prokocim	X		
30	Bieżanów Nowy	X	X	
31	Rżąka	X		

Most frequently, these situations occur simultaneously in housing estates. In addition to the already mentioned densifications within the perimeter of the housing estates, which occur nearly in all the analyzed housing estates, in 13 of them, a partial shift of their edge can be additionally observed, and seven share the neighborhood with new complexes of residential buildings. In four housing estates, all these phenomena were observed.

Considering the types and intensity of densifications, the housing estates that demonstrate the most far-reaching spatial changes are the Dywizjonu 303 housing estate (34.2%), the Oświecenia housing estate (30.6%), and Kurdwanów (30.0%) (Figure 7).

In the Dywizjonu 303 and Kurdwanów housing estates, intensive densifications with single buildings and clusters thereof take place within the perimeter of the housing estates, as well as in their outskirts, in some cases shifting the original edge of the housing estate, which has changed the original valuable urban composition. Furthermore, they are accompanied with areas of intensive residential development. In the housing estates in

Czyżyny (the Dywizjonu 303 housing estate and the II Pułku Lotniczego housing estate), unoccupied areas on the southern side of the runway have been adapted for the Nowe Czyżyny housing estate. Due to certain controversies and a significant influence of this project on the housing estates in Czyżyny, this case is described in more detail in Section 3.3.

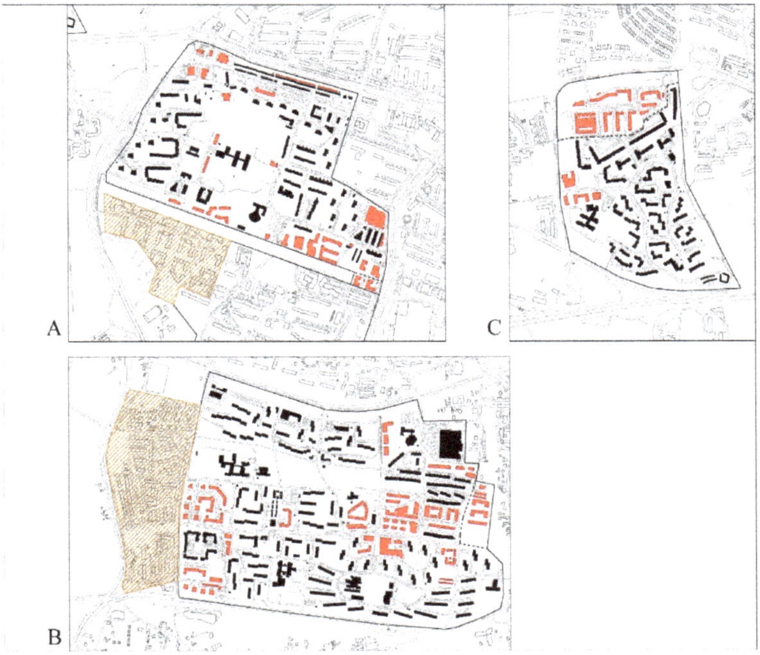

Figure 7. Cracow large-panel housing estates with the highest intensity of densification: (**A**)—the Dywizjonu 303 housing estate; (**B**)—the Kurdwanów housing estate; (**C**)—the Oświecenia housing estate. Color legend: black—developments from before 1995; red—densification—developments from after 1995; orange—other multi-family housing areas next to the large-panel housing estates.

In the Oświecenia housing estate, there are no single buildings added inside the housing estate; however, there are two distinct clusters of new buildings in the edge area. In the context of the formula of defining new buildings as belonging or not belonging to a housing estate as adopted in this paper, regarding the complex of buildings in the northern part of the housing estate as an extension of the Oświecenia housing estate instead of an independent project may seem debatable. This is due to the fact that this part of the development is generally regarded as officially belonging to this housing estate as its extension. It is worth pointing out that the southern part of the housing estate, with a characteristic meandering structure, has maintained its urban layout and there are no new additions between apartments blocks there.

The housing estates that have no or the lowest degree of densifications, i.e., Ceglana, Kombatantów, Ugorek, Tysiąclecia, and Piaski Nowe, demonstrate their original, nearly unchanged layout of the entire development. However, in the latter, a cluster of multi-family residential buildings has emerged adjacent to it, and it is still being extended.

3.3. Selected Examples of Spatial Devastation

The above considerations, presented in Sections 3.1 and 3.2, allow selecting examples in which adding new structures to large-panel housing estates recognized as urban heritage structures entails squandering of their potential by destroying their urban composition and other urban values.

The influence of this type of densifications can be considered according to the set of the following qualitative criteria, characterizing the basic aspects of the functioning of housing estates:

- Influence on the original urban layout of the housing estate and its legibility;
- Influence on the continuity of open public spaces and green areas and on the quality thereof;
- Influence on the amount of public open areas;
- Influence on the existing historical and cultural assets;
- Influence on the mixed-use development.

Ones of blatant examples in this respect are the aforementioned housing estates in Czyżyny: Dywizjonu 303 and II Pułku Lotniczego, which are located on both sides of the historic runway of the former airport Rakowice-Czyżyny (Figures 8 and 9). Aggressive measures undertaken by developers (level of densification: 34.2% and 19.2%) caused utter devastation of the post-military heritage of the area, which besides constituting a true loss in the scale of the city directly affects the adjacent large-panel housing estates. The new development is partially fenced off and almost monofunctional. This is an example of the densification process, which is aimed at private interest, not the common one, especially in the context of the neighboring large-panel housing estates [35].

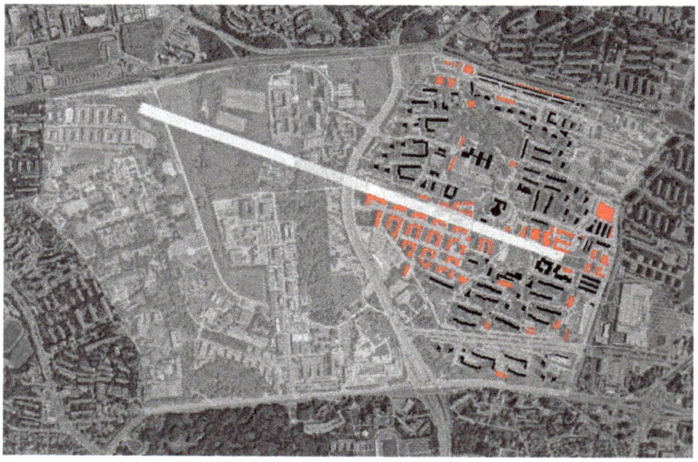

Figure 8. Densifications of housing estates in Czyżyny in the context of destruction of their main potential—the former runway. Based on an ortophotomap (www.maps.google.com; accessed on 5 December 2021). Color legend: white area—approximate area of the former airport Rakowice-Czyżyny with a specified runaway (white stripe); black buildings—developments from before 1995 (mostly within the scheme of the construction of the housing estates in Czyżyny); red—densification—developments from after 1995.

Similarly enough, metamorphoses observed in the Bohaterów Września housing estate and the Piastów housing estate in Mistrzejowice move toward a particularly unfavorable end. Devastating densifications (28.2% and 10.9%) take place predominantly on their northern and western outskirts, shifting and changing the character of their edges (Figures 10 and 11), which according to the original design were planned as a green buffer zone [42].

From the side of the Bohaterów Września housing estate, the green buffer zone is provided by Planty Mistrzejowickie park, and from the north, it is provided by areas of the former defense ring of the Cracow Fortress, with still preserved structures of the former forts: Mistrzejowice and Batowice, surrounded by abundant vegetation, once fulfilling camouflage purposes. The document 'Program of the Cracow Fortress Protection Plan' [52] describes this area as worth protecting.

The problems connected with the densification process described in the examples above are associated with—inter alia—their negative influence on the urban layout of these housing estates, breaking of the continuity of their public spaces, devastation of the historical and cultural potential, strengthening of monofunctionality, and appropriation of valuable green areas. Taking over green areas by aggressive investments constitutes a serious problem. Key green areas disappear and their continuity is broken, which causes irreversible destruction of these areas. It is a significant loss not only in the context of the housing estates themselves but the entire city, too. Urban values of the housing estates have been irreversibly destroyed.

(a) (b)

Figure 9. Character of an open space in the multi-family residential development in Czyżyny: (**a**) Arrangement of the area of the former runway of the Rakowice-Czyżyny airfield along with the residential development; (**b**) Character of a space in between the buildings. Photos: by author.

Figure 10. Densifications of housing estates in Mistrzejowice in the context of green areas of the fortress constituting their buffer zone and other green areas enveloping the housing estates (green). Color legend: black buildings—developments from before 1995 (mostly within the scheme of the construction of the Bohaterów Września i Piastów housing estates); red—densification—developments from after 1995. Based on an ortophotomap (www.maps.google.com, accessed on 5 December 2021).

(a) (b)

Figure 11. (**a**,**b**) Intensive multi-family residential development on the northern side of the Piastów housing estate. Photos: by author.

4. Discussion

One of the important reasons for undertaking the work was to draw attention to the growing problem of densification of large-panel housing estates with new buildings and at the same time destruction of Late Modernist urban structures, which are often recognized as heritage. Frequently, it is quite excessive and chaotic, and it constitutes a threat to their functioning as a valuable housing environment. Single buildings or clusters thereof built within the perimeter of housing estates as well as in their outskirts lead to the shifting of their edges' (Table 3) appropriate valuable open space, destroying their composition. Over time, such changes can lead to complete obliteration of the original design idea in the entire housing estate or a fragment thereof. In this way, the main potential of such housing estates, which results predominantly from their urban advantages, such as urban composition and its functional and spatial logic, large amounts of open public space, and abundance of greenery, is gradually annihilated.

The results presented herein demonstrated the possibility of obtaining essential information on this problem by quantitative determination of the degree of densification of large-panel housing estates. Upon the example of a comprehensive analysis of all (31) large-panel housing estates in Cracow, carried out on the basis of measurements and calculations of the additional built-up area performed on the basis of available materials (vector maps, satellite images, and planning materials), an average size of densification with new buildings was determined. For all the housing estates, this average is already 15.4%, whereas in some of them, it is very high—it has reached the value of nearly 35% (Table 1, Figure 3).

In light of the above, it is clear that the qualitative analysis may be more effective in the presence of quantitative data. Problems of threats and damages to the existing, already historic urban composition, caused by uncoordinated, chaotic additions to large-panel housing estates, incompatible with their initial plans, are illustrated upon selected examples—the housing estates in Czyżyny (Dywizjonu 303 and II Pułku Lotniczego; Figure 8) and selected housing estates in Mistrzejowice (Bohaterów Września and Piastów; Figure 10). The high density and chaotic arrangement of a new development caused significant spatial changes, which has a negative influence on the urban layout of these housing estates. They have lost valuable green areas, and the continuity of public spaces has been broken. Moreover, the densifications have led to devastation of the historical and cultural potential. As a result, the housing estates have lost some of their advantages irreversibly.

It proves that it is important to link quantitative data and qualitative criteria (e.g., influence of the densification on the urban layout, continuity of public spaces, resources of greenery, existing historical and cultural assets, mixed-use development) to obtain a proper characterization of the functioning of the estates.

According to the observations, a considerable share of the new buildings covers multi-family residential developments, which intensifies the monofunctionality of the housing estates. A similar effect can be caused by enclaves of such developments in the direct proximity of large-panel housing estates; hence, they have been taken into account in this study. Housing estates and their immediate surroundings need to be integrated within the urban context [53,54]. However, the study does not consider any detailed data relating to them, limiting solely to indicating their location. Likewise, the study does not focus on the quantitative determination of the function that emerges as new buildings appear. A quantitative functional and spatial analysis could be the next step, addressing transformations in housing estates in more detail. It has been found that despite the fact that prefabricated housing estates are a legacy of the Modernist segregation of functions and as such they are burdened with a number of issues, such as monofunctionality, monotony of development, repetitiveness of forms, and low quality of architecture of apartments blocks, they are often characterized by a functional and spatial logic and abundance of green areas. Frequently, the overall urban layout of a housing estate corresponds to local conditions, making use of natural advantages (the lay of the land, vicinity of valuable landscapes). The risks of losing these advantages caused by chaotic development may be mitigated to a certain extent by local spatial development plans. In the case of the housing estates in Cracow, such plans cover only 14 out of the 31 housing estates or complexes thereof investigated in this study (Table 2).

In this context, what deserves attention is an opinion that in the Study of Spatial Development Conditions and Directions in force [44], which provides the foundations for all the local spatial development plans and administrative decisions in this area, and consequently the grounds for building permits to be issued, the protection of valuable housing projects from that period is insufficient and ineffective [25]. In the scientific debate, there are more and more voices about the need to protect valuable urban structures of the second half of the 20th century [55–57]. Therefore, an interesting concept is the introduction of the term 'urban heritage structures' as a protection tool at the level of the local law (planning documents, study, local spatial development plans). These structures would comprise—among others—valuable 20th-century and contemporary projects of different scales and uses [2].

In the context of the results of studies and analyses presented in this paper, it should be concluded that some housing estates have lost some of their advantages irreversibly. Many housing estates still have such values, and it is important not to lose this potential.

However, it should be noted that following an idea of the 'compact city', many European planning strategies consider urban densification as one of the tools to promote sustainable urban development. One may agree with that opinion only when the densification results from a wider, comprehensive development plan and not from individual, uncoordinated decisions: not those aimed at private interest, but at the common one. It needs to be a thoroughly planned densification process [58], which is usually complex and slow [59]. Moreover, regeneration policies for the housing estates should be considered in view of a wider urban context of the city, which can be labeled as an outward-looking approach [60,61].

Furthermore, investing in such housing estates is a positive thing in itself, especially if new service facilities are introduced, which makes the functional program of the estate more diverse. Mixed use promotes sustainable urban development as well [62,63].

Therefore, even though quite paradoxically, it could be stated that up to a point, the current densifications protect housing estates from their functional degradation, at the same time, they lead to their spatial degradation and destroy their advantages. However, it is worth noticing that there is still a chance to improve the situation of housing estates by undertaking rational and consistent measures in the field of architecture and urban planning, which can change not only the living standards offered to residents but also the very image of housing estates.

The methodology proposed and verified within the scheme of the research presented in this publication can be used as a tool for analyzing the problem of densification of large-panel housing estates in other towns and cities.

5. Conclusions

The goal of the research presented in this paper is to demonstrate the scales and character of densifications of the large-panel housing estates built in the second half of the 20th century, which pose a threat of devastation of their urban layouts, which are often recognized as heritage.

The paper applies a quantitative and qualitative approach to present spatial transformations that take place within such housing estates. The study area comprised housing estates in Cracow, the main city of the Province of Małopolskie, which after filtering out certain local specificity demonstrates many tendencies and mechanisms characteristic for other big cities of Poland.

The paper illustrates the scale of the problem, subsequently classifying the ways in which these housing estates have been transformed in terms of their densifications. Two main tendencies have been observed: adding buildings or clusters of buildings within the perimeter of housing estates and in their outskirts, leading to shifting of their edges. It also points to the essence of functional and spatial changes in the direct vicinity of large-panel housing estates, consisting of building a new complex of multi-family residential buildings, demonstrating the potential negative influence on their functioning. On the basis of the analysis results, the paper demonstrates the threat of devastation of urban layouts of such housing estates and their main potential—green areas.

This paper constitutes a voice in a discussion devoted to the condition, but most of all to the future of large-panel housing estates, particularly in terms of their protection and stopping some unfavorable tendencies of urban destruction. It fills the gap in the existing studies focusing on this problem, particularly in the context of the ongoing debate on the significance of urban planning achievements in the second half of the 20th century and recognizing valuable urban layouts as heritage.

All measures leading to the development and improvement of the quality of large-panel housing estates are desirable. Nevertheless, they should constitute an element of a well-coordinated and comprehensive process, instead of a sum of one-off interventions, paying no or little heed to the existing context, particularly the ones that disregard public interest in favor of the private one. The study results presented herein indicate that chaotic densifications in such housing estates may reach a surprisingly big scale. Therefore, the presented method of analyzing the effects of this process in a systematic approach seems to be useful for controlling and rationalizing its propagation, in particular for revealing and eliminating the negative tendencies that emerge. The knowledge obtained this way may prove valuable in prognostic studies devoted to urban development, in the optimization of relevant formal and legal solutions, as well as in optimizing current management activities.

The research presented in this paper was limited to the problem of the percentage of the increase in the built-up area of these housing estates and the character of their densification as well as to demonstrating a threat of devastation of their urban layouts. Therefore, further research can develop the problem of densification and its effects on such housing estates in a more complex way. For instance, it is possible to study the impact of the densification process on the percentage of services and the mix of services, the degree of fragmentation of green areas, etc. The research topic can cover multidisciplinary issues, for example densification studies can be integrated with sociological aspects, such as residents' population, population by age, by income, and by nationality. It is recommended to conduct further research from the aspect of creating formal barriers to prevent the uncontrolled densification of large-panel housing estates.

The sustainable development of cities based on the concept of a compact city requires an appropriate density and intensity of development, which means that measures directed toward increasing this intensity within city limits are rational in many cases. Nevertheless, they need to be optimized, and they must not be implemented at the expense of existing heritage structures and their residents. This paper may constitute a source of information on tendencies of transformations occurring in large-panel housing estates as well as a source of inspiration to take actions aimed at the rationalization of these processes.

Funding: This research received no external funding.

Institutional Review Board Statement: Not applicable.

Informed Consent Statement: Not applicable.

Data Availability Statement: Data available in a publicly accessible repositories—they are all showed in the 'References' section.

Conflicts of Interest: The author declares no conflict of interest.

References

1. Corbusier, L. *Karta ateńska (orig. La Charte d'Athènes)*; Fundacja Centrum Architektury: Warszawa, Poland, 2017; Document Adopted at the IV Congrès Internationaux d'Architecture Moderne (CIAM) held in 1933.
2. Gyurkovich, M.; Szarata, A.; Zuziak, Z.K.; Ogrodnik, D.; Faron, A.; Ziobro, A.; Poklewski-Koziełł, D.; Suchoń, F.; Tota, P.; Sotoca, A.; et al. *Model Przestrzennej Struktury Krakowa. Perspektywa Planistyczna. Etap II Zdefiniowanie Modelu Przestrzennej Struktury Krakowa*; Reaserch project for BPP UM Kraków, Instytut Projektowania Urbanistycznego, Wydział Architektury; Politechnika Krakowska: Kraków, Poland, 2017.
3. Białkiewicz, A.; Żychowska, M.J.; Stelmach, B. Assets of Modern Culture: An Outline of Conservation Issues. *Wiadomości Konserw.* **2020**, *63*, 152–162. [CrossRef]
4. Sendi, R.; Kerbler, B. The Evolution of Multifamily Housing: Post-Second World War Large Housing Estates versus Post-Socialist Multifamily Housing Types in Slovenia. *Sustainability* **2021**, *13*, 10363. [CrossRef]
5. García-Pérez, S.; Oliveira, V.; Monclús, J.; Díez Medina, C. UR-Hesp: A methodological approach for a diagnosis on the quality of open spaces in mass housing estates. *Cities* **2020**, *103*, 102657. [CrossRef]
6. Dekker, K.; Van Kempen, R. Large housing estates in Europe: Current situation and developments. *Tijdschr. Econ. Soc. Geogr.* **2004**, *95*, 570–577. [CrossRef]
7. Díez Medina, C.; Monclús, J. Dealing with Mass housing estates legacy: The need of specific diagnoses from an urban design perspectives. In Proceedings of the 24th ISUF 2017—City and Territory in the Globalization Age, Valencia, Spain, 27–29 September 2017; Universitat Politċcnica Valċncia: Valencia, Spain, 2017; pp. 309–322. [CrossRef]
8. Szczerek, E. *Rewitalizacja Osiedli Wielkopłytowych a Ciągłość i Komplementarność Przestrzeni Publicznej Miasta*; Wydawnictwo Politechniki Krakowskiej: Kraków, Poland, 2018.
9. Van Beckhoven, E.; Van Kempen, R. Towards more social cohesion in large post-second world war housing estates? A case study in Utrecht, the Netherlands. *Hous. Stud.* **2006**, *21*, 477–500. [CrossRef]
10. Smagacz-Poziemska, M.; Bukowski, A.; Jabłońska, B. *Wykluczone Sąsiedztwa. Społeczne Aspekty Rewitalizacji w Przestrzeni w Wielkiego Miasta*; Instytut Socjologii Uniwersytetu Jagiellońskiego: Kraków, Poland, 2007.
11. Gorczyca, K. The Social Transformation of Large Housing Estates in Poland at the Turn of the 21st Century. *Czech Sociol. Rev.* **2016**, *52*, 861–892. [CrossRef]
12. Wassenberg, F. Large social housing estates: From stigma to demolition? *Neth. J. Hous. Environ. Res.* **2004**, *19*, 223–232. [CrossRef]
13. Brattbakk, I. Post-war large housing estates in Norway? Well-kept residential areas still stigmatised? *Neth. J. Hous. Environ. Res.* **2004**, *19*, 311–332. [CrossRef]
14. Racoń-Leja, K.; Kwiatkowski, K. W poszukiwaniu bezpiecznego osiedla—Eksperyment na krakowskim Nowym Bieżanowie. *Autoportret* **2005**, *4*, 28–29.
15. Jeżak, J.; Rydzik, W.; Węglowski, M. *Strategie Rehabilitacji Blokowisk*; Rębowska, A., Ed.; IRM: Kraków, Poland, 2006.
16. Högberg, L.; Lind, H.; Grange, K. Incentives for improving energy efficiency when renovating large-scale housing estates: A case study of the swedish million homes programme. *Sustainability* **2009**, *1*, 1349–1365. [CrossRef]
17. Jarczewski, W. Procedura badawcza i wyniki prac prowadzonych w ramach diagnozy. In *Rewitalizacja Miast Polskich—Diagnoza*; Ziobrowski, Z., Jarczewski, W., Eds.; IRM: Kraków, Poland, 2010; Volume 8, pp. 13–21.
18. Jarczewski, W. Skala degradacji miast w Polsce. In *Rewitalizacja Miast Polskich—Diagnoza*; Ziobrowski, Z., Jarczewski, W., Eds.; IRM: Kraków, Poland, 2010; Volume 8, pp. 57–64.
19. Raport—Stan Mieszkalnictwa w Polsce; Ministerstwo Rozwoju; Marzec 2020 (Report—The State of Housing in Poland; Ministry of Development; March 2020). Available online: https://www.gov.pl/web/rozwoj-technologia/raport-o-stanie-mieszkalnictwa (accessed on 2 November 2021).
20. Szafrańska, E. Transformations of large housing estates in post-socialist city: The case of Łódź, Poland. *Geogr. Pol.* **2014**, *87*, 77–93. [CrossRef]
21. Protić, I.B.; Mitković, P.; Vasilevska, L. Toward Regeneration of Public Open Spaces within Large Housing Estates–A Case Study of Niš, Serbia. *Sustainability* **2020**, *12*, 10256. [CrossRef]

22. Battisti, L.; Pille, L.; Wachtel, T.; Larcher, F.; Säumel, I. Residential Greenery: State of the Art and Health-Related Ecosystem Services and Disservices in the City of Berlin. *Sustainability* **2019**, *11*, 1815. [CrossRef]
23. Schwarz, N.; Haase, A.; Haase, D.; Kabisch, N.; Kabisch, S.; Liebelt, V.; Rink, D.; Strohbach, M.; Welz, J.; Wolff, M. How are urban green spaces and residential development related? A synopsis of multi-perspective analyses for Leipzig, Germany. *Land* **2021**, *10*, 630. [CrossRef]
24. Kopecká, M.; Szatmári, D.; Rosina, K. Analysis of urban green spaces based on sentinel-2A: Case studies from Slovakia. *Land* **2017**, *6*, 25. [CrossRef]
25. Gyurkovich, M.; Sotoca, A.; Szarata, A.; Szczerek, E.; Matusik, A.; Poklewski-Koziełł, D.; Suchoń, F. housing estates from the second half of the twentieth century as urban heritage structures: Example of housing estates in Mistrzejowice, Cracow. *Wiadomości Konserw.* **2021**, *65*, 54–65.
26. Temelová, J.; Slezáková, A. The changing environment and neighbourhood satisfaction in socialist high-rise panel housing estates: The time-comparative perceptions of elderly residents in Prague. *Cities* **2014**, *37*, 82–91. [CrossRef]
27. Kabisch, S.; Grossmann, K. Challenges for large housing estates in light of population decline and ageing: Results of a long-term survey in East Germany. *Habitat Int.* **2013**, *39*, 232–239. [CrossRef]
28. Kovács, Z.; Hegedűs, G. Gated communities as new forms of segregation in post-socialist Budapest. *Cities* **2014**, *36*, 200–209. [CrossRef]
29. Włodarczyk, M. (Ed.) *Krakowskie Osiedla Modernizmu lat 1945–1990. Wybrane Przykłady*; SARP Oddział Kraków: Kraków, Poland, 2014.
30. Tokajuk, A. Gry mieszkaniowe'—Wybrane problemy kształtowania zabudowy mieszkaniowej w Polsce XXI wieku. *Analiza krytyczna. Archit. Artibus* **2015**, *4*, 65–71.
31. Schneider-Skalska, G.; Lorek, A. The housing environment of Cracow—Selected areas, selected problems. In *Cracow: Selected Problems of the Urban Structure Evolution*; Gyurkovich, J., Matusik, A., Suchoń, F., Eds.; Wydawnictwo Politechniki Krakowskiej: Kraków, Poland, 2016; pp. 53–82.
32. Pancewicz, Ł. O potrzebie nowego modernizmu. *Autoportret* **2021**, *1*, 80–85.
33. Celińska-Janowicz, D. Druga młodość czy upadek? Warszawskie osiedle Służew nad Dolinką w okresie transformacji. *Studia Reg. Lokalne* **2010**, *1*, 89–104.
34. Masztalski, R.; Michalski, M. Podatność osiedli wielkopłytowych z drugiej połowy XX wieku we Wrocławiu na procesy rewitalizacyjne. *Bud. Arch.* **2018**, *17*, 101–108. [CrossRef]
35. Szczerek, E. Loss of Potential: Large-Panel Housing Estates—Czyzyny Case. *IOP Conf. Series Mater. Sci. Eng.* **2019**, *471*, 092034. [CrossRef]
36. Crisan, R.; Macri, Z.; Panait, A. (Eds.) *Visions on Mass-Housing Sustainable Rehabilitation*; Editura Universitară Ion Mincu: Bucuresti, Romania, 2011.
37. Gyurkovich, M. (Ed.) *Future of the City: Mass HOUSING Estates or Multifamily Housing Complexes? Eco Rehab 3 Cracow 2012*; Cracow University of Technology: Krakow, Poland, 2012.
38. Kristiánová, K. Post-socialist transformations of green open spaces in large scale socialist housing estates in Slovakia. *Procedia Eng.* **2016**, *161*, 1863–1867. [CrossRef]
39. Treija, S.; Bratuškins, U.; Koroļova, A. Urban densification of large housing estates in the context of privatisation of public open space: The case of Imanta, Riga. *Arch. Urban Plan.* **2018**, *14*, 105–110. [CrossRef]
40. Chmielewski, J.M.; Mirecka, M. *Modernizacja Osiedli Mieszkaniowych*; Oficyna Wydawnicza PW: Warszawa, Poland, 2007.
41. Jeżak, J.; Nejman, M.; Wierzchowski, M. *Wielokryterialna Analiza 19 Osiedli Zabudowy Blokowej Położonych na Terenie Gminy Miejskiej Kraków*; IRM: Kraków, Poland, 2011.
42. Karpińska, M.; Leśniak-Rychlak, D.; Wiśniewski, M. (Eds.) *Witold Cęckiewicz*; Instytut Architektury: Kraków, Poland, 2015; Volume I–II.
43. Fabiański, M.; Purchla, J. *Przewodnik po Architekturze Krakowa*; Wydawnictwo Literackie: Kraków, Poland, 2001.
44. Studium Uwarunkowań i Kierunków Zagospodarowania Przestrzennego Miasta Krakowa (2003), Uchwała RMK Nr XII/87/03 z dnia 16 kwietnia 2003 r. Zmiana Studium Uwarunkowa'n i Kierunków Zagospodarowania Przestrzennego Miasta Krakowa, t. I–III, załącznik do Uchwały nr CXII/1700/14 Rady Miasta Krakowa z 9 lipca 2014r., Biuro Planowania Przestrzennego UM Krakowa, Kraków 2014 (2003 Spatial Development Study, Resolution of the Cracow City Council of 16 April 2003 No. XII/87/03 together with the update vol. I–III dated on 9 July 2014). Available online: https://www.bip.krakow.pl/?bip_id=1&mmi=48 (accessed on 2 November 2021).
45. Seibert, K. *Plan Wielkiego Krakowa*; Wydawnictwo Literackie Kraków: Kraków, Poland, 1983.
46. Rehabilitacja Zabudowy Blokowej Osiedli w Krakowie. Available online: https://www.bip.krakow.pl/?dok_id=51896 (accessed on 2 November 2021).
47. Ilmurzyńska, K. Ewolucja przestrzenna Ursynowa Północnego. *PUA* **2018**, *1*, 203–218. [CrossRef]
48. Niezabitowska, E.D. *Metody i Techniki Badawcze w Architekturze*; Wydawnictwo Politechniki Śląskiej: Gliwice, Poland, 2014.
49. Lynch, K. *Good City Form*; The MIT Press: Cambridge, MA, USA, 1984.
50. MSIP Kraków: Miejski System Informacji Przestrzennej. Available online: https://msip.krakow.pl/ (accessed on 2 November 2021).
51. Biuletyn Informacji Publicznej. Miasto Kraków. Available online: https://www.bip.krakow.pl/ (accessed on 2 November 2021).

52. Myczkowski, Z.; Wielgus, K.; Chajdys, K.; Maj, A.; Mikulski, J.; Leonowicz, P.; Rymsza-Mazur, W.; Staniewska, A.; Środulska-Wielgus, J.; Latusek, K. *Program Planu Ochrony Twierdzy Kraków (Studium Parków Kulturowych i Stref Ochrony Krajobrazu Warownego)*; Reaserch project for WKiDN UM Kraków, Instytut Architektury Krajobrazu, Wydział Architektury; Politechnika Krakowska: Kraków, Poland, 2014.
53. Sotoca, A. (Ed.) *After the Project—Updating Mass Housing Estates*; II EcoRehab, Barcelona 2011; UPC: Barcelona, Spain, 2012.
54. Balla, R. A Sustainable Urban Concept for Housing Estates: Plot, Patch, Part. *Period. Polytech. Arch.* **2018**, *49*, 135–143. [CrossRef]
55. Węcławowicz-Gyurkovich, E. Podsumowanie ogólnopolskiej konferencji 'Ochrona dziedzictwa architektury polskiej II połowy XX wieku, Warszawa 16-17.XI.2016'. *Wiadomości Konserw.* **2017**, *49*, 7–14.
56. Gaczoł, A. Kryteria dotyczące wpisu obiektów architektonicznych z II połowy XX wieku do rejestru zabytków. Teoria i praktyka. *Wiadomości Konserw.* **2017**, *49*, 163–180.
57. Kadłuczka, A. Posłowie do Ogólnopolskiej Konferencji 'Ochrona dziedzictwa architektury i urbanistyki II połowy XX wieku'. *Wiadomości Konserw.* **2017**, *49*, 15–18.
58. Skovbro, A. Urban densification—A sustainable urban policy? In *The Sustainable City II*; Brebbia, C.A., Martin-Duque, J.F., Wadhwa, L.C., Eds.; WIT Press: Southampton, UK, 2002; pp. 517–527.
59. Puustinen, S. Urban densification as a challenge for city planning. Improving conditions and processes of infill developments. In Proceedings of the Conference: Plannord Conference in Kungliga Tekniska Högskolan, Stockholm, Sweden, 20–22 August 2015.
60. Hall, P. Regeneration Policies for Peripheral Housing Estates: Inward- and Outward-looking Approaches. *Urban Stud.* **1997**, *34*, 873–890. [CrossRef]
61. Muliuolytė, J. Rediscovering large scale housing estates in post socialist cities. *J. Arch. Urban.* **2013**, *37*, 51–58. [CrossRef]
62. Yang, H.; Fu, M.; Wang, L.; Tang, F. Mixed Land Use Evaluation and Its Impact on Housing Prices in Beijing Based on Multi-Source Big Data. *Land* **2021**, *10*, 1103. [CrossRef]
63. Maleki, M.; Zain, M.; Ismail, A. Variables communalities and dependence to factors of street system, density, and mixed land use in sustainable site design. *Sustain. Cities Soc.* **2012**, *3*, 46–53. [CrossRef]

MDPI
St. Alban-Anlage 66
4052 Basel
Switzerland
www.mdpi.com

Land Editorial Office
E-mail: land@mdpi.com
www.mdpi.com/journal/land

www.ingramcontent.com/pod-product-compliance
Lightning Source LLC
Chambersburg PA
CBHW040815250926
51850CB00007BA/1178
9783036585048